Polymer Nanocomposites

Polymer Nanocomposites: Fabrication to Applications offers readers an up-to-date interpretation of various polymeric nanocomposite materials and technologies via critical reviews. It covers developments and advancements in various nanomaterials, polymeric materials, biopolymers, and processes. It initiates from nanomaterial synthesis, fabrication, and characterization to the manufacturing aspect and feasible product applications of polymer-based nanocomposites. The prime focus is on polymer matrix nanocomposites and their future trends in the engineering sector.

Features:

- Explores synthesis, characterization, properties, fabrication/processing, and applications of polymer nanocomposite materials
- Elaborates on polymer manufacturing phase challenges using various control methods and statistical tools and modules
- Includes machining and micro (μ) machining investigation on the polymer nanocomposites
- Discusses modeling, simulation, and optimization of process parameters during the machining processes and applications of additive manufacturing
- Comprehends the significance of nanomaterials functionalizing synthetic fibrous and biocompatible composites

This book is aimed at researchers and graduate students in mechanical engineering, materials science, polymers, composites, and nanomaterials.

Polymer Nanocomposites
Fabrication to Applications

Edited by
Rajesh Kumar Verma, Shivi Kesarwani,
Jinyang Xu and J. Paulo Davim

CRC Press
Taylor & Francis Group
Boca Raton London New York

CRC Press is an imprint of the
Taylor & Francis Group, an **informa** business

First edition published 2024
by CRC Press
6000 Broken Sound Parkway NW, Suite 300, Boca Raton, FL 33487-2742

and by CRC Press
4 Park Square, Milton Park, Abingdon, Oxon, OX14 4RN

CRC Press is an imprint of Taylor & Francis Group, LLC

© 2024 selection and editorial matter, Rajesh Kumar Verma, Shivi Kesarwani, Jinyang Xu and J. Paulo Davim; individual chapters, the contributors

ISBN: 9781032381954 (hbk)
ISBN: 9781032381978 (pbk)
ISBN: 9781003343912 (ebk)

DOI: 10.1201/9781003343912

Typeset in Times
by codeMantra

Contents

Preface

Polymer nanocomposites are a novel class of composite materials in which one phase is made up of a polymer matrix, which can be thermosetting or thermoplastic, and the second phase is reinforcement having nanosize. One way to define nanocomposite is that when the reinforcement phase is on a nanosize scale (less than 100 nanometers), the developed composite is called nanocomposite. Polymer nanocomposites are widely used in multifunctional applications of aerospace, aircraft, biomedical, sensors, etc. due to their high synergistic effect and aspect ratio. The cost-effectiveness and durability are also the main factors for their uniqueness. In the current time of globalization, industries are placing more prominent efforts on customer satisfaction and the supply-to-demand ratio. Hence, every manufacturing/production unit develops a new quality of components to retain the required product quality and productivity. The utilization of carbon nanomaterials is an efficient approach to produce a strong polymeric matrix and fibrous composites. In this series, various exciting nanomaterials like carbon nanotube (CNT), carbon nano-onion, carbon nanosheet, carbon nanorod, graphene oxide (GO), reduced GO, and graphene nanoplatelets as well as several organic and inorganic nanomaterials are passing through the initial phase of exploration to boost fiber composites' properties. The practical functions of utilizing carbon nanomaterials/nanomaterials in the pristine thermoplastic or a thermoset polymer-based matrix/modified matrix are gaining research attention among material scientists and researchers. Nanomaterials are a category of diverged filler materials with multifunctional core molecules. To enhance the mechanical, electrical, and thermal characteristics, it is possible to end-functionalize the material, resulting in a product that can be cross-linked.

In prior state-of-the-art research explorations, polymer-based implants with nanoparticle-enhanced wear resistance and enhanced implant strength have been successfully demonstrated. Nanoparticles, as a result, have received a lot of attention in research works because of their large surface area and many features. Additionally, numerous studies have also shown that some nanocomposites made with a polymer matrix are more biocompatible and long-lasting than conventional materials/alloys.

Furthermore, recent advancements in polymer nanocomposites, a hybrid approach of fiber- and nanomaterial-reinforced nanocomposites, are popular in current trends and industrial applications. The formation of fiber composites, types of weaves, fiber orientation, thickness and nature, and shape and size of fibers affect the manufacturing principles. Studies on fiber nanocomposites require more attention to explore their highest benefits in society and trade interests. Limitations of conventional fiber composite production practices have persuaded scholars to delve into alternate fiber nanocomposite development. Fiber-reinforced plastic (FRP) is a type of composite material developed from the polymer matrix modified by fibers, usually glass, carbon, Kevlar, basalt, or aramid. By contrast, the matrix is typically an epoxy, vinyl ester, or polyester thermosetting plastic. FRPs are frequently used to develop the multifunctional components of aerospace and parts of other structural sectors. It is best suited for any type of design challenge that needs weight savings, accuracy, restricted tolerances, and streamlining functions in both production and operation.

For the consideration of manufacturing aspects, advancement in additive manufacturing technologies and rapid tooling are widely used. Component production in the mode of 3D printing technology has ensured the manufacturing of several complex structures and designs with the help of computer-aided design tools. Because of pressing environmental issues and inadequate physical-mechanical characteristics with low performance, polymer-based 3D printing materials are currently in use. The development of a sustainable high-performance polymer composite material is vastly needed in the current scenario. Fused deposition modeling (FDM), one of the mainly used 3D printing methods, is the most widely used method for manufacturing thermoplastic components. While evaluating machining and micro (μ) machining processes performed on the polymer nanocomposites, the proposed book focuses on the unified aim to evaluate conflicting machining performances such as material removal rate and surface roughness (roughness average), tool wear, tool damage issues, machining-induced defect, and damage analysis of the machined composite product. With the future use of FRPs in several products, the physical and mechanical characterization of these anisotropic and inhomogeneity composites has become a significant concern for the manufacturing industries.

This book offers readers an up-to-date interpretation of various polymeric nanocomposite materials and technologies via critical reviews. A leading expert in the field writes each chapter to ensure readers get a complete picture of developments and advancements in various nanomaterials, polymeric materials, biopolymers, and processes. The present work is related to the interdisciplinary research area and benefits materials science and engineering scholars. The proposed book will initiate from nanomaterial synthesis, fabrication, and characterization to the manufacturing aspect and feasible end-product applications of the polymer-based nanocomposites. The polymer manufacturing phase challenges will be elaborated using various control methods, statistical tools, and modules recently developed by eminent scholars. The exciting area of this book will be the prime focus on polymer matrix nanocomposites and their future trends in the engineering sector. The findings are charted and tabulated in this proposed book for better visualization.

Acknowledgments

This book was possible due to the help, support, and direction of several eminent people and scholars. First, our most profound gratitude goes to the Almighty, who endowed me with health, grace, wisdom, patience, and direction to complete this work. Our most heartfelt appreciation goes to Prof. Samsher, Honorable Vice Chancellor, Harcourt Butler Technical University (HBTU), Kanpur, India for their continuous support, stimulating environment toward research and constant direction for the timely completion of this book. We are extremely grateful to him for motivating me to think independently, good-naturedly, and solve numerous problems with perseverance. We are very gratified to Prof. J.P. Pandey, Honorable Vice Chancellor, Madan Mohan Malaviya University of Technology (MMMUT), Gorakhpur, India for their inspiration and motivation during this bookwork. We are fortunate to have their advice and support to express my thankfulness for their support and belief.

We are eternally obliged to the University Chancellor and Honorable Governor of Uttar Pradesh, India and the Government of Uttar Pradesh state for providing us with world-class infrastructure and technical support, which helped for finalizing this task.

The Lead editor would like to thank all the authors and eminent contributors for their continuous support in completing this task. We are thankful for the help of global research collaborators from the University of Aveiro, Portugal and Shanghai Jiao Tong University, China. We are very much thankful to research scholars of namely, Dr. Prakhar Kumar Kharwar, Dr. Jogendra Kumar, Mr. Devendra Kumar Singh, Mr. Kuldeep Kumar, Mr. Kaushlendra Kumar, Mr. Rahul Vishwakarma, Mr. Balram Jaiswal, and Mr. Virat Mani Vidhyasagar for their rigorous work and valuable findings in the polymer composites.

The editors are very gratified to the reviewers, editorial members, and team at CRC Press, Taylor & Francis, for their valuable time and effort in this book assignment. Their efforts were noteworthy in producing this book in its present-day form. We could not have accomplished this milestone without their constant and consistent advice, support, and collaboration.

About the Editors

Rajesh Kumar Verma is a professor in the Department of Mechanical Engineering, School of Engineering at Harcourt Butler Technical University, Kanpur, India. It is a renowned and old premier institute whose roots are as deep as its outlook. The Government Research Institute, Cawnpore was established in 1920 and was renamed "Government Technological Institute" in 1921. Finally, in 1926, it got the name we know today, "Harcourt Butler Technological Institute." As per Act No. 11 of 2016 by the Government of Uttar Pradesh, it has become a university, i.e., Harcourt Butler Technical University Kanpur, India. Prof. Verma received his Ph.D. (engineering) from Jadavpur University, Kolkata, India. He is a C.Eng. (mechanical engineering) of the Institution of Engineers (India). He is actively involved in teaching and research in waste management, fiber science, composites, nanomaterial, modeling, simulation, optimization, and manufacturing processes. Prof. Verma currently supervised/ongoing more than 13 master's and nine Ph.D. theses and published more than 135 research articles in peer-reviewed journals and conferences. The e-content and courses he developed are widely used in the Indian university system. He has completed/ongoing more than eight research and development (R&D) projects sponsored by various govt. agencies such as SERB-DST, AICTE, CST, MHRD, etc. He is now a reviewer of various international publishers. Moreover, Prof. Verma has been invited as keynote/invited speaker, section chair, and member for several international conferences with themes in mechanical and production engineering, emerging materials, and manufacturing science. He has developed various nanomaterials for structural applications and proposed hybrid optimization modules recognized by multiple peer-reviewed journal publishers.

Shivi Kesarwani is currently working as a senior research assistant in the Department of Mechanical Engineering at Madan Mohan Malaviya University of Technology, India. He is an associate member of the Institution of Engineers (India). His research interests are nanomaterial synthesis and development, composites machining, numerical modeling, and surface texturing. He has published over 30+ peer-reviewed articles in refereed international journals and conference proceedings. Additionally, he has filed two process and two design patents and has one Indian copyright-registered document. He has also utilized various nanomaterial-reinforced fibrous composites for structural applications and developed hybrid metaheuristic optimizing methodologies recognized by multiple peer-reviewed journals and book publishers.

Jinyang Xu is an Associate Professor and a Doctoral Supervisor of Mechanical Engineering at Shanghai Jiao Tong University, China. He received his Ph.D. (2016) in Mechanical Engineering from Arts et Métiers ParisTech, France. His research interests focus on composites machining, numerical modeling, micro/nano cutting, and surface texturing. He has published over 100 articles in highly-ranked JCR-referenced journals as a first/corresponding author, and edited 2 monographs, 5 book chapters, 6 int. conf. proceedings & 8 special issues. He now serves as the Editor-in-Chief of *Journal of Coating Science and Technology* (JCST), *International Journal of Precision Technology* (IJPTech), and *International Journal of Product Sound Quality* (IJPSQ). He is an Associate Editor of Proc. *Inst. Mech. Eng. Pt. E - J. Process. Mech. Eng.* (SCIE/EI), *Simulation–Transactions of the Society for Modeling and Simulation International* (SCIE/EI), and *Frontiers in Materials* (SCIE/EI). Presently, he is an Editorial Board Member of 7 SCI journals, including *Green Materials, Journal of Superhard Materials, International Journal of Aerospace Engineering, Coatings, Lubricants, etc.* He was honored with the prestigious IAAM Scientist Medal of the year 2020, and awarded the Shanghai Pujiang Scholar by the Shanghai Municipality in 2017. He is a fellow of IAAM and also a senior member of ASME and SCS.

J. Paulo Davim is a full professor at the University of Aveiro, Portugal. He is also distinguished as honorary professor in several universities/colleges in China, India, and Spain. He received his Ph.D. degree in mechanical engineering in 1997, M.Sc. degree in mechanical engineering (materials and manufacturing processes) in 1991, mechanical engineering degree in 1986 from the University of Porto (FEUP), the aggregate title (full habilitation) from the University of Coimbra in 2005, and D.Sc. (higher doctorate) from London Metropolitan University in 2013. He is a Senior Chartered Engineer by the Portuguese Institution of Engineers with an MBA and specialist titles in engineering and industrial management as well as in metrology. He is also Eur Ing by FEANI-Brussels and Fellow (FIET) of IET-London. He has more than 30 years of teaching and research experience in manufacturing, materials, mechanical, and industrial engineering, with a special emphasis on machining and tribology. He has also interest in management, engineering education, and higher education for sustainability. He has guided large numbers of postdoc, Ph.D., and master's students as well as has coordinated and participated in several financed research projects. He has received several scientific awards and honors. He has worked as an evaluator of projects for the European Research Council and other international research agencies as well as the examiner of Ph.D. thesis for many universities in different countries. He is the editor in chief of several international journals, guest editor of journals, book editor, book series editor, and scientific advisory for many international journals and conferences. Presently, he is an editorial board member of 30 international journals and acts as a reviewer for more than 100 prestigious Web of Science journals. In addition, he has also published as editor (and co-editor) more than 200 books and as author (and co-author) more than 15 books, 100 book chapters, and 500 articles in journals and conferences (more than 300 articles in journals indexed in Web of Science core collection/h-index 61+/12000+ citations, SCOPUS/h-index 65+/15000+ citations, Google Scholar/h-index 84+/25000+ citations). He has listed in World's Top 2% Scientists by Stanford University study.

Contributors

Bhagyashri Annaldewar
Department of Fibers and Textile Processing
    Technology
Institute of Chemical Technology (University
    Under Section-3 of UGC Act 1956)
Mumbai, India

Atul Bandyopadhyay
Department of Physics
University of Gour Banga
Malda, India

Bikash Chandra Behera
Department of mechanical Engineering
C. V. Raman Global University
Bhubaneswar, Odisha, India

Mohammed Berrada
Faculty of Sciences Ben M'Sick, Laboratory
    of Biomolecules and Organic Synthesis
    (BIOSYNTHO), Department of Chemistry
University Hassan II of Casablanca
Casablanca, Morocco

Jason T. Brantley
Applied Hypersonics and Space Systems
Kansas State University
Aerospace and Technology Campus, Salina,
    Kansas

Chandan R. S.
Department of Pharmaceutical Chemistry, JSS
    College of Pharmacy
JSS Academy of Higher Education and
    Research (JSSAHER)
Mysuru, India

Karan Chandrakar
Textile Engineering and Technology
Indian Institute of Technology-Delhi
New Delhi, India

Fouad Damiri
Faculty of Sciences Ben M'Sick, Laboratory
    of Biomolecules and Organic Synthesis
    (BIOSYNTHO),
Department of Chemistry
University Hassan II of Casablanca
Casablanca, Morocco

Apurba Das
Design Head-Freight
Titagarh Wagons Ltd
Kolkata, India

Kishore Debnath
Department of Mechanical Engineering
National Institute of Technology Meghalaya
Shillong, Meghalaya, India

M. Infanta Diana
Department of Physics, Dr. Mahalingam
    College of Engineering and Technology,
    Pollachi, India

Umang Dubey
Department of Production Engineering
National Institute of Technology
    Tiruchirappalli
Tiruchirappalli, Tamil Nadu, India

Hrishikesh Dutta
Centre for Additive Manufacturing
Chennai Institute of Technology
Chennai, India

Ergün Ekici
Faculty of Engineering
Çanakkale Onsekiz Mart University
Çanakkale, Türkiye

Vitthal L. Gole
Department of Chemical Engineering
Madan Mohan Malaviya University of
    Technology (MMMUT)
Gorakhpur, Uttar Pradesh, India

Prakash Chandra Gope
Mechanical Engineering Department
GBPUAT
Pantnagar, Uttarakhand, India

B. H. Jaswanth Gowda
Department of Pharmaceutics
Yenepoya Pharmacy College & Research
    Centre, Yenepoya (Deemed to be University)
Mangaluru, India

Harun Güçlü
Engineering Faculty, Applied Mechanics
    and Advanced Materials Research Group
    (AMAMRG) Laboratory, Automotive
    Engineering Department
Bursa Uludag University
Bursa, Türkiye

P. Adlin Helen
Luminescence and Solid-State Ionics
    Laboratory, Department of Physics
Bharathiar University
Coimbatore, India

Hemalatha Y. R.
Department of Pharmaceutical Chemistry
Vivekananda College of Pharmacy
Bengaluru, India

Akshay Jadhav
Department of Fibres and Textile Processing
    Technology
Institute of Chemical Technology (University
    Under Section-3 of UGC Act 1956)
Mumbai, India

Nilesh Jadhav
Department of Fibres and Textile Processing
    Technology
Institute of Chemical Technology (University
    Under Section-3 of UGC Act 1956)
Mumbai, India

Mark James-Jackson
Applied Hypersonics and Space Systems
Kansas State University
Aerospace and Technology Campus, Salina,
    Kansas

J. Jayaramudu
Polymer and Petroleum Group, Materials
    Sciences and Technology Division
CSIR-North East Institute of Science and
    Technology
Jorhat, Assam, India

Hemalata Jena
School of Mechanical Engineering
KIIT Deemed to be University
Bhubaneswar, India

Manoj Kumar Karnena
Department of Environmental Science
GITAM Institute of Science, GITAM (Deemed
    to be) University
Visakhapatnam, India

Amandeep Kaur
Bhaskaracharya College of Applied Sciences
University of Delhi
New Delhi, India

Madhavi Konni
Department of Humanities and Sciences
Malla Reddy Engineering College
Hyderabad, India

D. Lakshmi
Department of Physics
PSG College of Arts and Science
Coimbatore, India

Alisson Rocha Machado
Graduate Program
Pontifícia Universidade Católica do
    Paraná–PUC-PR
Curitiba, Brazil

Shikha Madan
Department of Applied Sciences and
    Humanities
Jamia Millia Islamia
New Delhi, India

Abhijit Maiti
Department of Polymer and Process
    Engineering
Indian Institute of Technology Roorkee
Saharanpur Campus, Saharanpur, Uttar
    Pradesh, India

Gurudas Mandal
Department of Metallurgical Engineering
Kazi Nazrul University
Asansol, India

Mohan S.
Department of Production Engineering
National Institute of Technology
    Tiruchirappalli
Tiruchirappalli, Tamil Nadu, India

Abhijit Mondal
ISR Group
Central Mechanical Engineering Research
    Institute
Durgapur, India

Ali Riza Motorcu
Faculty of Engineering
Çanakkale Onsekiz Mart University
Çanakkale, Türkiye

Murugabalaji V.
Department of Production Engineering
National Institute of Technology
    Tiruchirappalli
Tiruchirappalli, Tamil Nadu, India

Y. S. Negi
Department of Polymer and Process
    Engineering
Indian Institute of Technology Roorkee
Saharanpur Campus, Saharanpur, Uttar
    Pradesh, India

K. Panneerselvam
Department of Production Engineering
National Institute of Technology
    Tiruchirappalli
Tiruchirappalli, Tamil Nadu, India

Karthika Paul
Department of Pharmaceutical Chemistry
Vivekananda College of Pharmacy
Bengaluru, India

Department of Pharmaceutical Chemistry, JSS
    College of Pharmacy
JSS Academy of Higher Education and
    Research (JSSAHER)
Mysuru, India

M. Ramesh
Department of Production Engineering
National Institute of Technology
    Tiruchirappalli
Tiruchirappalli, Tamil Nadu, India

Shikha Rana
Department of Physics
Govt. College Sugh Bhatoli
Kangra, India

Department of Physics
Himachal Pradesh University
Shimla, India

Madhuparna Ray
Department of Polymer and Process
    Engineering
Indian Institute of Technology Roorkee
Saharanpur Campus, Saharanpur,
    Uttar Pradesh, India

Matruprasad Rout
Department of Production Engineering
National Institute of Technology
    Tiruchirappalli
Tiruchirappalli, Tamil Nadu, India

Sudesna Roy
School of Mechanical Engineering
KIIT (Deemed to be) University
Bhubaneswar, India

Rahul Samanta
Department of Metallurgical Engineering
Kazi Nazrul University
Asansol, India

Deba Kumar Sarma
Department of Mechanical Engineering
National Institute of Technology Meghalaya
Shillong, Meghalaya, India

P. Christopher Selvin
Luminescence and Solid-State Ionics
    Laboratory, Department of Physics
Bharathiar University
Coimbatore, India

Jyoti Sharma
Department of Chemical Engineering
Madan Mohan Malaviya University of
    Technology (MMMUT)
Gorakhpur, Uttar Pradesh, India

Ankur Shukla
Textile Engineering and Technology
Indian Institute of Technology-Delhi
New Delhi, India

Mohammad Ali Siddiqui
Department of Chemical Engineering
Madan Mohan Malaviya University of
    Technology (MMMUT)
Gorakhpur, Uttar Pradesh, India

Marcio Bacci da Silva
School of Mechanical Engineering
Federal University of Uberlandia
Uberlandia, Minas Gerais, Brazil

Rosemar Batista da Silva
School of Mechanical Engineering
Federal University of Uberlandia
Uberlandia, Minas Gerais, Brazil

Mahavir Singh
Department of Physics
Himachal Pradesh University
Shimla, India

Arijit Sinha
Department of Metallurgical Engineering
Kazi Nazrul University
Asansol, India

Kritika Singh Somvanshi
Department of Mechanical Engineering
ANDUAT
Ayodhya, Uttar Pradesh, India

Saurabh Tiwari
Department of Metallurgical and Materials
    Engineering
Indian Institute of Technology Roorkee
Roorkee, India

Unnikrishnan T. G.
Department of Production Engineering
National Institute of Technology
    Tiruchirappalli
Tiruchirappalli, Tamil Nadu, India

Abhishek Verma
Department of Polymer and Process
    Engineering
Indian Institute of Technology Roorkee
Saharanpur Campus, Saharanpur, Uttar
    Pradesh, India

Murat Yazıcı
Engineering Faculty, Applied Mechanics
    and Advanced Materials Research Group
    (AMAMRG) Laboratory, Automotive
    Engineering Department
Bursa Uludag University
Bursa, Türkiye

Abbreviations

AL (OH)$_3$	aluminum hydroxide
AL$_2$O$_3$	aluminum oxide
ANOVA	analysis of variance
SB$_2$O$_3$	antimony oxide
AFRP	aramid fiber-reinforced plastic
AI	artificial intelligence
BN	boron nitride
BNNS	boron nitride nanosheet
BUE	built-up-edge
CB	carbon black
CFRP	carbon fiber-reinforced polymer
CNFs	carbon nanofibers
CNT	carbon nanotube
CNRP	carbon nanotube-reinforced polymer
CA	cellulose acetate
CNCs	cellulose nanocrystals
CMCs	ceramic matrix composites
CMNC	ceramic matrix nanocomposites
CCD	charge-coupled device
CAGR	compounded annual growth rate
CT	computed tomography
DNA	deoxy ribose nucleic acid
DSC	differential scanning calorimetry
DWCNT	double-walled carbon nanotube
DLS	dynamic light scattering
DMTA	dynamical mechanical thermal analysis
EDM	electrical discharge machining
ECMM	electrochemical micromachining
ES	electrospinning
EVA	elliptic vibration assisted
ECM	extracellular matrix
FTP	feed per tooth
FRP	fiber-reinforced polymer
FE-SEM	field emission scanning electron microscopy
FD	fill density
FR	fire retardant
FCT	fixed circular tooling
FRT	fixed rhombic tooling
FST	fixed square tooling
FIB-SEM	focused ion beam–scanning electron microscopy
FDM	fused deposition modeling
GMR	giant magnetoresistance
GPa	gigapascal
GF	glass fiber
GFRP	glass fiber-reinforced polymer
GNP/GnP	graphene nanoparticle platelet
GO	graphene oxide

GPL	graphene platelet
HPAM	hydrolyzed polyacrylamide
HA	hydroxy apatite
nHA-PAA	hydroxyapatite polyacrylic acid
ITO	indium tin oxide
ICP-MS	inductively coupled plasma mass spectrometry
IPN	interpenetrating network
IoT	internet of things
LOC	laboratory on a chip
LH	layer height
LED	light-emitting diodes
LOI	limiting oxygen index
MG (OH$_2$)	magnesium hydroxide
MRI	magnetic resonance imaging
MRR	material removal rate
MALDI-MS	matrix-assisted laser desorption/ionization mass spectrometry
MMCs	metal matrix composites
MMNC	metal matrix nanocomposites
MOF(s)	metal organic framework(s)
MOS	metal oxides
MB	methylene blue
MEMS	micro-electromechanical systems
miRNAs	micro-RNAs
MUCT	minimum uncut chip thickness
MMM	mix-matrix membrane
MUC1	mucin 1 protein
MWCNT	multi-walled carbon nanotubes
NFs	nanofibers
NMT	nanoclay montmorillonite
NFC	nanofibrillated cellulose
nHA	nano-hydroxyapatite
NDPC	nanometric dispersed polymer composites
NO$_2$	nitrogen dioxide
OMLSs	organo-modified layered silicates
PP	phosphorous in polypropylene
PV	photovoltaic
PEO	plasma electrolytic oxidation
PBS	poly(butylene succinate)
PDADMAC	poly (diallyl methyl ammonium chloride)
PEO	poly(ethylene oxide)
PGA	poly(glycolic acid)
PHAS	poly(hydroxyalkanoates)
PHB	poly(ß-hydroxybutyrate)
PVOH	poly(vinyl alcohol)
PHBV	poly(3-hydroxybutyrate-co-3-hydroxyvalerate)
PLA	polylactic acid
PNC	polymer nanocomposites
PMMA	polymethyl methacrylate
PPS	polyphenylene sulfide
PEDOT: PSS	poly(3,4-ethylenedioxythiophene) polystyrene sulfonate
PHB	poly-3-hydroxybutyrate

PANI	polyaniline
PCL	polycaprolactone
PC	polycarbonate
PCD	polycrystalline diamond
PEA	polyester amide
PEEK	polyether ether ketone
PET	polyethylene terephthalate
PHA	polyhydroxyalkanoate
PLA	polylactic acid
PMCs	polymer matrix composites
PMNC	polymer matrix nanocomposite
PNC	polymer nanocomposite
POMs	polyoxometalates
PPy	polypyrrole
PS	polystyrene
PS/MWCNT	polystyrene/multi-walled carbon nanotube
PUR	polyurethane
PVA	polyvinyl alcohol
PVC	polyvinyl chloride
PS	print speed
PCBs	printed circuit boards
ROS	reactive oxygen species
rGO	reduced graphene oxide
RSM	response surface methodology
RCT	rotary circular tooling
STEM	scanning transmission electron microscopy
SCF	short carbon fiber
SIO$_2$	silicon dioxide
SWCNTs	single-walled carbon nanotubes
SERS	surface-enhanced Raman scattering
TGA	thermal gravimetric analysis
TMA	thermal mechanical analysis
TRGO	thermally reduced graphene oxide
TPS	thermoplastic starch
TFN	thin-film nanocomposite
3D	three-dimensional
TIO$_2$	titanium dioxide
TEM	transmission electron microscopy
2D	two-dimensional
UHMWPE	ultrahigh-molecular-weight polyethylene
UHR-SEM	ultrahigh-resolution SEM
UV	ultraviolet
UCT	uncut chip thickness
VOCs	volatile organic carbon
WEDG	wire electrodischarge grinding
ZNO	zinc oxide

SYMBOLS

a	amplitude of vibration of the tool in cutting direction, μm
b	amplitude of vibration of the tool in normal direction, μm

θ	angle of fiber orientation or chiral angle
A_f	area filled between grit projection and elastic membrane
A_{dull}	area of grit dulling
R_a	average surface roughness
VB_B	average width of the flank wear, μm
k_a	change in depth of cut
u_{chip}	chip creation energy
k_{chip}	chip creation function
p_{chip}	chip formation power
R	chip ratio (width-to-thickness)
t_h	chip thickness, mm
α	clearance angle of the tool
μ	coefficient of friction
c_c	constant
k_1	constant
M	constant
X	constant
Y	constant
l_c	contact length
p_c	contact stress in the contact zone
C	convexity of abrasive grit
KM	crater wear center distance, μm
KF	crater wear front distance, μm
KB	crater wear width, μm
h_{crit}	critical chip thickness
P^*	critical power at softening temperature
u^*	critical specific energy at softening temperature
t	cutting time, minutes
k_s	deflection of grinding wheel
k_w	deflection of workpiece
P	density
A	depth of abrasive incision
d	depth of cut, mm
Δ	difference in curvature
Z	distance of abrasive grits from surface to cutting depth
C_{ds}	distribution density of grits
a_o	dressing depth
a_d	dressing depth of cut
s_d	dressing feed
δ	dressing severity
$C_d(z)$	dynamic abrasive grit density
N	dynamic number of active grits
r, r_m	eccentricity constants
$A_{0, effective}$	effective wear land area
R_g	empirical constant
R_0	empirical constant
u_0	empirical constant
V_0'	empirical constant
ε	empirical constant, feed angle
d_e	equivalent diameter

F_{chip}	force (chip creation)
F_n	force (normal)
F_{plowing}	force (plowing)
F_{sliding}	force (sliding)
F_t	force (tangential)
F_{total}	force (total grinding)
K_{IC}	fracture toughness
P	grinding power
G	grinding ratio
d_s	grinding wheel diameter
v_s	grinding wheel velocity
H/H_v	hardness number
v_w	nanocomposite workpiece velocity
N_r	number of active grits per revolution
q_w	partitioned heat flux
p_{plowing}	plowing power
A_p	projected area of the grit
γ	rake angle of the tool
A_{sharp}	self-sharpening area
φ	shear angle
l_{sliding}	sliding length
p_{sliding}	sliding power
Q'_w	specific material removal rate
$C_s(z)$	static abrasive grit density
N_g	static number of active grits
$\dot{\gamma}$	strain rate
k_{plowing}	surface plowing function
k	thermal conductivity
α	thermal diffusivity
B	thermal parameter $\left(\dfrac{k\theta^m}{\beta\sqrt{\alpha}} \right)$
v	tool feed, m/min
f	tool vibration frequency, kHz
k_g	total deflection
v_{sh}	total kinematic volume generated by active grits
v_t	total volume of wheel engaged during cutting
σ_{uts}	ultimate tensile stress
t_0	uncut chip thickness, mm
h	undeformed chip thickness
V_f	volume fraction of fiber
V_m	volume fraction of matrix
KT	wear depth of the crater, µm
b	width of cut
d_w	workpiece diameter
E	Young's modulus
E_c	Young's modulus – composite
E_s	Young's modulus – grinding wheel

1 A Brief Overview of Polymer Composites and Nanocomposites

Akshay Jadhav, Bhagyashri Annaldewar, and Nilesh Jadhav

CONTENTS

DOI: 10.1201/9781003343912-1

1.1 INTRODUCTION

Polymers have restricted usage in the manufacture of products and structures all alone, since their properties are not extremely high when contrasted with different materials, for example, most metals. Their properties can be enhanced when the polymer system (matrix) is combined with reinforcing material (filler), which then produces a composite. In connection with matrix properties, the overall properties of the composite are strongly depended on the filler characteristics among which the size of the filler plays a dominant role. Polymer composites which have the filler size in the nanoscale regime generate the polymer nanocomposites (PNC) which is called as the nanotechnology. Apparently, the area of nanotechnology in polymer composites has been a very talked-about subject, and it does have the potential to make our livelihood better and the possibility to make this world a better place to live in [1].

In the area of nanotechnology, one of the most well-known areas for current research and innovations in the area of nanotechnology is of polymer nanocomposites (PNCs), and the examination field covers a wide range of topics which can be explored. This would incorporate nano-electronics, polymeric bio-nanomaterials, reinforced PNCs, nanocomposite-based drug delivery systems, etc. [2].

This chapter gives a brief overview of the polymer composites and nanocomposites. It also discusses their characteristics, synthesis methods, and applications in various fields.

1.2 POLYMER COMPOSITES

Composites have been known as the most innovative and compatible engineering materials which have been known to humans since ancient era. Composites have been discovered due to the research and development in the area of materials science and technology which in turn have given rise to these attractive, strong, and delightful materials. Composites are usually heterogeneous in nature, which are formed by mixing two or more components along with the fillers or by using reinforcing fibres in the presence of a compactable matrix. This matrix can be polymeric, metallic in nature, or ceramic in nature. This matrix is responsible for giving shape, good surface appearance, ecological tolerance, and complete resilience of the composites, whereas the fibre reinforcement is responsible for carrying the most of the structural loads, which gives macroscopic strength and stiffness. A composite material can offer excellent unique physical and mechanical properties as it associates the most necessary properties of its components while overpowering their minimum desirable properties. In today's current scenario, composites are applied in automobile industry, aerospace industry, and in other fields of engineering as they display excellent strength-to-weight and modulus-to-weight ratios. The fibre-reinforcing components used in the composites usually consist of thin continuous fibres or relatively segments. While using short segments, fibres which have high aspect ratio, that is, length-to-diameter ratio, are utilized. Generally, high-performance structural applications require continuous reinforced composites due to their specific strength (i.e., strength-to-density ratio) and stiffness (i.e., modulus-to-density ratio). The strength and stiffness of carbon fibre-reinforced composites are more superior than the conventional metal alloys. Also, based on the fibre alignment and orientation with the matrix, these polymer composites can be developed into materials that can give excellent structural and mechanical properties and are explicitly custom-made for a specific usage. Nowadays, concretes which are made of polymers are extensively utilized in the civil engineering to build structures. These polymer concrete structures are capable of resisting themselves in an extremely corrosive environment. Due to its extensively good strength-to-weight ratio and superior non-corrosive features of these reinforced polymer composites, they can be used to construct high-performance structures that are desirable and also cost-effective [3].

The main objective is to make a component that is strong and stiff and has a low density. A maximum percentage of commercial composites are prepared by using glass and carbon fibres as reinforcement materials in combination with thermoset resins like epoxy or polyester. Sometimes, thermoplastic polymers are also preferred as they remoulded after initial production. Also, there

```
                        Fibre reinforced composite
                                    |
          ┌─────────────────────────┴─────────────────────────┐
   Single layer composites                            Multilayered Composites
  (including composites having same                        (angle ply)
   orientation & properties in each layer)                     |
          |                                    ┌────────────────┴────────────────┐
   ┌──────┴──────┐                         Laminates                        Hybrids
Continous fibre reinforced    Discontinous fibre reinforced
     composites                        composites
        |                                    |
   ┌────┴────┐                          ┌────┴────┐
Unidirectional  Bidirectional      Random        Preferred
reinforcement   reinforcement    orientation    orientation
              (woven reinforcements)
```

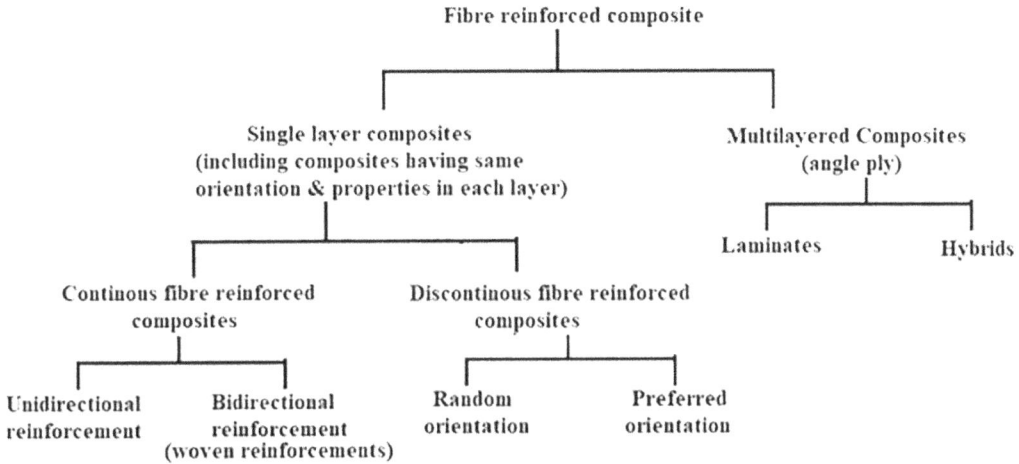

FIGURE 1.1 Classification of fibre-reinforced composites.

are other class of composites in which the matrix used is a metal or a ceramic. They are still in the developmental stage as the main problem to overcome is the high manufacturing costs. Fibre-reinforced composites have an extensive variety of applications in industries, like aircrafts, automobile industry, aerospace engineering, architectural structures, shipping industry, sports goods, and electrical products. Figure 1.1 shows the classification of reinforced composites.

1.3 TYPES OF POLYMER COMPOSITES

1.3.1 GREEN COMPOSITES

Currently, efforts are being made to develop a new class of complete biodegradable green composites. This is done by the combination of plant fibres with biodegradable natural resins. The most attractive part of the green composites is that they are environmentally friendly, completely biodegradable, and sustainable in nature. These show that they are truly 'green' in each and every aspect. These green composites can be fully disposed of or composted at the end of their life without damaging the environment. Figure 1.2 displays the classification of biodegradable polymers in four families. Apart from the fourth family which is naturally generated from the fossil origin, other polymers (from family 1 to 3) are attained from the renewable sources, that is, from biomass. The first family of biodegradable polymers consists of agriculturally based polymers (i.e., polysaccharides) which are obtained from biomass, whereas the second and third families consist of semisynthetic fibres usually obtained particularly by biofermentation method of agricultural residue or obtained via biologically modification of plants for example poly-hydroxyalkanoate PHA, and by chemical synthesis of monomers attained from biowaste for example polylactic acid (PLA). And lastly, the fourth family is of polyesters which are entirely synthesized from the petrochemical source for example polyester amide (PEA), polycaprolactone (PCL), and aliphatic or aromatic copolyesters. Today, a huge quantity of biopolymers are readily available in the market. These biodegradable polymers have a wide range of varieties and properties and can therefore have the potential to compete with synthetic polymers in various industrial fields, for example, packaging [4].

The next important category of biocomposites is created from agriculturally based polymer matrices which largely focus on starch-based materials. Plasticized starch, which is called as thermoplastic starch (TPS), is attained due to the disruption and plasticization process of natural starch, along with water and an addition of a plasticizer, for example, polyol. This is usually attained by applying external thermal and mechanical energies in a nonstop extrusion process. The major

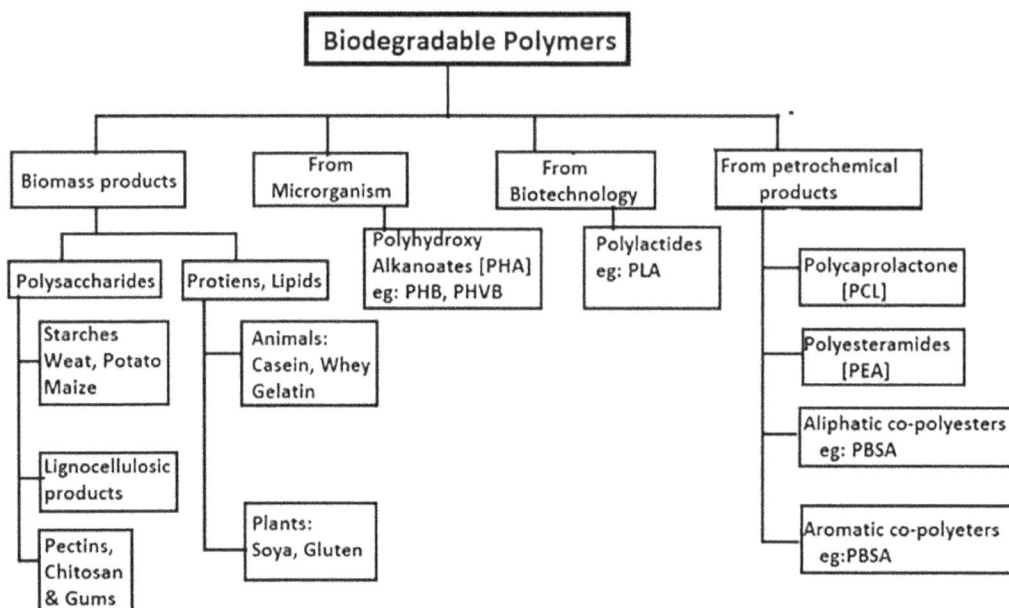

FIGURE 1.2 Classification of biodegradable polymers.

drawback of the TPS is that it possesses good hydrophilic behaviour and poor mechanical properties when compared with conventional polymers. The overall properties of TPS usually attain equilibrium after some weeks. Therefore, to overcome these drawbacks, TPS is frequently associated by other chemical components.

In today's current scenario, green composites have been efficiently utilized and applied in a wide range of mass-produced consumer products which have short life cycle, or applied in 'use and throw' products, or interim use products before disposal.

Polylactic acid polymers are very crystalline in nature with a high melting point. It is a hydrophobic polymer as it possesses CH_3 side groups. Therefore, polylactic acid is used in combination with plant fibres such as jute, hemp, and kenaf to produce biodegradable green composites. Also, innovative biodegradable polymer films were prepared from polylactic acid and chitosan. These composite films presented boundless benefit in avoiding the superficial growth of mycotoxinogen stains due to its antifungal activity. But due to poor physicochemical properties, it dramatically hinders its further application in packaging materials. Green composites, which are composed of regenerated cellulose (lyocell) fabric, recyclable polyesters (poly-(butylene succinate) (PBS)-, poly (3- hydroxybutyrate-co-3-hydroxyvarelate) (PHBV)-, and poly (lactic acid) (PLA)-based composites), have shown good biodegradability, which is the most important need towards sustainability. Apart from these materials, vegetable oil-based resins and starch are also utilized as a matrix to prepare green composites [5].

Soy protein has also been used in the preparation of biodegradable green composites. Soybeans provide more than 60% of the oils and fats utilized in food and majority of protein feed. Soybeans generally consist of 20% oil and 40%–50% protein. They comprise distinct groups of proteins (polypeptides) that differ in molecular dimensions and are embraced with almost 37% of nonpolar and non-reactive amino acids deposits, whereas 58% consist of polar and reactive residues. Therefore, modifications are implemented in order to improve the water solubility and reactivity of the soy protein which can be further utilized in preparation of plastic materials and biomaterials. Soy protein plastics are injection-moulded in different compositions. Soy plastic green composites are prepared on compression moulding machine. These soy plastics are reinforced with natural fibres and can be applied in various engineering fields.

Similarly, biofibre natural rubber composites are also manufactured extensively. The mechanical and structural properties of biofibre-reinforced rubber composites comprise improved strength along with better bonding at higher fibre loading percentage, improved modulus, diminished elongation at break, significantly enhanced creep resistance over particulate-filled rubber, enhanced hardness, and significant enhancement in cut resistance, and it also improves tear resistance and simultaneously puncture resistance. Biodegradation of vulcanized rubber is achievable, even though it is tough as a result of inter-associations of the poly(*cis*-1,4-isoprene) chains, which in turn result in decreased water retention rate as well as decreased gas permeability of the composite material. Natural rubber has been extensively reinforced with coir fibre, sisal fibre, oil palm fibre, kenaf fibre, isora fibre, bagasse fibre, jute fibre, and so on.

The main obstacle in the commercialization of green composites is the high price of biodegradable matrices. Biodegradable resins are very costly when compared to the commonly available resins. Low-cost production of oils for resin production by biotechnology process could be surely of great help in advancing with their commercialization. The production costs can be further minimized by fast, improved, and more effective processing [5].

1.3.2 HYBRID COMPOSITES

Hybrid composites are comprised of various fibres into a single matrix. The overall behaviour of the hybrid composites depends upon the addition of the individual fibres which have distinct properties, and this results in further satisfactory equilibrium amongst the advantages and disadvantages. The importance of utilizing a hybrid composite is that it consists of more than one type of fibres, as this helps either of the fibre to complement each other, which lacks essential properties. As a result, equilibrium in price and performance via appropriate material design could be achieved. The overall properties of the hybrid composites are mostly dependent on the fibre content, type of fibre, amount of fibres combined together, fibre orientation, fibre to matrix bonding, and alignment of the fibres. Also, in hybrid composites, the overall strength is reliant on the failure strain of the individual fibres. Utmost hybrid properties are attained when the individual fibres are in highly stressed state. A positive or a negative hybrid effect is defined as a deviation of a certain property such as structural or mechanical from the rule of hybrid mixture.

The hybrid effect was utilized to depict the phenomenon of an obvious synergistic enhancement in the physiochemical properties of a composite comprising of at least two kinds of fibre. The choice process of the fibre type that is selected for the fabrication of hybrid composite isn't totally set by the reason behind hybridization, prerequisites forced on the material, or the plan being created and arranged for the particular composite. The issue of picking the sort of feasible fibres and their properties is of major significance while planning, designing, and creating hybrid composites. The fruitful utilization of hybrid composites is not at all self-contented entirely by the mechanical, physical, and chemical stability properties of the fibre/matrix system.

Generally, hybrid biocomposites are created by blending of a natural fibre with synthetic fibres in a single matrix. Usually, hybridization performed by utilizing glass fibre gives a strategy to improvise mechanical properties of the natural fibre composites. The results of this method stress hinge on the planning, designing, and lastly development of the composite material. Fabrication of cellular biocomposite cores from flax or hemp fibres with unsaturated polyester was hybridized with jute, glass, and carbon fabrics. Characterization of the composite material showed further developed strength, moisture absorption stability, and good stiffness, whereas flexural test results in laboratory-scale plates exhibited upgraded mechanical behaviour. Therefore, hybrid biofibre-based composites were found to deliver an economic and ecologically friendly substitute to synthetically engineered composites [5].

1.4 APPLICATION OF POLYMER COMPOSITES

Fibre-reinforced composites have been utilized in almost all fields of engineering. Stalks of sunflower and poplar wood mixed in different ratios of urea formaldehyde adhesive have been used to manufacture three-layered particle boards. Also, cellulose fibres have been fabricated with soy oil-based resin to prepare sheets which have been used to fabricate roof structures. It was found that the composite sheets provided good stiffness and strength as per the required roof construction standard. Also, flame-retardant composites have been prepared by loading phosphorous in polypropylene (PP) or polyurethane (PUR) composites comprising waste biofillers and recycled polyol. Both filler and matrix proved to be very effective and participated in the mechanism of flame retardancy. Epoxidized soy bean oil along with hollow keratin fibres has been fabricated to prepare low dielectric constant material which are utilized in the application of electronic appliances. These low-cost vegetable oils have the potential to substitute the dielectrics used in circuit boards and microchips in the field of constantly increasing electronics materials. Apart from this, plastic/wood composites have been extensively used in the manufacturing of docks, decks, window frames, and moulded panel components. Statistics show that around 460 million pounds of wood/plastic composites were manufactured in 1999 which drastically increased to 750 million pounds till 2002. Also, in civil construction industry, bamboo fibre has been utilized as a reinforcement in the structural concrete elements. Biocomposites have shown a tremendous amount of applications in the furniture market, even though biofibre in construction industry has been known long before. Currently, the utilization of natural fibres as a reinforcement material in technical applications is taking place mostly in the packaging and automobile sectors [5].

1.4.1 Nanotechnology

Dimensional stability plays a very significant role in defining the properties of material. The nanostructure of a composite material has always been the significant aspect considering the improvement of novel properties and in monitoring the structure at the nanolevel. Nanotechnology is very profoundly encouraging field of today's world, and would be considered to thoroughly rebuild the technological applications in the various fields of organic and inorganic materials, semiconductors, biotechnology, and energy storage [6].

The term nanotechnology assigns to the building and comprehending of nanoscaled materials and gadgets. It can be defined as 'Study and preparation of particles having the size in range of 1–100 nm' [7]. Nanotechnology is useful in development of innovative products having unique features in the various fields such as chemistry, pharmacy, and textile [8].

1.4.2 Nanomaterials

1.4.2.1 Classification of Nanomaterials

Nanostructures are grouped based on their dimensions. Figure 1.3 shows different nanomaterials based on their dimensions [9].

- Zero-dimensional (0D) nanomaterial
  These nanomaterials are characterized by nanodimensions in all the three directions. Metallic nanoparticles and quantum dots are the examples of this type of nanostructures.
- One-dimensional (Quasi 1D) nanomaterial
  These kinds of nanomaterials have one dimension outside the nanometre range. Nanotubes and nanorods are the examples of this type of nanomaterial.
- Two-dimensional (2D) nanomaterial
  This kind of nanomaterials has two dimensions outside the nanometre range. Nanocoatings and nanosheets belong to this type.
- Three-dimensional (3D) nanomaterial
  If the materials have either an overall size in the millimetre or micrometre range, so far exhibit nanometric features, or prepared using the nanosized building blocks, they can be termed as '3D nanomaterials' [10].

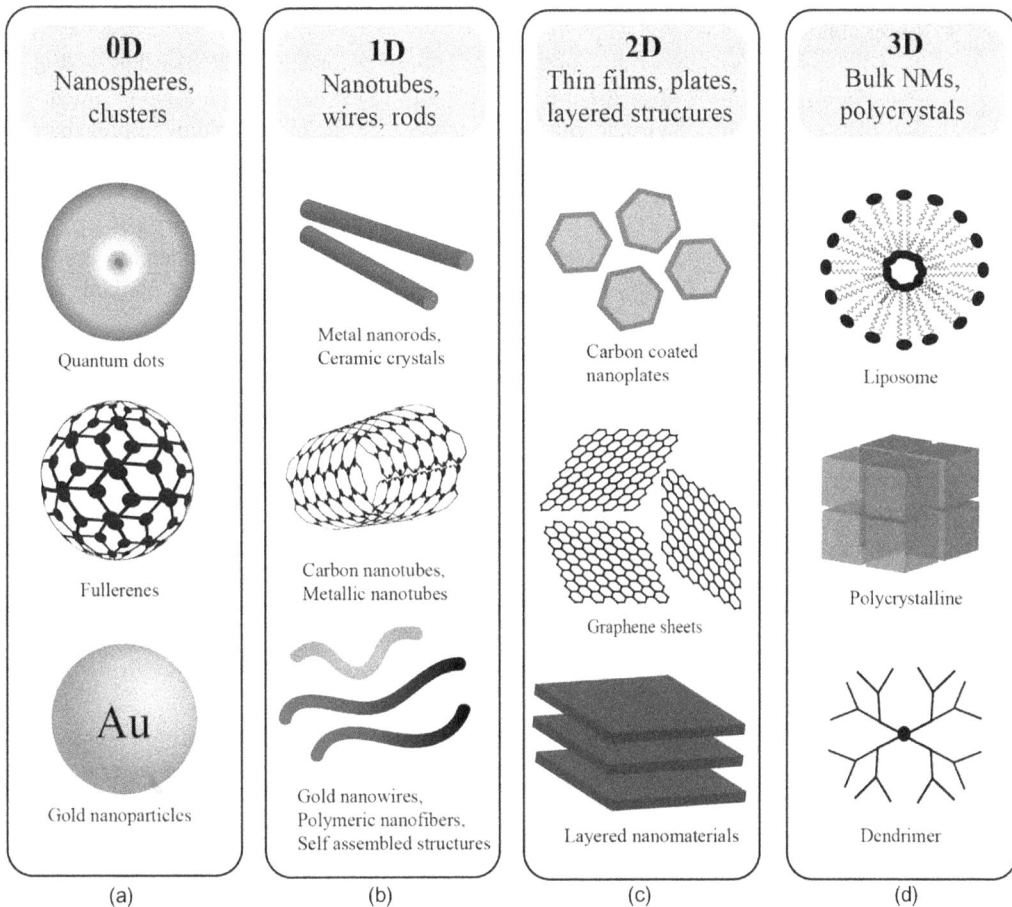

FIGURE 1.3 Classification of nanomaterials: (a) zero-dimensional, (b) one-dimensional, (c) two-dimensional, and (d) three-dimensional.

1.4.3 SYNTHESIS OF NANOPARTICLES

Nanoparticles can be synthesized using three different methods, namely, physical, chemical, and biological, and are described below:

- **Physical synthesis**
  - Evaporation–condensation

    In this method, the synthesis of nanoparticles is done by utilizing a tube furnace at atmospheric pressure. Precursor is allowed to vaporize into carrier gas, and subsequent condensation takes place in furnace to form the nanoparticles [11].
  - Mechanical milling

    In this technique, synthesis of nanoparticles is done in solid state with the help of ball milling apparatus. Aim of this process technique is to decrease particle size and mixing the particles in different phases.
  - Mechanochemical synthesis

    In this technique, nanoparticles are synthesized by chemical reaction which is carried out using mechanical energy. During the synthesis, chemical precursors such as blends of chlorides, metals, or oxides react with each other by ball milling or thermal treatment process which produces nanoparticles dispersed in a matrix. Further, the matrix is removed using solvent washing in order to get purified nanoparticles [12].

- Laser ablation

    This method produces the nanoparticles having the narrow size distribution and thus widely used. It involves the use of laser pulse to raise the temperature of precursor material up to boiling point which in turn produces the plasma plume of precursor material. Further, plasma plume undergoes the condensation, and nanoparticles are obtained [13].

- **Chemical synthesis**

  Chemical synthesis methods are based on the reduction of metal ions under favourable conditions which results in formation of nanoparticles. Some of the chemical methods are mentioned below:

    - Chemical reduction

        It is a widely known and simple method to prepare metal nanoparticles. In this method, nanoparticles are synthesized by the reduction of metal precursor salts using a suitable reducing agent either at room temperature or at higher temperature. Sodium borohydride, hexamethylenetetramine, sodium citrate, etc. are used as reducing agents. To avoid agglomeration of nanoparticles in solution, stabilizers such as polyvinyl pyrrolidone and sodium dodecyl sulphate are used [14,15].

    - Microemulsion process

        Microemulsion is a clear solution composed of water, oil, surfactant, and cosurfactant. During the nanoparticle's synthesis, microemulsion helps to react water-soluble precursor with oil-soluble reducing agent by holding them together [16].

    - Hydrothermal/solvothermal

        In case of hydrothermal method, nanoparticles can be synthesized at high temperature and high pressure using water and a precursor which is insoluble in water under normal conditions. Solvothermal method is similar to hydrothermal method except that it involves the use of solvent instead of water, and reaction is carried out under supercritical conditions [17].

    - Sonochemical method

        In this method, synthesis of metal nanoparticles is carried out by applying strong ultrasonic radiations (having frequencies in the range of 20 kHz to 10 MHz) to the metal precursor solution. During the process, cavitation takes place which results in formation of nanoparticles. It is one of the simplest methods used in preparation of nanoparticles [9].

    - Sol–gel process

        Sol–gel method includes the development of inorganic systems via the generation of a colloidal suspension (sol) and gelation of a sol to form a diphasic system having both solid and liquid phases (gel). Nanocoating on fabrics is widely synthesized by sol–gel process. Metal alkoxides are used as a precursor for sol–gel synthesis of nanoparticles. Out of various metal alkoxides, silicon alkoxides [Si (OR)$_4$] are generally used. In a sol–gel process, precursor undergoes hydrolysis and condensation by reacting with water in the presence of a catalyst (mineral acid/base) as shown in Figure 1.4 [14,16].

- Biological methods

  Preparation of nanoparticles by a chemical method is found to be non-eco-friendly and costly. Hence, several researchers recommended the eco-friendly and economical approaches for nanoparticles synthesis. Nowadays, several plant extracts and microorganisms have been utilized as substitutes for synthetic reducing agents as well as stabilizers for nanoparticles synthesis [17].

$$RO-\underset{\underset{OR}{|}}{\overset{\overset{OR}{|}}{Si}}-OR + H_2O \xrightarrow{\text{Hydrolysis}} RO-\underset{\underset{OR}{|}}{\overset{\overset{OR}{|}}{Si}}-OH + HOR$$

$$2\ RO-\underset{\underset{OR}{|}}{\overset{\overset{OR}{|}}{Si}}-OH \xrightarrow{\text{condensation}} RO-\underset{\underset{OR}{|}}{\overset{\overset{OR}{|}}{Si}}-O-\underset{\underset{OR}{|}}{\overset{\overset{OR}{|}}{Si}}-OR + H_2O$$

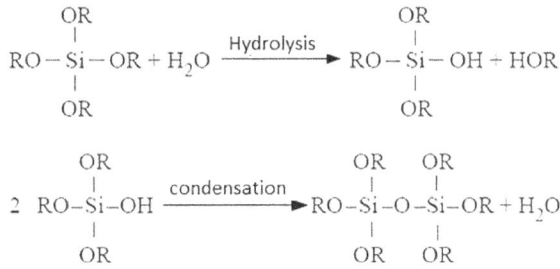

FIGURE 1.4 Mechanism of sol–gel process.

1.5 POLYMER NANOCOMPOSITES

Polymer nanocomposites are defined as materials in which nanoscopic inorganic particles, at least in one dimension, are distributed in the organic polymer matrix which improves its overall properties such as chemical and mechanical. PNCs address another option in contrast to conventionally filled polymers. In view of their size, the dispersion of fillers in the nanocomposites displays extraordinarily enhanced properties with respect to other polymers or their conventional composites. The properties which are enhanced due to these filler dispersions are tensile strength, flexural strength, conductivity, and thermal properties as well as decreased flammability [18].

1.5.1 SYNTHESIS OF POLYMER NANOCOMPOSITES

Nanocomposites usually consist of two sections: the first phase is a continuous process, and the second phase consists of a discontinuous process. PNCs comprise a polymer matrix which acts as a continuous phase, which has a size under 100 nm on the nanoscale reinforcing itself in dispersed phase. Therefore, it is critical to take into consideration that nanoparticles that are added into the continuous phase not only act as a reinforcement but are also responsible for modifying more than one property of the continuous phase. Contingent upon the type of nanomaterials utilized, other performance characteristics result [19].

Polymers that are used in the nanocomposite fabrication can be classified as follows: Naturally occurring biopolymers are usually obtained from proteins and carbohydrates. Protein-based polymers consist of casein, soya protein, corn zein, wheat gluten, gelatine whey protein, collagen, whereas carbohydrate-based polymers consist of cellulose, agar, chitosan, starch, alginate, and carrageenan. Also, chemically synthesized biopolymers are poly (l-lactide) (PLA), poly (glycolic acid) (PGA), poly(e-caprolactone) (PCL), poly (vinyl alcohol) (PVOH), poly (butylene succinate), and more. Also, microbial polyesters are (3-hydroxybutyrate-co-3-hydroxyvalerate) (PHBV), poly (ß-hydroxybutyrate) (PHB), poly (hydroxyalkanoates) (PHAS), and more. Apart from biodegradable polymers, non-biodegradable polymers like polyamide, nylon, polyolefins, polyethylene terephthalate, polyurethane, etc. are also utilized in nanocomposite fabrications [20].

1.5.2 CLASSIFICATION OF POLYMER NANOCOMPOSITES (PNCS)

PNCs are for the most part classified into two groups: intercalated or exfoliated. This classification usually depends on the condition of diffusion of the nanoparticles in the polymer matrix. The nanomaterials in the PNCs keep up with their unique crystallographic structures; however, the distance between the interlayers, sheets, or planes is more when compared to original form, as the nanomaterials are intercalated by the polymer chains and because of which a sandwich-like structure is created.

After the nanomaterials lost their unique design structures, the hap-hazard or disorder prearrangement originates in the polymer matrix, hence labelled as exfoliated PNCs. Once more, contingent upon the dimension of nanomaterials integrated into the polymer matrix, they are classified into three groups, namely, zero-dimensional, one-dimensional, and two-dimensional PNCs [21–23].

Polymer composites are also categorized into thermoplastic and thermosetting polymers nanocomposites which are based on the thermal properties and the behaviour of the polymer in the nanocomposites. Polymers which acquire shapes on being remoulded after heating and cooling are called as thermoplastic polymers. These polymers can be cured repetitively and reversibly time after time again. This repetitive heating and remoulding process is possible as there is no covalent bonding involved, whereas thermosetting polymers are those polymers which behave exactly the opposite of thermoplastics. Once they are heated, moulded, and cooled, they cannot be remoulded time after time again under any circumstances. They possess a three-dimensional covalent-bonded structure which makes the remoulding process difficult [24].

Classification of PNCs is based upon the class of polymers utilized, which is usually constructed on the key linkages existing in the constructions alike virgin polymers. These usually consist of different types of PNCs such as polyester, polycarbonate, polyurethane, polyamide, olefinic, silicone, polyacrylate, diene, and also cellulosic nanocomposites.

Further, based on the structure of the polymer matrices, these can be further classified into dendritic PNCs and linear polymer-based nanocomposites, particularly dendrimer-based and hyperbranched PNCs. Also, hyper-branched polymers are described via polymer structure that comprises numerous divided ends connected to their molecular structures by moderately short chains whose patterns are uneven, whereas the single molecules usually possess different molar mass and divarication degrees. Dendrimers are monodispersing macromolecules that comprehend symmetric branching units assembled around a small molecule or a linear polymer core [25].

1.6 PREPARATION OF POLYMER NANOCOMPOSITES

A high-performance PNC is developed through thorough design, formulation technique, proper material selection, and composite manufacture method. The most important part is the selection of a polymer matrix along with the filler material. Also, the best process of production should be determined based on the end-usage requirement, and lastly nanocomposite is prepared. The designer faces a real challenge during the application process as he has to choose the appropriate polymers and filler materials from an enormous variety of components available today. The designer has to choose a polymer and a filler which are compatible with each other and their adhesion and ease of incorporating filler into the polymer along with proper dispersion, which will make the design of the PNC a rather complex procedure. In order to formulate a desired PNC, the following factors have to be taken into consideration: (a) Its end application, (b) Functionality, (c) Required properties, (d) Environmental necessities, and (e) Economical and lastly processing parameters. These situations are responsible in constraining the design of PNCs which is based on rationality, tailored properties, and functionality. Some of the nanocomposite fabrication processes are as follows: (a) solution mixing, (b) in situ polymerization, (c) melt intercalation, and (d) template synthesis [26]. Fabrication process of nanomaterial-based nanocomposite is shown in Figure 1.5.

1.6.1 SOLUTION MIXING

Solution mixing is a technique in which a homogeneously mixed polymer solution along with a perfectly dispersed nanomaterial is designed by the process of mechanical stirring along with ultra-sonication. In this method, the nanomaterial is subjected to swelling followed by dispersion in appropriate solution by appropriate homogenization process. Simultaneously, polymer solution which is formed separately in the solvent is miscible with the nanomaterial dispersed solution. Further, removal of unwanted liquid or solvent from the homogeneous medium mixture is done

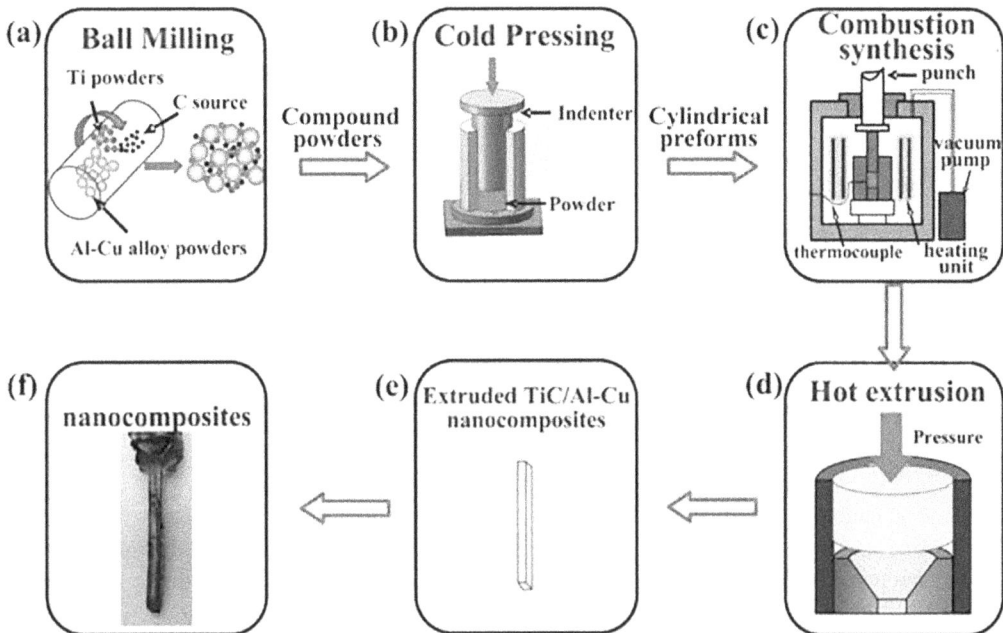

FIGURE 1.5 Fabrication process of nanomaterials and nanocomposites.

by precipitation or evaporation of the nanomaterial which is dispersed in the polymer matrix. The process of mixing between the nanomaterials and polymer chains usually relies on the thermodynamics of blending and the interaction amongst the components of the subsequent nanocomposite. Homogenous diffusion of nanomaterials in the polymer matrix is attained when interactions between nanomaterial and polymer chains are greater than interactions between nanomaterial and polymer solvent. Strong interactions of polymer nanomaterial indeed help the polymer chains enter the structure of the nanomaterials, whereas the chain molecules are adsorbed on the superficial area of nanomaterial, thus creating the anticipated nanocomposite. Also, intercalated and exfoliated nanocomposites are prepared by this process, dependent on the way of interactions and distribution in between the components. Agitation can be attained by shear mixing, magnetic stirring, reflux, or the most commonly used method of sonication [27–28].

1.6.2 MELT INTERCALATION

Melt intercalation is the common standard methodology for synthesizing thermoplastic-based PNCs. It includes strengthening the polymer matrix at high temperatures, addition of the filler, and lastly blending the composite mixture to accomplish uniform conveyance. It enjoys the benefit of being harmless to the ecosystem due to the absence of solvent usage. Furthermore, it is thought of as viable with modern industrial processes, for example, injection moulding and extrusion which makes it more advantageous to use and, hence, more conservative and economical. Notwithstanding, the high temperatures utilized in the process can harm the surface alteration of the filler [29].

1.6.3 IN SITU POLYMERIZATION TECHNIQUE

This is the very common technique for polymer synthesis to produce nanocomposites. This method generally contains the blending of monomer and nanoparticles in a compatible solvent, just as it is said that monomers are intercalated with nanoparticles, further followed by polymerization with a compatible reagent or free radical initiator to produce PNC. The metal precursor is utilized to change over nanoparticle to produce metal or metal oxide particles inside the polymer matrix. This process assists to synthesize multidimensional complex distinct structures, with totally different properties from the original precursors [30].

1.6.4 TEMPLATE SYNTHESIS

In this technique, a template is utilized to make nanocomposite materials of specific shape for instance layered, hexagonal shape, and so on. The water-dissolvable polymer acts as a template for the development of layers. This process is generally utilized for the synthesis of mesoporous materials, however, less produced for the development of layered silicates [31,32].

1.7 COMMONLY USED NANOFILLERS IN POLYMER NANOCOMPOSITES

1.7.1 CARBON-BASED NANOFILLERS

- Carbon nanotubes
  Carbon nanotubes (CNTs) are generally ultra-thin carbon fibres having measurements in nanometre and micrometre. The design of CNT comprises enlisted graphite sheet, in which the carbon molecules are arranged and conveyed in a honeycomb lattice in planar hexagonal. Notwithstanding the remarkable electrical and conductive properties, CNTs are additionally associated with phenomenal mechanical properties. The integration of CNTs in the polymer matrix has been considered as a technique to get nanocomposite materials with electrical properties and unrivalled thermomechanical properties.

1.7.2 GRAPHENE

Graphene has ignited gigantic attention in various research facilities all over the world because of its uncommon electrical, physical, and excellent chemical properties. Graphene can be created from graphite by various techniques, like thermal expansion of chemically intercalated graphite, micro-mechanical exfoliation of graphite, chemical vapour deposition, and chemical reduction technique for graphene oxide.

1.7.3 LAYERED NANOCLAYS

Nanoclays have a place with a different class of materials largely manufactured by using layered silicates or clay minerals along with fractions of metal oxides and natural organic matter. Generally, clay minerals are hydrous aluminium phyllosilicates, here and there with different measures of magnesium, iron, alkali metals, alkaline earth metals, and others cations. Examples of some of the clays are saponite, montmorillonite, laponite, sepiolite, vermiculite, and hectorite. From these clays, montmorillonite is broadly utilized in PNCs, as a result of its enormous accessibility, notable intercalation/exfoliation chemistry, large superficial area, and reactivity.

There are numerous works published which have displayed the distribution of exfoliated clays in polymer, which leads to incredible upsurge in toughness, flame-retardant properties, and barrier properties [33].

1.7.4 NANO-METALS AND NANO-METAL OXIDES

Insertion of metal nanoparticles into polymer lattices addresses a basic method for utilizing the upsides of nanoparticles. Metal nanoparticles ordinarily work on the mechanical properties of polymers, because of the intrinsic properties of nanosized metals like enormous surface region and high

TABLE 1.1

Effect of Nanomaterials on Nanocomposites

Nanomaterials	Effect on Nanocomposites
Silicon dioxide (SiO_2), aluminium oxide (Al_2O_3), or titanium dioxide (TiO_2)	Enhances protection from mechanical impacts and to decrease wear
Titanium dioxide (TiO_2)	UV protection for plastics
Magnesium hydroxide ($Mg(OH)_2$), Aluminium hydroxide ($Al(OH)_3$), or Antimony oxide (Sb_2O_3)	Fire resistance
Nano-silver	Antimicrobial properties

modulus. Likewise, the development of a solid interfacial/interphase region between polymer lattice and nanoparticles facilitates the accomplishment of huge mechanical properties. By and large, the thermal properties of PNCs upsurge, in light of the fact that the nanoparticles act as barrier to volatile compounds. Numerous polymers are electrically insulated; however, the conductive metal nanoparticles give an adequate conductivity in polymer/metal nanocomposites. The conductive nanocomposites can be used in different applications like rechargeable battery, shield, electrode, and sensor [34].

Silicon dioxide (SiO_2), aluminium oxide (Al_2O_3), and titanium dioxide (TiO_2) are the most often utilized nano-oxides in polymer composites, chiefly to enhance protection from mechanical impacts and to decrease wear. Heat resistance can likewise be enhanced by utilizing nano-oxides. TiO_2 nanoparticles can likewise act as UV protection for plastics. FRs should be added to plastics to give fire resistance. In recent years, the utilization of nanoscale FRs, for example, ultrafine magnesium hydroxide ($Mg(OH)_2$), aluminium hydroxide ($Al(OH)_3$), or antimony oxide (Sb_2O_3), has been explored. Nano-silver has antimicrobial properties and can be utilized in plastics, for instance to produce food-packaging films, or containers to safeguard food from getting spoiled [35]. Table 1.1 shows the effect of nanomaterial on nanocomposite.

1.8 PROPERTIES OF POLYMER NANOCOMPOSITES

Polymer nanocomposites have uncovered obviously the benefits that nanomaterial-added substances can give in contrast with both their conventional fillers and base polymer.

Polymer nanocomposites have the following advantages [33,36–37]:

- They are light in weight than traditional composites since high levels of degree of strength and stiffness are acknowledged with undeniably less high-density material.
- Their thermal and mechanical properties are greater.
- Their barrier properties are enhanced when compared with neat polymer.
- Optical clarity improved in contrast to conventionally prepared polymers.
- Enhanced flame retardancy and reduced smoke emissions.
- Improved electrical conductivity.
- Improved chemical resistance.

1.9 APPLICATIONS OF POLYMER NANOCOMPOSITES

1.9.1 FOOD PACKAGING

Food packaging acts a major job to keep up with the sanitation and food quality at the time of stockpiling and while transportation, and to delay the period of food items utilized, by safeguarding it from microbial contamination, and physicochemical and environmental risks. Paper, paperboard, plastic, glass, and metal materials that are primarily involved in packaging of different food varieties are paperboards, paper, glass, plastic, and metal. Likewise, a blend of these materials can be utilized to successfully satisfy the necessary functions more efficiently [38].

In this way, food-packaging materials based on PNCs should give protection from solutes, moistures, water vapours, and gases, as well as acting as transporters of active chemicals which show explicit properties as packaging materials pointed towards improving overall quality of the food and drawing out time span of usability by hindering the improvement of fungus and microorganisms in the food products [20].

1.9.2 CURRENT ADVANCES IN THE PACKAGING FIELD

   I. Enhanced packaging process which proposes enhanced mechanical properties, upgraded hindering properties against gases, water, toughness, temperature, and moisture dependability, etc.
   II. Active and bioactive food-packaging materials usually give antioxidant and antimicrobial properties. This can be achieved with the assimilation of active or bioactive mixtures into matrix utilized in current packaging materials, otherwise via the use of coatings with the referenced functionality over physiochemical modifications on the surface. Application of active packaging is mostly implemented in the pharmaceutical, food, and consumer goods industries in order to further enhance and develop safety, and increase shelf life and quality of packaging food varieties. Coating is beneficial due to the fact that the physiochemical properties of the packaging materials are protected practically flawlessly by utilizing a small quantity of active agent which helps to confer viability and subsequently cost is also less.
   III. Utilization of smart and intelligent packaging materials is a very promising research area in active packaging. This is generally developed by the utilization of nano-biosensors which can specify superiority of food products, of nano (bio), shift to release preservatives, and nanocoatings as antioxidant, antifungal, antimicrobial, barrier coatings, external stimuli reactive materials, and self-cleaning food contact surfaces. Smart inks such as nanoparticles and reactive nanolayers permit appreciation at nanoscale. Printed tags are used to show time, temperature, freshness, humidity, pathogen, and integrity. Clever packaging might display several constraints, for instance oxygen, temperature, moisture, and pH of packaged foodstuffs [39].

The consolidation of antimicrobial mixtures into food-packaging materials has received extensive consideration, as it could assist with controlling the development of microorganisms. It is especially attractive because of the adequate underlying trustworthiness and hindrance properties conferred by the polymeric matrix, and antimicrobial properties by compounds embedded inside the packaging materials. Because of the high surface-to-volume proportion, nanomaterials can create more duplicates of biological particles when contrasted with respect to their microscale counterparts, which prompts a more noteworthy effectiveness. Nanomaterials are adaptable materials that can be utilized in more than one way, like antimicrobials, antibiotic carriers, or growth inhibitors. The quintessential nanocomposites utilized as antimicrobials in food-packaging materials depend on silver, zinc oxide (ZnO), copper oxide (CuO), titanium dioxide (TiO_2), etc., which are notable for its strong killing properties for many microorganisms [40].

Clays and other silicate materials are extremely stable, apparently non-toxic in nature, and are always obtainable at very cheaply, which therefore make them very attractive as filler material. This is the reason that nanoclay-based PNCs are studied widely and applied in food packaging for the past two decades [41]. In Figure 1.6, a schematic of the gas molecule diffusion through (a) a polymer-only barrier and (b) a polymer composite barrier is shown. Unvaryingly nanoplates decrease the permeability by increasing the resistance through tortuosity.

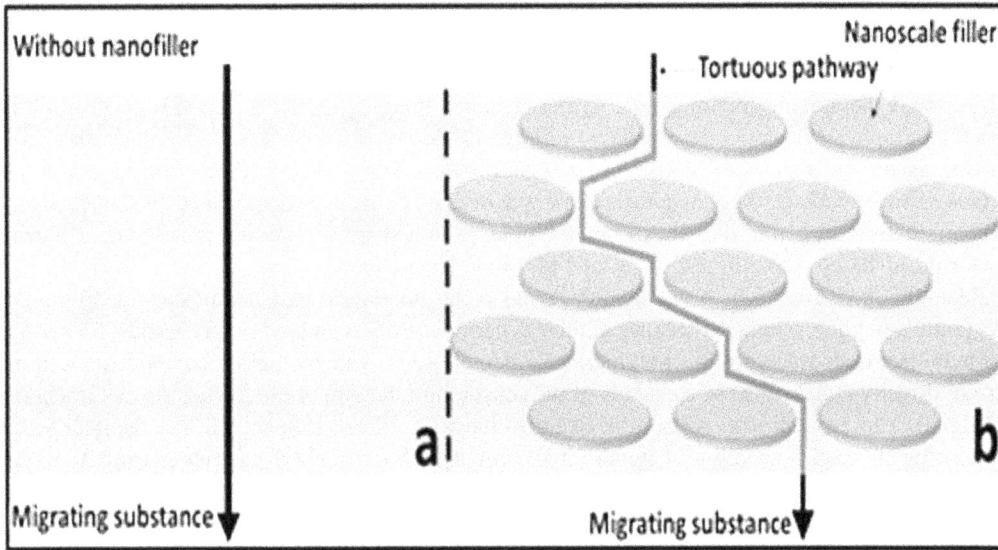

FIGURE 1.6 Schematic of gas molecule diffusion.

1.9.3 APPLICATION OF POLYMER NANOCOMPOSITES IN HEALTHCARE SECTOR

The field of nanomaterial is an important prospect in various fields related to drug delivery, tissue engineering, biosensors, wound healing, prosthetic materials, and in coating of medical implants and devices. Use of biomaterials is a technique which helps in taking part alone or together with a complex system simultaneously controlling the interaction with living system with respect to therapeutic and diagnostic procedure. Bone tissues are now engineered in such manner that they can be used with biocompatible scaffold alone, or can be used in combination along with different cells or bioactive parameters as replacement for bone grafting. Alternatively, as a replacement of traditional methods utilized for the treatment of fractured bone for prostheses and grafting, biodegradable and biocompatible polymers were developed as a 3D approach for bone generation in tissue engineering. The recent advances of such composite material have resulted in one of the turning points having great performance in architectural design for modern orthopaedic and prosthetic devices [42–44].

Nanocomposites have generally been explored as scaffolding materials on the grounds of that the polymers utilized as matrix are effectively processable in correlation with metals and ceramics, apart being biocompatible, biodegradable, and bioresource. Furthermore, concerning bone cells' cooperation with nanostructured materials, consideration of certain nanoparticles in polymers improves biological behaviour of the scaffolds. Because of their nanoscale dimensions, nanoparticles with bigger surface area bring about excellent mechanical behaviour, which, when combined with polymer, are fit for mirroring the properties possessed by the natural bone. For example, for repairing bone fractures, epoxide carbon fibre composites are used [45,46].

Because of superior physical, chemical, and biocompatible properties, the utilization of PNCs has been acquainted to secure the maximum healing properties with minimum side effects. The significant job of nanotechnology, particularly nanocomposites, has been perceived in drug delivery systems. Thus, segregation of these types of drug delivery methods should be active on the target site without any adversative consequences on human body tissues, organs, or cells. PNCs have been testified to change the analysis and treatment methodologies of assorted diseases through designated and precise discharge of drug. For example, nanocomposites prepared with polylactic acid and calcium phosphate nanoparticles are used in tissue engineering. Also, nanocomposite prepared from polyethylene glycol along with nano-graphene oxide is used in drug delivery; similarly, chitosan along with nano-graphene oxide is used in tissue engineering and biosensing medical instruments.

Lately, polymeric nanocomposites have arisen in the area of bioimaging as an innovative class of fluorescent probes. PNCs offer noteworthy enhancements in the area of biodistribution and pharmacokinetics of bioimaging probes. These nanocomposites have been created as a differentiation agent for assorted bioimaging techniques (e.g., fluorescence imaging, MRI, X-ray, CT positron emission tomography). Magnetic nanoparticle-reinforced PNCs have been accounted as successful contrasting agents for improving MRIs of tissues and organs. Point-of-care (POC) testing guarantees improved and speedy medical diagnosis choices at or close by the site of the event. POC gadgets are fundamentally micro- and biosensors, for example, (a) stat strips for glucose monitoring, (b) urine test strips, (c) blood troponin test strips, and (d) blood lactate meters.

Recently, biosensors which are manufactured using polymeric nanocomposites have attained substantial attention primarily because of their synergistic effects and hybrid properties. Biosensors are manufactured by compiling polymeric nanocomposites with biological components and are further directly attached to the electrode of the sensor with the aim of modifying the electrochemical signal. The growth of biosensors is likely to have significant influence in the medical sector concerning the rapid detection of human and animal pathogens, food product examination, and environmental monitoring. Mostly, nanomaterials like quantum dots, gold nanoparticles, and nanographene oxide-based nanocomposites are used in biosensing materials [47,48].

1.9.4 Aerospace Applications

Nanocomposites are also applied in aerospace applications. The main requirement is to achieve lightweight structure, but apart from this property, high performance of the nanocomposite materials plays a very important role in the aerospace structures, for example, aircraft interiors, cockpit, equipment enclosures, coatings, crew gear, nozzles, space durable mirrors, and solar array substrates. Nanocomposites that are generally used in aerospace are nano-carbon–carbon composites which act as base materials for space shuttles, brake lining, and brake disc material for civil and military aircraft. Also, zirconia-based nanocomposites are used in thermal protection of turbo engines. These nanocomposite materials often offer fire resistance and chemical stability apart from the benefit of low operating cost because of their light weight. Generally, aerospace structures are subjected to extremely diverse environments that embrace differences in temperature and moisture. These materials are always exposed to flammable liquids such as jet fuel, hydraulic fluid, and de-icing fluid. Therefore, nanomaterial coating should be able to resist ultraviolet exposure due to sun, lightning strikes, and erosion [44,45].

1.9.5 Automobile Application

Nanocomposites play a major role in the manufacturing of automobiles. A lot of wear and tear occurs in automobile parts. Therefore, nanocomposite is the ideal material used in manufacturing of automobile parts because of their high fracture resistance. These nanocomposites are utilized to provide enormous strength to the parts where high efficacy is required. In the United States, vehicles contain more nanocomposites, and the most commonly used are CNTs in nylon blends which act as a shield against the static electricity of fuel system. In 2001, Toyota company started using nanocomposites in bumpers of cars which made the overall car 60% lighter in weight and also additionally made the car two times more scratch-proof and dent-resistant. The most commonly used nanocomposites in the automotive field are carbon nanotubes (CNTs). Also, ceramic matrix nanocomposites (CMNC) and metal matrix nanocomposites (MMNC) are used in the automotive field [49].

1.10 CONCLUSION

This chapter have clearly demonstrated the fundamentals of polymer composites and their nanocomposites. The concept of polymer composites has now been of great significance because of the advance applications required in today's modern engineering world. Biofibres like jute, hemp, kenaf, flax, okra, sisal, etc. have now found their applications in a wide range of industries. The main industry where these polymer composites have found increasing usage is the automobile industry, predominantly in its interior applications.

The concept of nanocomposites has allowed technologically inspiring method for fabricating innovative and advanced polymer materials. This has permitted the fabrication of different PNCs displaying an extensive range of fascinating and exclusive properties including thermal, mechanical, electrical, and barrier properties. The particle size, shape, dispersion, orientation, and volume dispersion of nanofillers affect the properties of PNCs. These innovative classes of materials are not only exceptional with respect to their structure and properties, but they are also used in an extensive series of applications. Composites with improved properties are extensively utilized in structural, industrial, coating, thermoelectric, biomedical, and photovoltaic applications. The possibility of using natural resources and the fact of being environmentally friendly have also presented new opportunities for applications.

REFERENCES

1. Delides, C. G. "Everyday life applications of polymer nanocomposites." *Technological Educational Institute of Western Macedonia* (2016): 1–8.
2. Paul, D. R., and L. M. Robeson. "Polymer nanotechnology: Nanocomposites." *Polymer* 49, no. 15 (2008): 3187–3204.
3. Hariharan, A. B. A., and H. P. S. Abdul Khalil. "Lignocellulose-based hybrid bilayer laminate composite: Part I-Studies on tensile and impact behavior of oil palm fiber-glass fiber-reinforced epoxy resin." *Journal of Composite Materials* 39, no. 8 (2005): 663–684.
4. Khan, A., R. A. Khan, S. Salmieri, C. Le Tien, B. Riedl, J. Bouchard, G. Chauve, V. Tan, M. R. Kamal, and M. Lacroix. "Mechanical and barrier properties of nanocrystalline cellulose reinforced chitosan-based nanocomposite films." *Carbohydrate Polymers* 90, no. 4 (2012): 1601–1608.
5. Jadhav, N. C., and A. C. Jadhav. "Synthesis of acrylate epoxidized rice bran oil (AERBO) and its modification using styrene and Shellac to study its properties as a composite material." *Polymer Bulletin* 80, no. 5 (2022): 5023–5045.
6. Dubey, U., S. Kesarwani, and R. K. Verma. "Incorporation of graphene nanoplatelets/hydroxyapatite in PMMA bone cement for characterization and enhanced mechanical properties of biopolymer composites." *Journal of Thermoplastic Composite Materials* (2022): 08927057221086833.
7. Tolle R, Nunn P, Maynard T, Baxter D. Lloyd's of London report on nanotechnology: recent development, risks and opportunities. London, England, United Kingdom, Lloyd's of London Emerging Risk Team (2007).
8. Kathirvelu, S. "Nanotechnology applications in textiles." *Indian Journal of Science and Technology* 1, no. 5 (2008): 1–10.
9. Kalpana, V. N., and V. Devi Rajeswari. "A review on green synthesis, biomedical applications, and toxicity studies of ZnO NPs." *Bioinorganic Chemistry and Applications* 2018 (2018).
10. Suresh, S. "Semiconductor nanomaterials, methods and applications: A review." *Nanoscience and Nanotechnology* 3, no. 3 (2013): 62–74.
11. Abou El-Nour, K. M. M., A. Eftaiha, A. Al-Warthan, and R. A. Ammar. "Synthesis and applications of silver nanoparticles." *Arabian Journal of Chemistry* 3, no. 3 (2010): 135–140.
12. Umer, A., S. Naveed, N. Ramzan, and M. Shahid Rafique. "Selection of a suitable method for the synthesis of copper nanoparticles." *Nano* 7, no. 05 (2012): 1230005.
13. Kesarwani, S., and R. K. Verma. A critical review on synthesis, characterization and multifunctional applications of reduced graphene oxide (rGO)/Composites. *Nano* 16, no. 09 (2021): 2130008.
14. Guzmán, M. G., J. Dille, and S. Godet. "Synthesis of silver nanoparticles by chemical reduction method and their antibacterial activity." *International Journal of Chemical and Biomolecular Engineering* 2, no. 3 (2009): 104–111.
15. Swathy, B. "A review on metallic silver nanoparticles." *IOSR Journal of Pharmacy* 4, no. 7 (2014): 2250–3013.
16. Yu, C. H., K. Tam, and E. S. C. Tsang. "Chemical methods for preparation of nanoparticles in solution." *Handbook of Metal Physics* 5 (2008): 113–141.
17. Byrappa, K., and T. Adschiri. "Hydrothermal technology for nanotechnology." *Progress in Crystal Growth and Characterization of Materials* 53, no. 2 (2007): 117–166.
18. Ismail, W. N. W. Sol–gel technology for innovative fabric finishing—a review. *Journal of Sol-Gel Science and Technology*, 78, no. 3 (2016): 698–707.

19. Coêlho, R. A. L., H. Yamasaki, E. Perez, and L. B. de Carvalho Jr. "The use of polysiloxane/polyvinyl alcohol beads as solid phase in IgG anti-Toxocara canis detection using a recombinant antigen." *Memórias do Instituto Oswaldo Cruz* 98 (2003): 391–393.
20. Iravani, S., H. Korbekandi, S. V. Mirmohammadi, and B. Zolfaghari. "Synthesis of silver nanoparticles: Chemical, physical and biological methods." *Research in Pharmaceutical Sciences* 9, no. 6 (2014): 385.
21. Zhang, X., W. Fan, and T. Liu. "Fused deposition modeling 3D printing of polyamide-based composites and its applications." *Composites Communications* 21 (2020): 100413.
22. Bustamante-Torres, M., D. Romero-Fierro, B. Arcentales-Vera, S. Pardo, and E. Bucio. "Interaction between filler and polymeric matrix in nanocomposites: Magnetic approach and applications." *Polymers* 13, no. 17 (2021): 2998.
23. Idumah, C. I., A. Hassan, and D. E. Ihuoma. "Recently emerging trends in polymer nanocomposites packaging materials." *Polymer-Plastics Technology and Materials* 58, no. 10 (2019): 1054–1109.
24. Karak, N. "Fundamentals of nanomaterials and polymer nanocomposites." In Niranjan Karak (Ed.) *Nanomaterials and Polymer Nanocomposites*, pp. 1–45. Elsevier, Amsterdam, Netherlands, 2019.
25. Idumah, C. I., A. Hassan, and A. Chioma Affam. "A review of recent developments in flammability of polymer nanocomposites." *Reviews in Chemical Engineering* 31, no. 2 (2015): 149–177.
26. Youssef, A. M. "Polymer nanocomposites as a new trend for packaging applications." *Polymer-Plastics Technology and Engineering* 52, no. 7 (2013): 635–660.
27. Abbasi, E., S. F. Aval, A. Akbarzadeh, M. Milani, H. T. Nasrabadi, S. W. Joo, Y. Hanifehpour, K. Nejati-Koshki, and R. Pashaei-Asl. "Dendrimers: synthesis, applications, and properties." *Nanoscale Research Letters* 9, no. 1 (2014): 1–10.
28. Majka, T. M., and K. Pielichowski. "Functionalized clay-containing composites." In Krzysztof Pielichowski and Tomasz M. Majka (Eds.) *Polymer Composites with Functionalized Nanoparticles*, pp. 149–178. Elsevier, Amsterdam, Netherlands, 2019.
29. Fischer H. "Polymer nanocomposites: from fundamental research to specific applications." *Materials Science and Engineering: C* 23, no. 6–8 (2003):763–772.
30. Rane, A. V., K. Kanny, V. K. Abitha, and S. Thomas. "Methods for synthesis of nanoparticles and fabrication of nanocomposites." In *Synthesis of Inorganic Nanomaterials*, pp. 121–139. Woodhead publishing, 2018.
31. Fawaz, J., and V. Mittal. "Synthesis of polymer nanocomposites: Review of various techniques." In V. Mittal (Ed.), *Synthesis Techniques for Polymer Nanocomposites*, pp. 1–30. Wiley-VCH Verlag GmbH & Co. KGaA, Weinheim, 2014.
32. Shameem, M. M., S. M. Sasikanth, R. Annamalai, and R. Ganapathi Raman. "A brief review on polymer nanocomposites and its applications." *Materials Today: Proceedings* 45 (2021): 2536–2539.
33. Müller, K., E. Bugnicourt, M. Latorre, M. Jorda, Y. E. Sanz, J. M. Lagaron, O. Miesbauer et al. "Review on the processing and properties of polymer nanocomposites and nanocoatings and their applications in the packaging, automotive and solar energy fields." *Nanomaterials* 7, no. 4 (2017): 74.
34. Camargo, P. H. C., K. G. Satyanarayana, and F. Wypych. "Nanocomposites: Synthesis, structure, properties and new application opportunities." *Materials Research* 12 (2009): 1–39.
35. de Oliveira, A. Dantas, and C. A. G. Beatrice. "Polymer nanocomposites with different types of nanofiller." *Nanocomposites-Recent Evolutions* (2019): 103–104. IntechOpen, London. doi: 10.5772/intechopen.81329
36. Zare, Y., and I. Shabani. "Polymer/metal nanocomposites for biomedical applications." *Materials Science and Engineering: C* 60 (2016): 195–203.
37. Pavlicek, A., and F. Part. "Polymer Nanocomposites-Additives, properties, applications, environmental aspects (NanoTrust-Dossier No. 052en–February 2020."
38. Giannelis, E. P. "Polymer-layered silicate nanocomposites: Synthesis, properties and applications." *Applied Organometallic Chemistry* 12, no. 10–11 (1998): 675–680.
39. Tyagi, M., and D. Tyagi. "Polymer nanocomposites and their applications in electronics industry." *International Journal of Electrical and Electronics Engineering* 7, no. 6 (2014): 603–608.
40. Turan, D., G. Gunes, and A. Kilic. "Perspectives of bio-nanocomposites for food packaging applications." In M. Jawaid and S. Swain (Eds.), *Bionanocomposites for Packaging Applications*, pp. 1–32. Springer, Cham, 2018.
41. Vasile, C. "Polymeric nanocomposites and nanocoatings for food packaging: A review." *Materials* 11, no. 10 (2018): 1834.
42. Luna, J., and A. Vílchez. "Polymer nanocomposites for food packaging." In Rosa Busquets (Ed.), *Emerging Nanotechnologies in Food Science*, pp. 119–147. Elsevier, Amsterdam, Netherlands, 2017.

43. Sarfraz, J., T. Gulin-Sarfraz, J. Nilsen-Nygaard, and M. K. Pettersen. "Nanocomposites for food packaging applications: An overview." *Nanomaterials* 11, no. 1 (2020): 10.
44. Idumah, C. I. "Progress in polymer nanocomposites for bone regeneration and engineering." *Polymers and Polymer Composites* 29, no. 5 (2021): 509–527.
45. Kumar, S., M. Nehra, N. Dilbaghi, K. Tankeshwar, and K. H. Kim. "Recent advances and remaining challenges for polymeric nanocomposites in healthcare applications." *Progress in Polymer Science* 80 (2018): 1–38.
46. Krasno, S., and K. Swathi. "A review on types of nanocomposites and their applications." *International Journal of Advance Research, Ideas and Innovations in Technology* 4, no. 6 (2018): 235–236.
47. Hassan, T., A. Salam, A. Khan, S. U. Khan, H. Khanzada, M. Wasim, M. Q. Khan, and I. S. Kim. "Functional nanocomposites and their potential applications: A review." *Journal of Polymer Research* 28, no. 2 (2021): 1–22
48. Hassan, M., A. K. Deb, F. Qi, Y. Liu, J. Du, A. Fahy, Md. A. Ahsan, S. J. Parikh, and R. Naidu. "Magnetically separable mesoporous alginate polymer beads assist adequate removal of aqueous methylene blue over broad solution pH." *Journal of Cleaner Production* 319 (2021): 128694.
49. Wen, Y., J. Yuan, X. Ma, S. Wang, and Y. Liu. "Polymeric nanocomposite membranes for water treatment: A review." *Environmental Chemistry Letters* 17, no. 4 (2019): 1539–1551.

2 Nano-Engineered Polymer Matrix-Based Composites

Madhuparna Ray, Abhishek Verma, Abhijit Maiti, and Y. S. Negi

CONTENTS

2.1 INTRODUCTION

Polymer composites are reinforced materials where the polymer functions as a matrix resin in which the reinforcing fillers are incorporated, and it penetrates the bundles of reinforcing material and bonding to it. At least two components require to make a composite: the matrix or the continuous phase and the reinforcement, which can also be referred to as the dispersed phase. Metals, ceramics, and other polymers may be used in the composite either as a matrix or as a reinforcement. The matrix for polymer composites has traditionally been a mixture of thermosetting and thermoplastic resins. The various systems are carefully blended to create a hybrid material with more beneficial structural, mechanical, chemical, thermal, and other functional features that cannot be attained by any constituents alone [1]. Typically, the matrix is a less hard phase and more ductile. It shares a load with the dispersed phase and holds a dispersed phase with it. Uniform dispersion of reinforcing filler is desired to obtain superior properties in the final composite [2].

DOI: 10.1201/9781003343912-2

Blends and composites vary, so the two primary components in composites can still be identified as individual parts, whereas they cannot be in blends [3]. Interestingly, the most significant polymeric composites can be found in nature. Mammal connective tissues are among the most sophisticated polymer composites ever created, with collagen, a fibrous protein that serves reinforcement. It performs both functionalities for soft and rigid connective tissues. Wood is a natural composite of cellulose fibers embedded in a lignin matrix. For the construction of buildings, straw and mud were the most basic man-made composite materials. Our roads are paved with either asphalt concrete or Portland cement with steel and aggregate reinforcement, which is the most prominent example of a composite application. Another illustration of composite material is reinforced concrete. High-performing advanced composites are fabricated using fiber reinforcements made of the epoxy polymer matrix. Composites made of graphite, Kevlar, and boron are a few examples. Traditional uses of advanced composites in the aerospace sector today also find their essential benefits in other sectors like environmental pollution control, construction, automobiles, bulletproof clothing, weapons, and energy devices.

Nanocomposites are another kind of heterogeneous multiphase material that differ from traditional composites in the sense that the fillers here usually have at least a one dimension in the nanoscale range, that is, less than 100 nm. A polymer nanocomposite (PNC) is a polymer-based mixture of two or more materials in which the dispersed phase is composed of a nanomaterial. This provides some special advantages due to the extremely high aspect ratios of nanoparticles that causes some surprising change in properties even when embedded at low content into the matrix. It has been demonstrated that some nanocomposite materials are 1000 times more durable than their bulk components. Due to the few atoms per particle in fillers containing nanoparticles, they may have distinct properties from the bulk material and also displays strong interactivity with the matrix phase. The existence of a very large nanoparticle–matrix interfacial area and molecular interactions are thought to be key factors for the unique physical and mechanical properties that arise in nanocomposites.

Most commercially available composites are made with polymer matrix substances which are known as a resin solution. Depending on the initial basic materials, a wide variety of polymers can be used to prepare nanocomposite. Polyester, vinyl ester, polyether ether ketone (PEEK), epoxy, polyimide, phenolic, polyamide, polypropylene, PAN, cellulose acetate, and other materials are among the most popular. Further, lower cost, good chemical resistance, easier processability, and accessible fabrication techniques lead to increased demand for polymer composites. Different kinds of nanoparticles, such as graphene, clays, nanocellulose, carbon nanotubes, halloysite, and metal–organic frameworks (MOFs), have been explored by numerous studies to obtain nanocomposites with different polymers. Since the mechanical and thermal properties are closely related to the morphology of the end product, the assessment of the nanofiller dispersion in the polymer matrix is crucial and should preferably be uniform. Three different nanocomposite morphologies are feasible depending on the nanoparticles' degree of separation, such as conventional composites (or microcomposites), intercalated nanocomposites, and exfoliated nanocomposites. Phase separation occurs in polymeric composites when a polymer cannot intercalate effectively between the silicate layers, and its properties are similar to those in conventional polymeric composites. A well-ordered multilayer morphology with intercalated layers of clay and polymer is produced, leading to an intercalated structure in which a single (and occasionally more than one) polymer chain is inserted between two silicate layers. An exfoliated structure is created when the silicate layers are evenly and thoroughly scattered across a continuous polymer matrix. Because the polymer matrix and nanoparticles have a wide surface area of contact, maximum enhancement in properties can be achieved in the case of exfoliated nanocomposites. This is one of the primary distinctions between ordinary composites and nanocomposites (Figure 2.1) [4].

In this chapter, we discuss the uses of different nanofillers like graphene, carbon nanotubes, and porous nanomaterials like zeolites and MOFs to obtain superior polymer composite. They are being extensively used to fabricate advanced composites in the current time for several high-end applications like gas separation, water purification, catalysis, drug delivery, fuel cells, energy devices, and other biomedical applications.

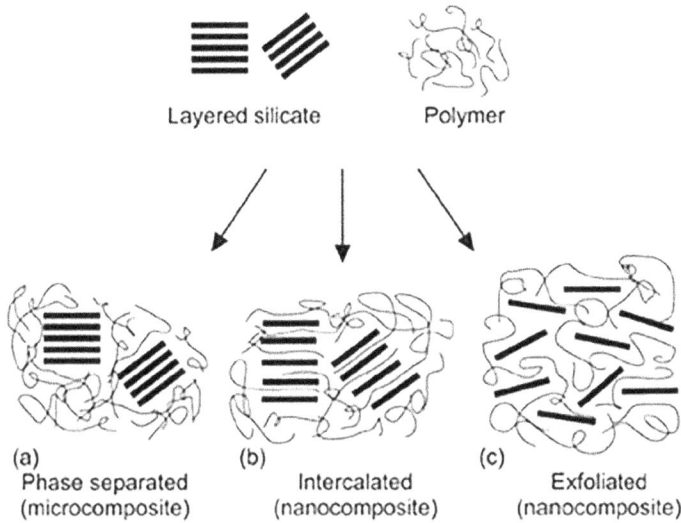

(a)
Phase separated
(microcomposite)

(b)
Intercalated
(nanocomposite)

(c)
Exfoliated
(nanocomposite)

FIGURE 2.1 Different kinds of polymer composites morphologies [5]. (Reprinted with permission.)

2.2 NANOFILLERS: STRUCTURE, PROPERTIES, APPLICATIONS

2.2.1 GRAPHENE

In many disciplines, graphene, a 2D allotrope of carbon, has been the focus of investigation and sparked enormous interest. Numerous businesses, including the composite industry, have benefited since its discovery. Because of its remarkable electrical, chemical, optical, and mechanical properties, it is found to be a potential candidate as a nanofiller. Although a significant challenge remains in its large-scale commercial synthesis, numerous researchers have tried new strategies to produce massive amounts of high-quality graphene sheets. Graphene has a unique structure in which the carbon atoms bonded in sp^2 hybridization are arranged hexagonally, giving rise to a layered (planar, lamellar) structure.

The distance between each carbon atom is 0.192 nm and between the planes is 0.332 nm. The physical characteristics of the two recognized types of graphite, hexagonal and rhombohedral, are quite similar. The lateral planes of the lamellar structures are subject to substantially larger forces than between the planes themselves. Each lattice contains three strong linkages that combine to form a sturdy hexagonal structure. The vertical bond is primarily responsible for graphene's electrical conductivity. Each lattice contains three strong linkages that combine to form a sturdy hexagonal structure (Figure 2.2a). The vertical bond is primarily responsible for graphene's electrical conductivity to the lattice plane. The stability of graphene is a result of the tightly packed carbon atoms and hybridization of the sp^2 orbitals, which are the combination of the orbitals s, p_x, and p_y that form the σ-bond. The final p_z electron sets up the π-bond. The π-π bands formed due to the hybridization of the π bonds of graphene give rise to the exceptional properties of graphene that allows for free-moving electrons and superior electrical properties (Figure 2.2b) [6,7].

2.2.2 GRAPHENE-BASED NANOCOMPOSITES AND APPLICATIONS

Many desired characteristics of graphene include high strength and elastic modulus of 1 TPa that gives good dimensional stability. It also demonstrates high electrical and thermal conductivity, high aspect ratio, and high gas impermeability are essential features for its use as a filler in polymer composites. Incorporating graphene at a modest volume fraction can significantly enhance the characteristics of

(a)

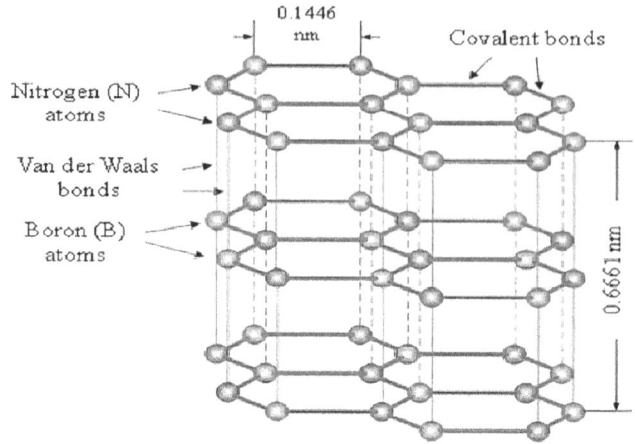

(b)

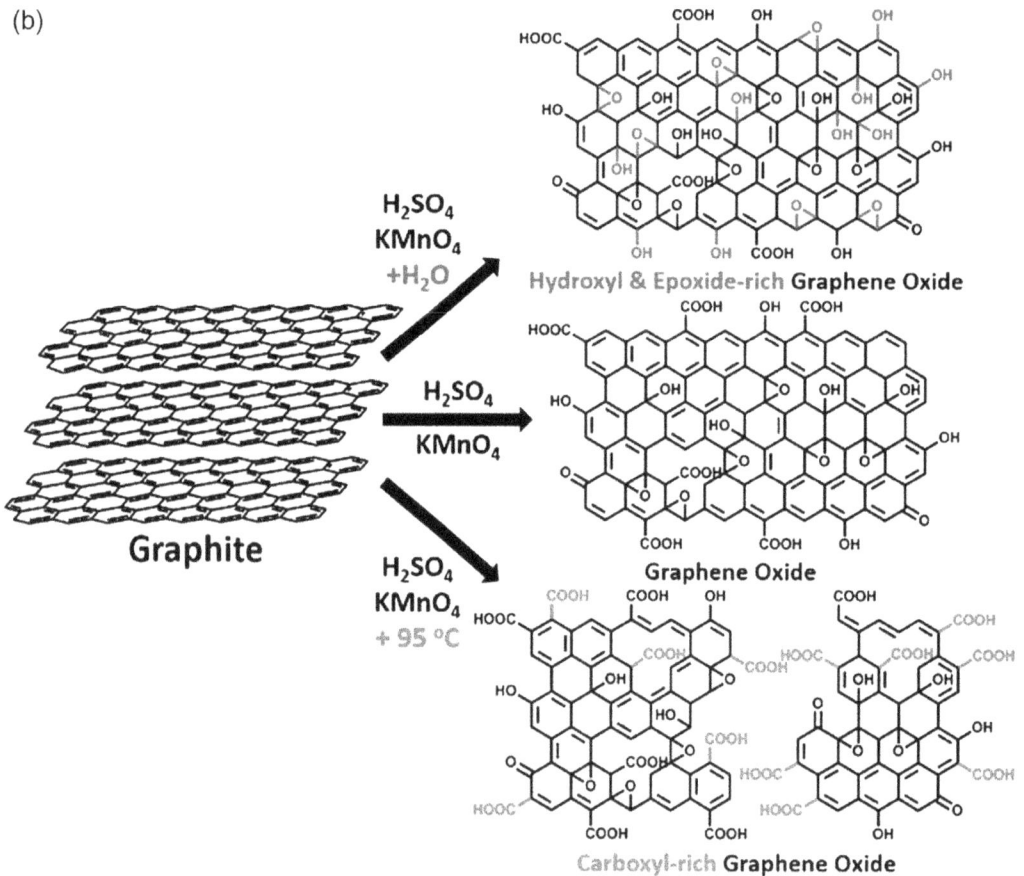

FIGURE 2.2 (a) Structure of graphene and (b) synthesis of graphene oxide from graphene [8]. (Reprinted with permission.)

polymers. The surface-to-volume ratio of graphene is also higher. Graphene-based nanocomposites are found to have significant potential as energy storage materials. For instance, (a) graphene-based nanocomposites have shown to achieve better performance in lithium-ion batteries because of their high power and energy density as well as a quick recharging capability in hydrogen-based fuel cells; (b) graphene as an electrode material enhances electrocatalytic activity; and (c) composites based on graphene are being used in photovoltaics for their low resistivity and noteworthy electrical conductivity and high charge carrier and its mobility at room temperature ($10,000\,cm^2$/V-s). Different polymers such as epoxy, HDPE, polystyrene, PMMA, and nylon have been explored to fabricate PNCs with graphene as the dispersed phase. Graphene/PNCs are primarily prepared using graphene oxide (GO), reduced GO, or thermally reduced graphene oxide (TRGO) as fillers. Table 2.1 describes the various reported polymer/graphene composites for myriad applications.

TABLE 2.1
Description of Different Polymer/Graphene Composites

Composition	Brief Description	Role of Graphene	Application	Reference
Polyaniline/ graphene/epoxy	GO nanosheets modification with Polyaniline via in situ polymerization and introduced in an epoxy matrix.	Platelet-like structure of graphene-based materials aids helps to improve the properties of corrosion and weathering resistance of coatings. A significantly high aspect ratio equips them to act as a physical barrier against corrosive and humidity agents.	A high-performing epoxy nanocomposite film have superior capability of antioxidant and anticorrosion as well as UV blocking.	[9]
PMMA/ GOPMMA/ GO-Fe$_3$O$_4$	In situ polymerization is employed for growing PMMA on GO surface.		Nanocomposite for Malachite green dye adsorption	[10]
Polyamide 12/RGO	Polyamide granules and RGO nanofiller were melt-compounded after drying and then hot-pressed to obtain square sheets.	It acts as a conductive nanofiller for polymer matrices, helping overcome the problems associated with the weak electrical conductivity of usual polymers.	Novel electroactive nanocomposites.	[11]
(PEG–NH$_2$-rGO)/ polyimide amine-terminated poly (ethylene glycol)	Graphene oxide, a graphene precursor, was initially functionalized by amine-terminated polyethylene glycol (PEG–NH$_2$) to prepare GO–PEG. It is further reduced using N$_2$H$_4$·H$_2$O to con vert it to modified reduced graphene oxide (rGO)–PEG. The synthesized material was then introduced into polyimide (PI) during in situ polymerization and a thermal reduction process to prepare its composites.	In this study, the brittleness of polyimide could be reduced by modifying rGO with long chain PEG and increase in material flexibility. Also due to low electrical conductivity of PI the static electrons may accumulate heavily on the surface, causing breakdown due to the high voltage. Modified rGO significantly enhances the properties of when made into a composite with this polymer.	These nanocomposites have many applications in the electronics and aerospace industries.	[12]
Polypropylene (PP)/graphene nanoplatelet (GnP)	Solid state mixing followed by melt blending.	Incorporation of GnP to develop composites increased storage modulus in DMA. Particularly smaller GnP particle increases crystallization temperature in DSC analysis with an increase in thermal stability and degradation temperature.	Structural components for automobiles, industries.	[13]

(Continued)

TABLE 2.1 (*Continued*)
Description of Different Polymer/Graphene Composites

Composition	Brief Description	Role of Graphene	Application	Reference
Nafion/graphene oxide	After synthesis of graphene oxide by modified Hummers method, GO, and platinum functionalized GO were used to prepare composite membranes using the solution casting method.	Nafion/GO composite membrane showed a drastic increment in water uptake than neat Nafion, which enhanced the fuel cell performance under low humidity conditions and helped to improve thermal and dimensional stabilities.	Polymer electrolyte membrane fuel cell (PEMFC)	[14]
Poly urethane/ graphene nanocomposites	Waterborne polyurethane (WPU) and covalently conjugated hydroxyl-functionalized GO wrapped in P-N flame retardants have been prepared by in situ strategies.	On comparison with pristine WPU, total heat released and peak heat release rate significantly reduced by 18.6% and 39.2%, respectively. Moreover, due to the uniformly dispersed graphene nanosheet demonstrating the catalytic charring performance of P-N flame retardant improvement, the lamellae-blocking effect, and ensured better fire safety. Also, WPU/fGO composite revealed a 139% increment in tensile strength and simultaneously maintained greater elongation at break (almost 857.5%)	Fire-retardant composites	[15]
Polyethylene terephthalate/ graphene nanocomposites	PET/graphene nanocomposites can be fabricated by melt compounding.	Morphological analysis displayed a fine dispersion in the PET matrix that led to significant improvement in the electrical conductivity of PET, which causes a sharp transition from an electrical insulator to a semiconductor with a low separation threshold of 0.47 vol%. At only 3 vol% graphene, electrical conductivity as high as 2.11 S/m was reported. Unique features of graphene, such as high aspect ratio and surface area with good compatibility in PET matrix, contributed to achieving the mentioned improved characteristics in the composite than pure polymer.	EMI shielding	[16]

2.2.3 Carbon Nanotubes

Carbon nanotubes have been the center of focus for many years due to their extraordinary and fascinating properties. Although the applicability of CNT as a bulk material faces severe restrictions, using it as a nanofiller in suitable matrices to develop innovative structures with remarkable properties has been explored in depth. This material has been strategically combined with many other materials such as thermosets, thermoplastics, and elastomers to obtain hybrid materials that can cooperate to build desirable characteristics in the end product. CNTs are the stiffest known material, according to recent studies, and they buckle elastically (vs. fracture) under significant flexural or compressive strains, showing their potential in advanced composites [14].

2.2.3.1 Structure of CNTs

Carbon nanotubes can be considered hollow cylinders obtained by rolling graphene sheets. They usually fall into two classes—single-walled carbon nanotube (SWCNT), which consists of only one sheet of graphene, and multiwalled carbon nanotube (MWCNT), in which layers of graphene are rolled up coaxially to form concentric arrangements of nanotubes as shown in Figure 2.3a. A nanotube is a single molecule comprising atoms in millions and lengths of which can be tens of micrometers, whereas diameters are less than 0.7 nm.

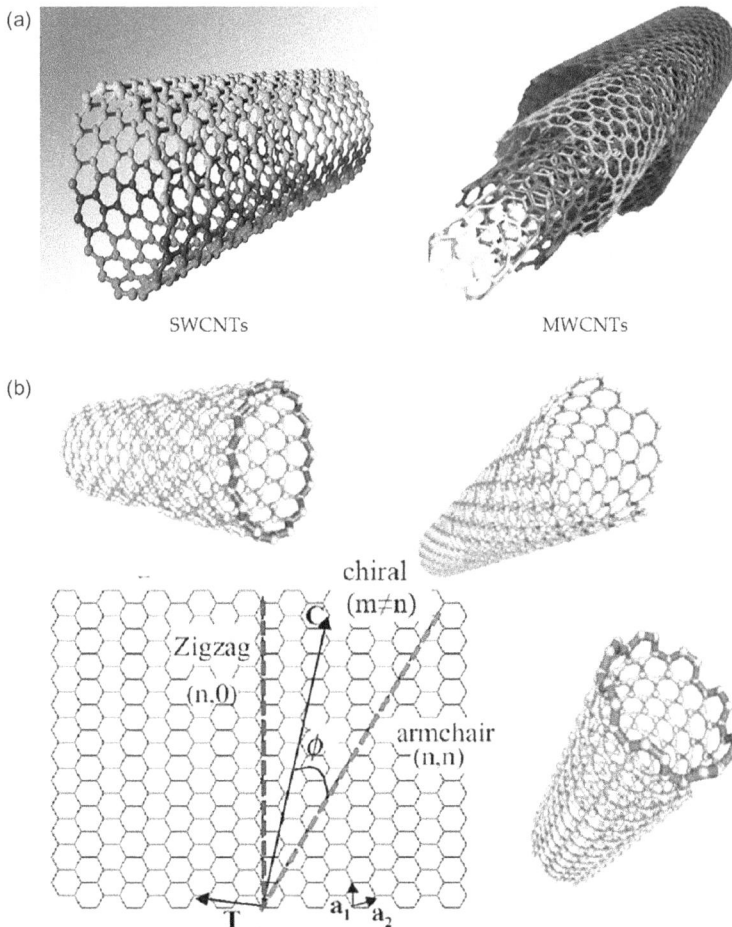

FIGURE 2.3 (a) Single-walled CNT and multiwalled CNT [17] and (b) Different structures of SWCNT based on the rolling axis of graphene sheet [18]. (Reprinted with permission.)

In SWCNTs generally, ten atoms constitute the cylinder's circumference, but fascinatingly, it is only one atom thick [19]. Immanent characteristics usually are large aspect ratio near about 1000; henceforth, these allotropes of carbon can be considered 1-dimensional. In MWCNTs, stacking SWCNTs inside one another gives rise to larger structures with outer diameters that should be less than 15 nm. The graphene sheets can be rolled up in different combinational geometries based on which CNTs can further be divided into subcategories—zigzag, armchair, and chiral as shown in Figure 2.3b [20]. The vector of the rolling axis of the graphene sheets relative to the radius of the hollow closing cylinders and hexagonal graphene arrays gives rise to such combinations. As we see, the rolling vector (C_h) is equal to the sum of integral multiples of graphene vector lattices as described below:

$$C_h = na_1 + ma_2$$

where n and m are integers and a_1 and a_2 are unit vectors of graphene lattice. For equal chiral indices ($n=m$), the armchair conformation is taken up by nanotube, with a chiral angle of 30°. When either of the integers is zero (n, 0) or (0, m), the SWCNT is called zigzag, and the chiral angle is 0° (achiral nanotubes). For chiral SWCNTs, the integers are unequal ($n \neq m$), the nanotube is called chiral, and its angle is $0° < \theta < 30°$. The following equation determines the nanotube diameter (D):

$$D = a\sqrt{n^2 + m^2 + nm}$$

where a denotes

$$a = a_{C-C}\sqrt{3} = 1.44 \text{Å}$$

chiral angle (θ) is described as

$$\cos\theta = \frac{(2n + m)}{2\sqrt{(n^2 + m^2 + nm)}}$$

2.2.3.2 Properties of CNTs

- CNTs have very low densities, sometimes as low as 1.3 cm³/g (almost one-sixth that of steel), but they have Young's moduli approximately equivalent to 1 TPa, higher than carbon fibers, and nearly five times greater than steel. They are the most robust materials discovered due to interlocking carbon to carbon covalent bonds.
- On exploring tensile strength or breaking strain for a CNT, it went up to 63 GPa, which is around 50 times higher than steel. Even the lower-quality CNTs display tensile strength in several GPa, additionally, considerable chemical and environmental endurance and superior thermal conductivity (~3000 W/m/K, comparable to diamond). Coupled with their lightweight properties, they are potential candidates for high-performing applications.
- Although nanotubes are strong, they are highly elastic and not easily deformed under force. Since they are one large molecule, weak spots such as crystal grain boundaries are absent, helping them preserve their inherent strength.
- The extraordinary electronic properties (electrical conductivity comparable to copper) of CNTs arise from the ballistic transport of electrons like a wave within the tube according to the rules of quantum mechanics with negligible resistance [21]. Depending on the conformation taken up by the nanotube while rolling the sheets, its electrical properties can be that of a metal (for armchairs) or semiconducting (for zigzag and chiral kinds).

2.3 APPLICATIONS IN POLYMER-BASED NANOCOMPOSITES

Due to the chemical inertness and contaminants like fullerenes, amorphous carbon, or metal catalysts, CNTs' wettability and dispersion in the polymer matrix are impeded. Different kinds of modifications and functionalization are carried out in the carbon nanotubes, broadly classified into two categories—covalent and non-covalent to utilize the full potential of CNTs as nanofillers and formulate advanced composites.

2.3.1 COVALENT FUNCTIONALIZATION

Functional groups such as −COOH, −NH$_2$, and phenolic by addition reactions are attached to the sidewalls to obtain stable and uniform dispersions of CNTs in polymer matrices. The addition of polar groups helps to form stable dispersions in polar solvents. Moreover, groups like carboxyl or amines help to create different functionalization by polymer molecules through the grafting technique, which gives superior polymer composites.

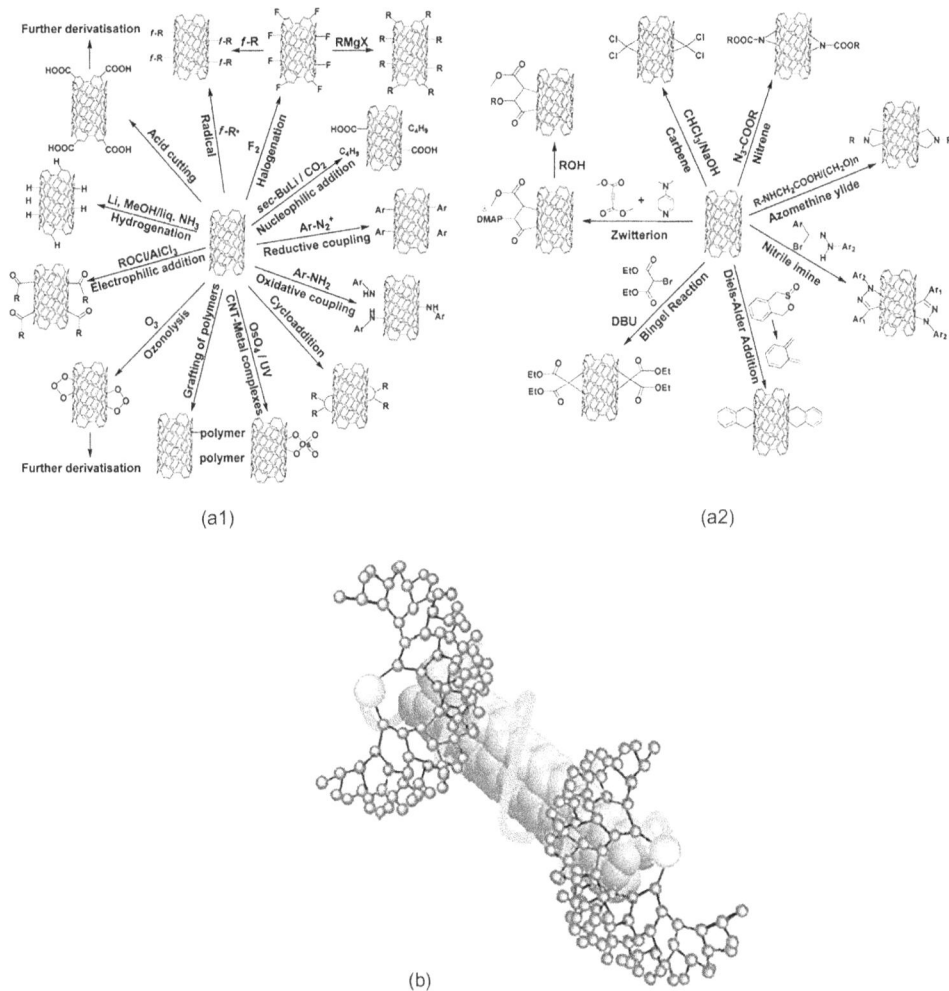

(a1)   (a2)

(b)

FIGURE 2.4   (a$_1$) Covalent functionalization of CNT and (a$_2$) Covalent functionalization of CNT by cycloaddition reaction [22] and (b) Non-covalent functionalization of CNT surface by dendritic copolymer—polyglycerol–poly (ethylene glycol)–polyglycerol [23]. (Reprinted with permission.)

TABLE 2.2

List of Various Polymer/CNT Composites That Have Been Explored for Different Applications

Composition	Brief Description	Role of CNTs	Application	Reference
Epoxy/CNT	The standard calendaring technique is employed to fabricate.	Increment in stiffness, Young's modulus, fracture toughness, and strain to failure at only 0.1 wt.% of CNT addition.	Aerospace, radar-absorbing material, EMI shielding,	[24]
PMMA/CNT	Ultrasonication and precipitation of respective solutions.	High EMI efficiency in the terahertz (THz) frequency range is demonstrated equally in oxidized and unoxidized MWCNT/PMMA at 0.25–2 wt.%.	As innovative materials for nano-electronics and also nanophotonics.	[25]
PET/CNT	PET resin compounded with CNTs by the twin-screw extruder.	At 4 wt.% of CNTs in PET, the volume electrical resistance is 12 times lower than pure PET. Also, the cloth women from this composite helped in achieving excellent antistatic property.	Conducting woven fabrics.	[26]
Polyurethane/CNT	Electrospinning	On incorporation of CNTs at optimal content, the biological properties such as clotting time, in vivo stability, adhesion, and chemical and physical properties such as Young's modulus, toughness, and water absorption are enhanced.	Biomaterials are designed for tissue engineering, bone tissue engineering, cardiovascular, tendons ligaments, and heart valve graft parts.	[27]

2.3.2 Non-covalent Functionalization

Non-covalent functionalization is preferred when an alteration in the structure of CNTs is not desired for specific application purposes. This technique becomes beneficial for keeping intact the original electrical and mechanical properties of these nanostructures. Here, molecular interactions such as van der Waals attraction - π-π, CH-π between the matrix molecules and walls of dispersed CNT occur. The different techniques explored in this aspect involve polymer wrapping, plasma treatment, or surfactant-assisted dispersions.

2.4 ZEOLITES AND METAL–ORGANIC FRAMEWORKS

Massive environmental challenges are faced globally due to industries' toxic gases, heavy metals dumping in soil and water bodies, and dyes in water through everyday activities. We are in dire need of commercially viable cutting-edge technologies to overcome such issues that are affecting our health and daily life. In this regard, PNC-based superior membranes for water purification and gas adsorbents to purify the air are being widely investigated and have been commercially successful in many cases. A new generation of nanocomposite membranes constructed by combining porous materials such as zeolite, molecular sieves, MOFs, and versatile, low-cost polymers to develop mixed-matrix membranes (MMM) are being intensively explored. The superior gas separation features, selectivity of porous materials, tuneability, and specific functionalization, have opened the doors for widespread applications of these novel materials, including separation, water purification, gas storage, drug delivery, and biomedical applications. An advantage of porous materials is the selective permeation and transport pathways for desirable components so that various strategies can be defined as per the needs of an application.

2.4.1 Zeolites

These porous aluminum silicate materials are widely applied in gas separation techniques. Zheng et al. reported polyether-block-amide (Pebax), a thermoplastic elastomer that has been modified with NaY zeolite to form mixed-matrix membranes by simple solution casting followed by solvent evaporation technique [28]. The membranes showed good polymer filler compatibility, defect-free surface, improved CO_2 permeability, and better CO_2/N_2 separation performance. Another study investigated the effect of –COOH group on zeolite to obtain efficient separation of gas mixtures such as CO_2/N_2 and CH_4/N_2 [29]. The base material is comprised of PEBAX and a permeable PES (polyether sulfone) support layer. The results showed that carboxyl-modified zeolites (NaX), when incorporated to fabricate mixed-matrix membranes (MMM), showed improved CO_2 selectivity due to increased sorption and excellent separation abilities. The nano-zeolites help rigidity of the polymer chains of the matrix due to continuous hydrogen bonding interactions that arise with the –COOH group, thus preventing compaction and plasticization of the membranes. Zeolite ZSM-5(also called MFI referring to the framework) belongs to the pentasil family of the same forms 2D MFI nanosheets used as filler in PEBAX matrix for separation of CO_2/CH_4 mixture [30]. The MMM shows a high increment in CO_2/CH_4 separation, almost by 63.5% in CO_2 permeability and 76.4% in CO_2/CH_4 selectivity compared to pure polymer. These nanofillers' ultrathin and straight channels have a sheet-like morphology with large cavities of 0.54×0.56 nm (along b axes) and also possess CO_2 adsorption preferentially over methane. These factors, coupled with shorter diffusion paths, increase the CO_2 permeation. Simultaneously, the high aspect ratio of these nanosheets increases the filler compatibility with the polymer interface, improving mechanical properties and adding barriers selectively to make the methane molecules locomote through more tortuous diffusion routes among filler nanosheets.

2.4.2 Metal–Organic Framework

Metal–organic frameworks are composite of metals and organics with very high pore volume and high surface area, including uniform pore sizes, which is very useful in the separation process, gas, and energy storage processes, to increase the electrical conductivity of materials, adsorption of pollutants, drug delivery, and sensor applications [31]. It is a complex compound of metal ions or atoms and organic linkers. Trivalent or bivalent aromatic carboxylic acid or nitrogen-containing aromatics are used to form a framework with aluminum, copper, chromium, zinc, and zirconium metals [32]. Preparation of MOFs by a solvothermal or hydrothermal process generally takes several hours or days to complete the reaction. Researchers have also investigated several other short-duration methods to prepare MOF, including electrochemical, mechanochemical, microwave-assisted, and sonochemical [33]. For the preparation of MOF, there are several organic linkers are widely used, such as H_3BTC (1,3,5-benzenetricarboxylic acid), H_2BDC-$(OH)_2$; 2,5-dihydroxy-1,4-benzenedicarboxylic acid), H_2BDC (1,4-benzenedicarboxylic acid), (H_2BDC-NH_2; 2-amino-1,4-benzenedi-carboxylic acid, and TPA (Terephthalic acid).

2.4.2.1 Solvothermal Synthesis

In the solvothermal synthesis process, metal salts and their corresponding organic linkers are dissolved in a specific solvent and sealed in vials in a conventional heating environment for a particular time. After crystallization, the solution is removed for solvent exchange and washing, and the prepared crystals are the synthesized MOF, as shown in Figure 2.5 [34].

2.4.2.2 Microwave-Assisted Synthesis

Due to the shortest preparation time and fast crystallization, the microwave-assisted process is widely used to prepare MOF under isothermal conditions. The advantages of the microwave-assisted technique include narrow particle size distribution, phase selection, and facile morphology control [35]. In the microwave-assisted process, a suitable solution of linker and metal ions is placed in Teflon tubes, and a sealed tube is placed in the microwave oven at a certain temperature and power for a specific time (Figure 2.6). There are many MOFs that have been prepared by microwave irradiation method, such as Cr-MIL-101 [36], MOF-5, HKUST-1 [37], IRMOF-3(H2BDC-NH$_3$), Fe-MIL-101-NH$_2$, and ZIF-8 as described in Figure 2.7 [35–39].

2.4.2.3 Sonochemical Synthesis

A suitable solution of linker and metals ion has been placed in the horn-type pyrex reactor under ultrasonic radiation. Due to the presence of sonic waves. formation of a bubble starts, and when it collapses, it releases some energy, which causes an increase in temperature and pressure up to 1000 bar, resulting in the formation of MOF crystals as in Figure 2.7 [40]. This technique can reduce the

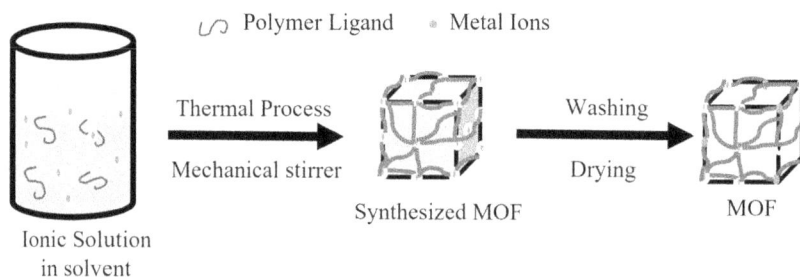

FIGURE 2.5   Solvothermal synthesis of MOF [33].

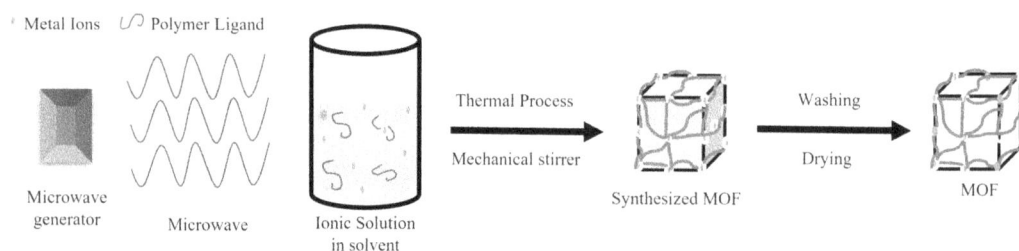

FIGURE 2.6   Microwave-assisted process to synthesize MOF [33].

MOF production time and crystal size significantly smaller than the solvothermal process [41]. This technique, such as MOF-5, prepare several MOFs by using NMP as a solvent, HKUST1 using DMF/EtOH, and Mg-MoF-74 using triethylamine (TEA), IRMoF-9, PCN-6, etc. [39,41,42].

2.4.2.4 Electrochemical Synthesis

In this process, metal ions are provided by the anode to react with an organic linker in the conducting salt medium. BASF's researchers first adopted this technique for preparing HKUST-1. It took 150 minutes at 12–19-volt electric potential using copper plates with H_3BTC in methanol solvent, Figure 2.8 [43]. Recently researchers have discovered many MOFs from this technique, such as Al-MIL-53-NH_3, ZIF-8, Al-MIL-3, and HKUST-1 [44].

2.4.2.5 Mechanochemical Synthesis

This technique involves breaking intramolecular bonds by using mechanical energy and then producing ions allowed to react with organic linkers in the dry condition at room temperature to form MOF nanocrystals. A tiny amount of solvent during the mechanical stress can boost the reaction of the metal ion with organic linkers because solvent provides the mobility space to move the particle from one place to another, Figure 2.9 [31]. The preparation of ZIF-8 MOF is a typical example of a mechanochemical synthesis process [45].

2.4.3 MORPHOLOGY AND CRYSTAL SIZE OF MOFS

There are many process parameters such as temperature profile, compositional parameters, additives, reverse microemulsion, and other process parameters that define the size and morphology of the crystals.

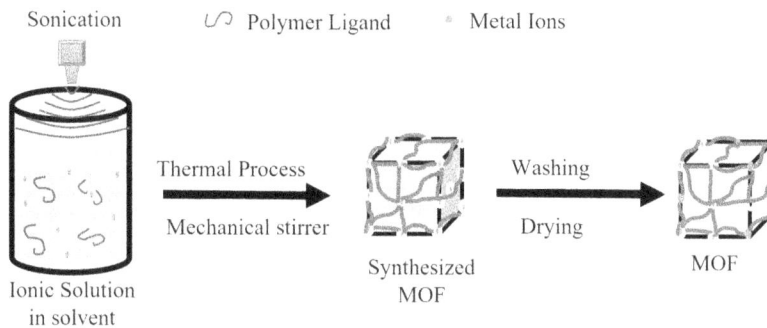

FIGURE 2.7 Sonochemical process for MOF preparation [33].

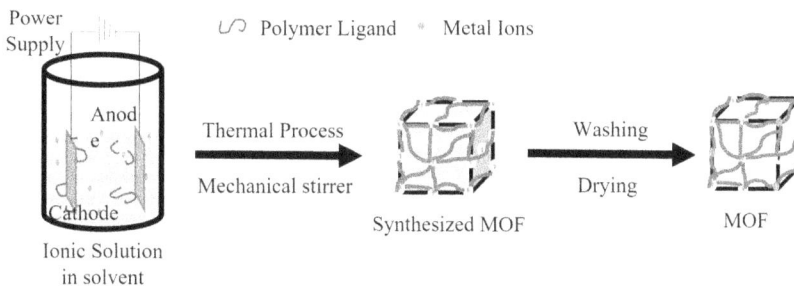

FIGURE 2.8 Electrochemical synthesis process for MOF [33].

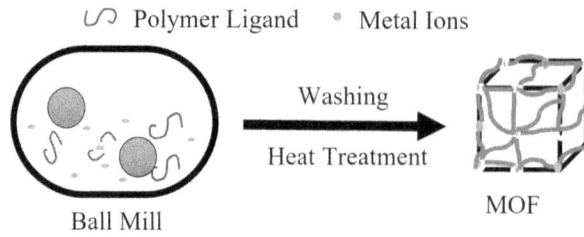

FIGURE 2.9 Mechanochemical process for MOF synthesis [33].

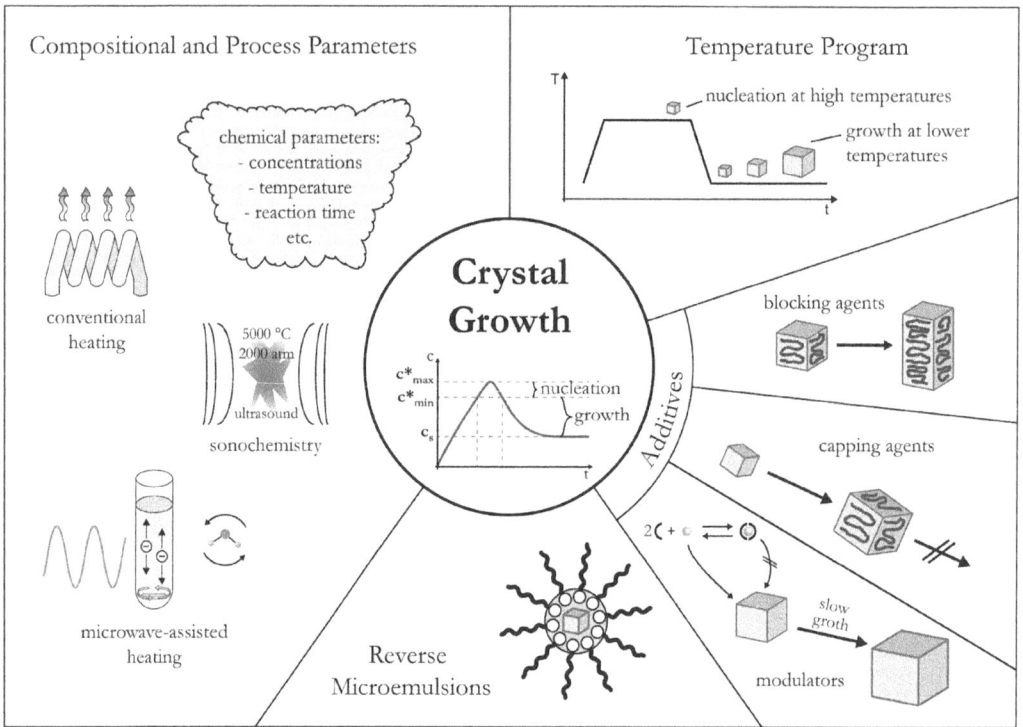

FIGURE 2.10 Summary of controlling parameters to the growth of size and morphology of MOF crystal [34]. (Reprinted with permission.)

Compositional parameters, including the type of solvent, the concentration of reactant, pH, source of metal ions, and molar ratio, can be advantageous in controlling the size and morphology of MOF crystals. Temperature, pressure, time of processing, and source of heating are the process parameters that also greatly influence the size and morphology of the MOF's crystals. The temperature profile can control the rate of crystal growth and relative nucleation. Additives like molecules blocking or capping agents also play a significant role in maintaining the size of the crystals. Microemulsions are a mixture of water, hydrophobic liquid, and surfactant. When organic liquid is a continuous phase and water is a dispersed phase, the mixture is known as reverse microemulsions, Figure 2.10. These reverse microemulsions can control crystal size by changing the molar ratio of surfactant and continuous phase [34].

TABLE 2.3
MOF/Polymer Nanocomposites Explored in Recent Times

Composition	Fabrication Technique	Role of MOF	Application	References
UiO-67/polyurethane	Ultrasonication of MOF nanofillers in polymer solution for thorough dispersion and then casting the solution and evaporation on glass plates for developing composite films.	60-fold enhancement in the adsorption capacity was demonstrated after loading the polymer matrix (PU) with MOF (UiO-67) crystals as nanofillers.	Gas separation, Air purification, as drug delivery device materials for ocular therapeutics.	[46]
ZIF-8/polyaniline	The composites were prepared via one-pot interfacial polymerization and characterized by electrochemical techniques such as cyclic voltammetry and electrical impedance spectroscopy.	High power density and current density could be obtained, paving the pathway to fabricating novel capacitor devices.	Lightweight energy storage.	[47]
Ionic liquid/MOF (Cu-BTC)/PEBAX MMM	1-ethyl-3-methylimidazolium acetate, an ionic liquid, was loaded into Cu-BTC MOF by wet impregnation technique and further incorporated into the PEBAX membrane.	The novel modified MOF nanofiller increased the CO_2 permeability by 2.5 times and CO_2/N_2 selectivity by 5.6 times compared to the pure PEBAX membrane.	CO_2/N_2 separation	[48]
Poly(N-isopropylacrylamide) (PNIPAM)/ UiO-66-NH$_2$	MOF was synthesized solvothermal and was post synthetically modified with the polymer to generate precisely controlled "On–Off" properties based on temperature stimulus.	PNIPAM modification MOF's surface generates thermosensitive behavior and releases encapsulated guest molecules in MOFs by switching between the open state at low temperature because of the coiled conformation of PNIPAM. The globule collapses when the temperature is raised, and the closed state is achieved.	Controlled release of different guest molecules such as resorufin, caffeine etc. in solutions.	[49]
ZIF-8/MWCNT/ epoxy	MWCNT walls were modified with MOF particles to act as curing agents and nanofiller.	Imidazole-based MOFs readily initiate the ring opening of epoxy resins for curing. Additionally, it acts as a superior compatibilizer for the dispersion of MWCNTs. Also, the MOF modification helps in low dielectric loss, high CTE, and high tensile strength.	NO_2 adsorption from the air, high dielectric constant polymer nanocomposites for energy storage devices and microelectronics.	[50]

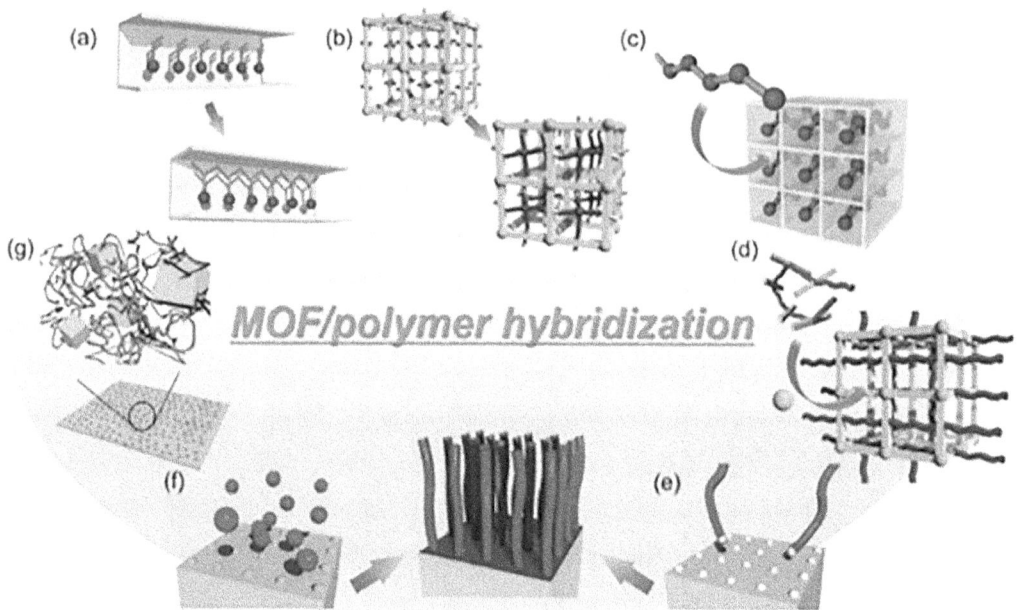

FIGURE 2.11 Various kinds of MOF-based polymer composites [51]. (Reprinted with permission.)

2.4.4 MOF-BASED POLYMER NANOCOMPOSITES

Table 2.3 gives an account of how MOF has been used as a filler in polymer matrices to design novel materials. In Figure 2.11, different types of polymeric compositions of MOF are also explained.

2.5 CONCLUSION

In this chapter, we discussed some of the nanofillers currently employed to design innovative architectures for addressing various challenges in modern times. Graphene, carbon nanotubes, and layered silicates are well-known materials. Scientists and engineers have explored them extensively to generate high-performing composites for gas separation, air purification, automobiles, aerospace, military, and construction. New porous nanomaterials like MOFs are a booming area of research to cater to the need of the present era. The synthesis of the MOF by using solvothermal, microwave-assisted, electrochemical, sonochemical, and mechanochemical process has also been described. It was found that due to the high surface area and pore volume, MOFs can be utilized to design novel composite materials for water purification, air purification, capacitors, fuel cells, gas separation, and nanocoatings covered in this chapter. Although much research is being done on these materials, several persisting issues hinder their widespread applicability, like difficulties in large-scale processing and scalability, defects in crystal grain boundaries, and extended synthetic techniques. Future research should focus on overcoming limitations to gain advantages and significant benefits from these materials to increase the utilization in various applications.

REFERENCES

1. R. Hsissou, R. Seghiri, Z. Benzekri, M. Hilali, M. Rafik, and A. Elharfi, "Polymer composite materials: A comprehensive review," *Compos. Struct.*, vol. 262, p. 113640, 2021, doi: 10.1016/J. COMPSTRUCT.2021.113640.
2. V. V. Kumar, G. Balaganesan, J. K. Y. Lee, R. E. Neisiany, S. Surendran, and S. Ramakrishna, "A review of recent advances in nanoengineered polymer composites," *Polymers (Basel).*, vol. 11, no. 4, pp. 1–20, 2019, doi: 10.3390/polym11040644.

3. L. Yu, K. Dean, and L. Li, "Polymer blends and composites from renewable resources," *Prog. Polym. Sci.*, vol. 31, no. 6, pp. 576–602, 2006, doi: 10.1016/J.PROGPOLYMSCI.2006.03.002.

4. S. Sinha Ray and M. Okamoto, "Polymer/layered silicate nanocomposites: A review from preparation to processing," *Prog. Polym. Sci.*, vol. 28, no. 11, pp. 1539–1641, 2003, doi: 10.1016/j.progpolymsci.2003.08.002.

5. H. M. C. de Azeredo, L. H. C. Mattoso, and T. H. McHugh, *Nanocomposites in Food Packaging: A Review*, Rijeka: IntechOpen, 2011, pp. 57–58, Ch. 4.

6. G. Yang, L. Li, W. B. Lee, and M. C. Ng, "Structure of graphene and its disorders: A review," *Sci. Technol. Adv. Mater.*, vol. 19, no. 1, pp. 613–648, 2018, doi: 10.1080/14686996.2018.1494493.

7. S. Kesarwani and R. K. Verma, "A critical review on synthesis, characterization and multifunctional applications of reduced graphene oxide (rGO)/composites," *Nano*, vol. 16, no. 09, p. 2130008, 2021, doi: 10.1142/S1793292021300085.

8. P. P. Brisebois and M. Siaj, "Harvesting graphene oxide-years 1859 to 2019: A review of its structure, synthesis, properties and exfoliation," *J. Mater. Chem. C*, vol. 8, p. 1517, 2020, doi: 10.1039/c9tc03251g.

9. S. Amrollahi, B. Ramezanzadeh, H. Yari, M. Ramezanzadeh, and M. Mahdavian, "Synthesis of polyaniline-modified graphene oxide for obtaining a high performance epoxy nanocomposite film with excellent UV blocking/anti-oxidant/ anti-corrosion capabilities," *Compos. Part B Eng.*, vol. 173, p. 106804, 2019, doi: 10.1016/J.COMPOSITESB.2019.05.015.

10. M. Rajabi, K. Mahanpoor, and O. Moradi, "Preparation of PMMA/GO and PMMA/GO-Fe_3O_4 nanocomposites for malachite green dye adsorption: Kinetic and thermodynamic studies," *Compos. Part B Eng.*, vol. 167, pp. 544–555, 2019, doi: 10.1016/J.COMPOSITESB.2019.03.030.

11. A. Dorigato and A. Pegoretti, "Novel electroactive polyamide 12 based nanocomposites filled with reduced graphene oxide," *Polym. Eng. Sci.*, vol. 59, no. 1, pp. 198–205, 2019, doi: 10.1002/PEN.24889.

12. L. Ma, G. Wang, and J. Dai, "Preparation of functional reduced graphene oxide and its influence on the properties of polyimide composites," *J. Appl. Polym. Sci.*, vol. 134, no. 30, p. 45119, 2017, doi: 10.1002/APP.45119.

13. M. Ajorloo, M. Fasihi, M. Ohshima, and K. Taki, "How are the thermal properties of polypropylene/graphene nanoplatelet composites affected by polymer chain configuration and size of nanofiller?" *Mater. Des.*, vol. 181, p. 108068, 2019, doi: 10.1016/J.MATDES.2019.108068.

14. T. W. Odom, J. L. Huang, P. Kim, and C. M. Lieber, "Structure and electronic properties of carbon nanotubes," *J. Phys. Chem. B*, vol. 104, no. 13, pp. 2794–2809, 2000, doi: 10.1021/jp993592k.

15. P. Zhang, P. Xu, H. Fan, Z. Sun, and J. Wen, "Covalently functionalized graphene towards molecular-level dispersed waterborne polyurethane nanocomposite with balanced comprehensive performance," *Appl. Surf. Sci.*, vol. 471, pp. 595–606, 2019, doi: 10.1016/J.APSUSC.2018.11.235.

16. H. Bin Zhang, et al., "Electrically conductive polyethylene terephthalate/graphene nanocomposites prepared by melt compounding," *Polymer (Guildf)*., vol. 51, no. 5, pp. 1191–1196, 2010, doi: 10.1016/J.POLYMER.2010.01.027.

17. A. Zeeshan, N. Shehzad, M. Atif, R. Ellahi, and S. M. Sait, "Electromagnetic flow of SWCNT/MWCNT suspensions in two immiscible water- and engine-oil-based newtonian fluids through porous media," *Symmetry 2022*, vol. 14, no. 2, p. 406, 2022, doi: 10.3390/SYM14020406.

18. C. M. Tîlmaciu and M. C. Morris, "Carbon nanotube biosensors," *Front. Chem.*, vol. 3, no. OCT, pp. 1–21, 2015, doi: 10.3389/fchem.2015.00059.

19. N. Saifuddin, A. Z. Raziah, and A. R. Junizah, "Carbon nanotubes: A review on structure and their interaction with proteins," *J. Chem.*, 2013, doi: 10.1155/2013/676815.

20. W. Khan, R. Sharma, and P. Saini, "Carbon nanotube-based polymer composites: Synthesis, properties and applications," *Carbon Nanotub. - Curr. Prog. Their Polym. Compos.*, 2016, doi: 10.5772/62497.

21. G. Dresselhaus, M. S. Dresselhaus, and R. Saito, *Physical Properties of Carbon Nanotubes*. Tokyo: World Scientific Publishing Company, 1998.

22. H.-C. Wu, X. Chang, L. Liu, F. Zhao, and Y. Zhao, "Chemistry of carbon nanotubes in biomedical applications," 2009, doi: 10.1039/b911099m.

23. P. Bilalis, D. Katsigiannopoulos, A. Avgeropoulos, and G. Sakellariou, "Non-covalent functionalization of carbon nanotubes with polymers," 2014, doi: 10.1039/c3ra44906h.

24. F. H. Gojny, M. H. G. Wichmann, U. Köpke, B. Fiedler, and K. Schulte, "Carbon nanotube-reinforced epoxy-composites: Enhanced stiffness and fracture toughness at low nanotube content," *Compos. Sci. Technol.*, vol. 64, no. 15, pp. 2363–2371, 2004, doi: 10.1016/J.COMPSCITECH.2004.04.002.

25. J. MacUtkevic, et al., "Multi-walled carbon nanotubes/PMMA composites for THz applications," *Diam. Relat. Mater.*, vol. 25, pp. 13–18, 2012, doi: 10.1016/J.DIAMOND.2012.02.002.

26. Z. Li, G. Luo, F. Wei, and Y. Huang, "Microstructure of carbon nanotubes/PET conductive composites fibers and their properties," *Compos. Sci. Technol.*, vol. 66, no. 7–8, pp. 1022–1029, 2006, doi: 10.1016/J. COMPSCITECH.2005.08.006.

27. Z. Eivazi Zadeh, A. Solouk, M. Shafieian, and M. Haghbin Nazarpak, "Electrospun polyurethane/ carbon nanotube composites with different amounts of carbon nanotubes and almost the same fiber diameter for biomedical applications," *Mater. Sci. Eng. C*, vol. 118, p. 111403, 2021, doi: 10.1016/J. MSEC.2020.111403.

28. Y. Zheng, Y. Wu, B. Zhang, and Z. Wang, "Preparation and characterization of CO_2-selective Pebax/ NaY mixed matrix membranes," *J. Appl. Polym. Sci.*, vol. 137, no. 9, p. 48398, 2020, doi: 10.1002/ APP.48398.

29. M. S. Maleh and A. Raisi, "CO_2-philic moderate selective layer mixed matrix membranes containing surface functionalized NaX towards highly-efficient CO_2 capture," *RSC Adv.*, vol. 9, no. 27, pp. 15542–15553, 2019, doi: 10.1039/C9RA01654F.

30. Q. Zhang, M. Zhou, X. Liu, and B. Zhang, "Pebax/two-dimensional MFI nanosheets mixed-matrix membranes for enhanced CO_2 separation," *J. Memb. Sci.*, vol. 636, p. 119612, 2021, doi: 10.1016/J. MEMSCI.2021.119612.

31. J. L. C. Rowsell, E. C. Spencer, J. Eckert, J. A. K. Howard, and O. M. Yaghi, "Chemistry: Gas adsorption sites in a large-pore metal-organic framework," *Science (80-.).*, vol. 309, no. 5739, pp. 1350–1354, 2005, doi: 10.1126/science.1113247.

32. J. R. Long and O. M. Yaghi, "The pervasive chemistry of metal-organic frameworks," *Chem. Soc. Rev.*, vol. 38, no. 5, pp. 1213–1214, 2009, doi: 10.1039/b903811f.

33. Y. R. Lee, J. Kim, and W. S. Ahn, "Synthesis of metal-organic frameworks: A mini review," *Korean J. Chem. Eng.*, vol. 30, no. 9, pp. 1667–1680, 2013, doi: 10.1007/s11814-013-0140-6.

34. N. Stock and S. Biswas, "Synthesis of metal-organic frameworks (MOFs): Routes to various MOF topologies, morphologies, and composites," *Chem. Rev.*, vol. 112, no. 2, pp. 933–969, 2012, doi: 10.1021/ cr200304e.

35. T. Jin, et al., "Microwave synthesis, characterization and catalytic properties of titanium-incorporated ZSM-5 zeolite," *Res. Chem. Intermed.*, vol. 33, no. 6, pp. 501–512, 2007, doi: 10.1163/156856707782565804.

36. V. I. Isaeva and L. M. Kustov, "Microwave activation as an alternative production of metal-organic frameworks," *Russ. Chem. Bull.*, vol. 65, no. 9, pp. 2103–2114, 2016, doi: 10.1007/s11172-016-1559-9.

37. M. Schlesinger, S. Schulze, M. Hietschold, and M. Mehring, "Evaluation of synthetic methods for microporous metal-organic frameworks exemplified by the competitive formation of $[Cu_2(btc)_3(H_2O)_3]$ and $[Cu_2(btc)(OH)(H_2O)]$," *Microporous Mesoporous Mater.*, vol. 132, no. 1–2, pp. 121–127, 2010, doi: 10.1016/j.micromeso.2010.02.008.

38. R. Banerjee *et al.*, "High-throughput synthesis of zeolitic imidazolate frameworks and application to CO_2 capture," *Science (80-.).*, vol. 319, no. 5865, pp. 939–943, 2008, doi: 10.1126/science.1152516.

39. F. Bigdeli, H. Ghasempour, A. Azhdari Tehrani, A. Morsali, and H. Hosseini-Monfared, "Ultrasound-assisted synthesis of nano-structured Zinc(II)-based metal-organic frameworks as precursors for the synthesis of ZnO nano-structures," *Ultrason. Sonochem.*, vol. 37, pp. 29–36, 2017, doi: 10.1016/j. ultsonch.2016.12.031.

40. M. Y. Masoomi and A. Morsali, "Sonochemical synthesis of nanoplates of two Cd(II) based metal-organic frameworks and their applications as precursors for preparation of nano-materials," *Ultrason. Sonochem.*, vol. 28, pp. 240–249, 2016, doi: 10.1016/j.ultsonch.2015.07.017.

41. A. Mirzaie, T. Musabeygi, and A. Afzalinia, "Sonochemical synthesis of magnetic responsive $Fe_3O_4@$ TMU-17-NH_2 composite as sorbent for highly efficient ultrasonic-assisted denitrogenation of fossil fuel," *Ultrason. Sonochem.*, vol. 38, pp. 664–671, 2017, doi: 10.1016/j.ultsonch.2016.08.013.

42. A. Mehrani, A. Morsali, Y. Hanifehpour, and S. W. Joo, "Sonochemical temperature controlled synthesis of pellet-, laminate- and rice grain-like morphologies of a Cu(II) porous metal-organic framework nano-structures," *Ultrason. Sonochem.*, vol. 21, no. 4, pp. 1430–1434, 2014, doi: 10.1016/j. ultsonch.2014.01.011.

43. S. N. Kane, A. Mishra, and A. K. Dutta, "Preface: International conference on recent trends in physics (ICRTP 2016)," *J. Phys. Conf. Ser.*, vol. 755, no. 1, 2016, doi: 10.1088/1742-6596/755/1/011001.

44. U. Mueller, M. Schubert, F. Teich, H. Puetter, K. Schierle-Arndt, and J. Pastré, "Metal-organic frameworks - Prospective industrial applications," *J. Mater. Chem.*, vol. 16, no. 7, pp. 626–636, 2006, doi: 10.1039/b511962f.

45. C. Mottillo and T. Friščić, "Advances in solid-state transformations of coordination bonds: From the ball mill to the aging chamber," *Molecules*, vol. 22, no. 1, 2017, doi: 10.3390/molecules22010144.

46. J. Gandara-Loe, B. E. Souza, A. Missyul, G. Giraldo, J.-C. Tan, and J. Silvestre-Albero, "MOF-based polymeric nanocomposite films as potential materials for drug delivery devices in ocular therapeutics," *ACS Appl. Mater. Interfaces*, vol. 12, pp. 30189–30197, 2020, doi: 10.1021/acsami.0c07517.

47. A. P. M. Udayan, O. Sadak, and S. Gunasekaran, "Metal-organic framework/polyaniline nanocomposites for lightweight energy storage," *ACS Appl. Energy Mater.*, vol. 3, no. 12, pp. 12368–12377, 2020, doi: 10.1021/ACSAEM.0C02376.

48. N. Habib, O. Durak, M. Zeeshan, A. Uzun, and S. Keskin, "A novel IL/MOF/polymer mixed matrix membrane having superior CO_2/N_2 selectivity," *J. Memb. Sci.*, vol. 658, p. 120712, 2022, doi: 10.1016/J. MEMSCI.2022.120712.

49. S. Nagata, K. Kokado, and K. Sada, "Metal-organic framework tethering PNIPAM for ON-OFF controlled release in solution," *Chem. Commun*, vol. 51, p. 8614, 2015, doi: 10.1039/c5cc02339d.

50. C. Liu, et al., "High dielectric constant epoxy nanocomposites based on metal organic frameworks decorated multi-walled carbon nanotubes," *Polymer (Guildf).*, vol. 207, 2020, doi: 10.1016/J. POLYMER.2020.122913.

51. A. Kirchon, L. Feng, H. F. Drake, E. A. Joseph, and H.-C. Zhou, "From fundamentals to applications: A toolbox for robust and multifunctional MOF materials," *Chem. Soc. Rev*, vol. 47, p. 8611, 2018, doi: 10.1039/c8cs00688a.

3 Electrospun Polymer Nanofibers in Catalysis for Pollutant Removal

Madhavi Konni and Manoj Kumar Karnena

CONTENTS

3.1 INTRODUCTION

One-dimensional nanomaterial has gained much prominence owing to its application in energy, catalysis, and membrane filtration. Many researchers have recently researched the same topic using various nanofibers (NFs) of various compositions. NFs have unique properties like high surface areas and porosities (Thenmozhi et al., 2017; Bölgen and Vaseashta, 2021; Karnena et al., 2020). Electrospun nanofibers have much scope for a wide variety of applications in science as they are easily modified and cost-effective (Shin et al., 2012; Konni et al., 2022). Additionally, stable structures are obtained by avoiding the particles' aggregation, and the catalyst's performance is expected to be stable during their performance in the reactions (Munnik et al., 2015). Over the past few decades, several researchers switched their attention from nanoparticles (NPs) to NFs. The electrospinning (ES) process is the only method for preparing the NFs in the larger scale synthesis compared to the other available methods. Even though the ES techniques are not new as they were started in the 1930s, they came to notice in only the 20th century as the instrumentation is advanced and working parameters are testified. The literature shows that several electrospun nanofibers were fabricated and used in various applications after 2000 and continuously grew. The basic principle involved in these techniques is electrostatic interaction—the peristaltic pump pushes the solution via a high-voltage needle, and the hand ejects the fibers and collects them on the collector.

DOI: 10.1201/9781003343912-3

3.2 HISTORY OF THE ELECTROSPUN NANOFIBERS

The electrospray techniques were started in the 1890s and later transformed into ES techniques in 1900. A brief timeline of ES techniques is shown in Figure 3.1. It is worth identifying the research advances after 2010, which dealt mainly with the applications of electrospun nanofibers in the allied fields using the polymer nanomaterials and metal oxide (MO) nanocomposites for doping and modifying NFs.

Further, from their review, Konni et al. (2022) revealed that electrospun nanofibers are widely used for catalytic applications to degrade environmental pollutants. In a few cases, inorganic/organic/mixtures are used for catalyst preparation as they act as supporting materials or catalysts during reactions. For obtaining specific geometrical structures, organic compound matrixes are used as templates; different nanomaterials are used in some cases during the catalytic process to immobilize the compounds in solid phases (Tucker et al., 2012). If the materials obtained have a high surface area and porosity with dimensional networks, these materials are widely researched for numerous scientific applications. The immobilization of nanomaterials on electrospun nanofibers improves the thermal stability, selectivity, durability of NFs and activity of the electrospun nanofibers (Figure 3.2).

In addition to the porosity and surface areas, these materials also offer unique properties. The surface area and porosity improve the catalytic transport activities and kinetics as they provide access to the NFs. The low strength of fluid flow allows adaptability to geometry with any shape (Xue et al., 2017). The fiber structure will not allow nanomaterial aggregation or leaching within the frameworks. The electrospun nanofibers have another salient feature, i.e., they can be used in batch/continuous flow systems and are easy to recover from the reactor. The catalysts might replace the nano compounds to reduce the limitations during the catalyst recovery. Further, catalytic material loading and mechanical properties limit the NF-type catalyst for many applications (Duru Kamaci and Peksel, 2021). Many researchers attempted to develop and design a simple construction and process the conditions to acquire extremely active nanostructures for the catalytic process. In the current chapter, we briefly discussed electrospun nanofibers, their preparation, catalyst types, and the application areas of ES in catalysis to remove pollutants (Filiz, 2020).

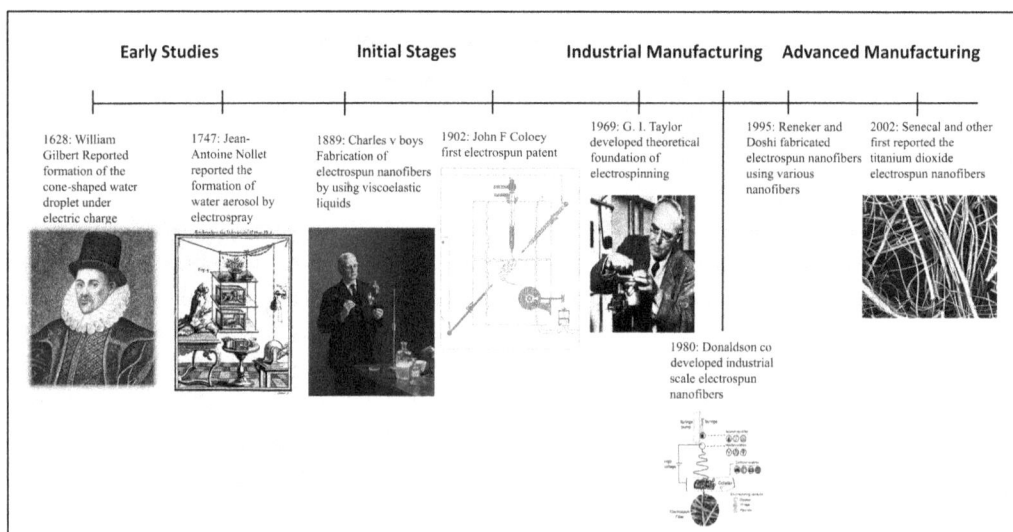

FIGURE 3.1 A brief timeline of the electrospinning nanofibers in science (Konni et al., 2022).

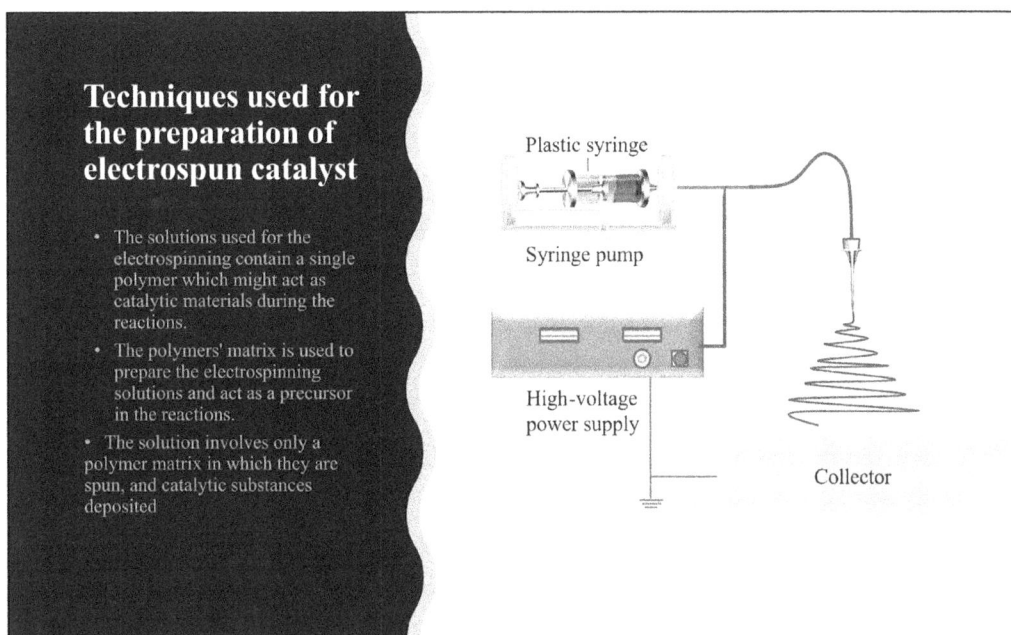

Techniques used for the preparation of electrospun catalyst

- The solutions used for the electrospinning contain a single polymer which might act as catalytic materials during the reactions.
- The polymers' matrix is used to prepare the electrospinning solutions and act as a precursor in the reactions.
- The solution involves only a polymer matrix in which they are spun, and catalytic substances deposited

Plastic syringe

Syringe pump

High-voltage power supply

Collector

FIGURE 3.2 Techniques used for the preparation of electrospun catalysts.

3.3 TYPE OF CATALYSTS

The catalyst types are classified based on the inorganic or organic templates and MO supporters as shown in Figure 3.3.

3.3.1 SUPPORTER CATALYST

Organic-type NFs can be considered carriers for the catalytic NPs as they support the inorganic types of NFs. The immobilization of NPs mainly depends on the reactions and kinds of NF materials (ceramics, carbon, or polymers). New methods like wet impregnation (Soukup et al., 2014), metallization (Hong et al., 2008; Guo et al., 2017), liquid-phase deposition (Liu et al., 2017), reduction by in situ methods (Moreno et al., 2015; Yang et al., 2016), drop coating methods (Niu et al., 2017), photo deposition method (Qin et al., 2017), and impregnation are considered to be prominent compared to conventional methods for catalytic loading and supporting. According to the literature available, catalytic NPs are homogenous and distributed on NF surfaces during catalytic activity and enhance stability and regeneration capabilities. The one-step ES techniques are used to prepare organic supports like porous carbon fibers, cellulose, composite nanofibers, and N-doped NFs with metals by controlled macroscopic structures (Wang et al., 2017). Even though many techniques are available to synthesize NFs, ES techniques are considered efficient as they produce NFs with uniform diameters (Xu et al., 2018).

Different experimental methods can prepare carbon nanofiber(CNF)-type supports; ES after thermal treatment and catalytic particle loading is prominent. Polyvinyl alcohol, 2,6-dimethylphenol polymer, poly(1-phenylethene), and polyacrylonitrile are used as fibers. Inorganic NF support is one of several researchers' research interests for many reactions. Supporting catalyst materials like silicon dioxide (Kang et al., 2010; Wen et al., 2015), titanium dioxide (Formo et al., 2008; Someswararao et al., 2021), selenium dioxide, and zirconium dioxide (Zhou et al., 2020) are

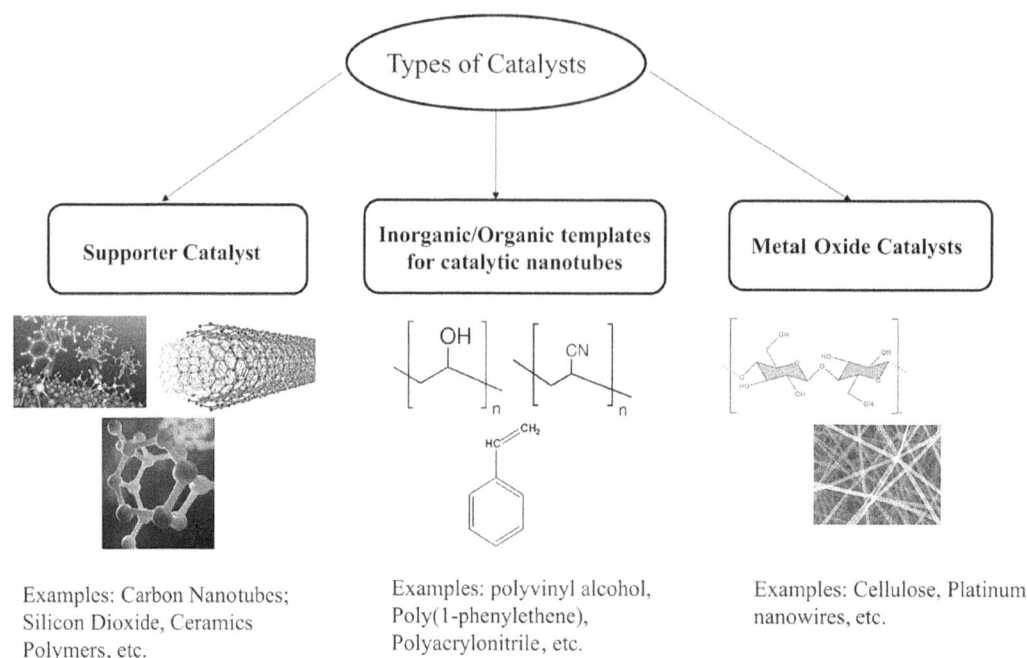

FIGURE 3.3 Types of catalysts (Filiz, 2020).

intensively studied like the organic NF supporters. The temperature, laboratory conditions, etc. should be considered and not ignored during the synthesis and formation of NFs with unique features like surface areas, stability, and crystal lattice.

3.3.2 TEMPLATES

Several organic or inorganic materials are used as a template for catalysts; the ES techniques have been used widely to synthesize catalytic NFs. The material used as a template is removed by the chemical/thermal treatment methods to obtain only nanotube forms. Organic polymers like polyvinyl alcohol, poly(1-phenylethene), and polyacrylonitrile are used as templates to prepare catalytic nanotubes (Hosseini et al., 2017). The ES techniques are one of the approaches for synthesizing catalytic nanotubes. Likewise, tin oxide is used by several researchers for catalyst preparation (Chen et al. 2017). Ochanda and Jones (2005) stated that counterpart compounds in the catalytically activated reactions locate on the nanotube's surface or hollow channels. Further, Chen et al. (2017) revealed that these approaches enhance the interactions between the reactants and catalytically activated counterparts on the inner and outer surfaces of the tubes.

3.3.3 METAL OXIDES

Nanomaterials like nanorods, NFs, and nanotubes of metals enhanced the surfaces' surface-to-volume ratios and active sites for the reactions. The utilization of metal nanostructures alone as a material for catalytic reactions in the energy applications field might be helpful as these materials have high conductivity and electron transfer; thus, these structures reduce the drawbacks associated with the MO-type catalysts (Ochanda and Jones, 2005). Active reaction sites of the trigger, present on the surfaces of the nanomaterials, leach and might cause corrosion with time (Kim et al., 2010; Choi et al., 2008). ES is a conventional technique used to prepare fibers from the solutions of the polymers, even though the metal precursors and polymer mixtures are

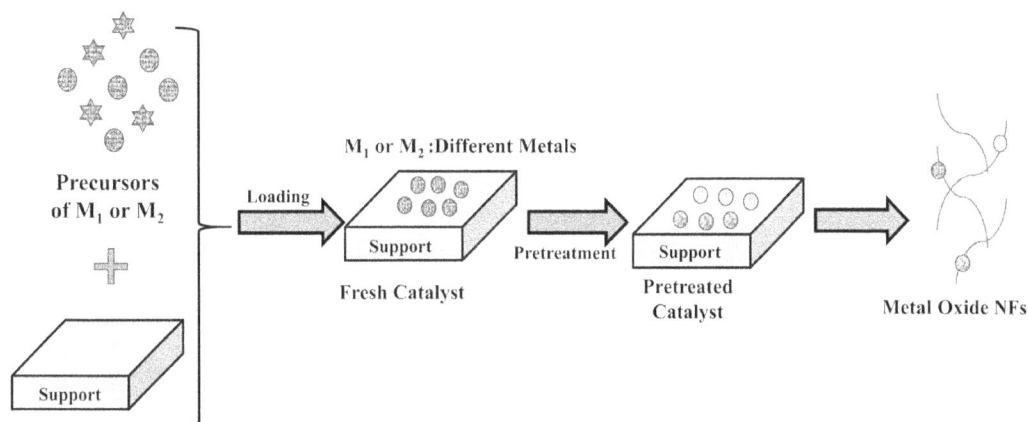

FIGURE 3.4 Precursors of metal interaction with catalyst supporters for NF formation.

used to develop hybrid metal–polymer NFs. By subjecting the polymer to precursor solutions for thermal treatment, MO NFs are synthesized (Figure 3.4). Desired structures of the compounds are attained by this process in single or multiple steps, and the heating environment plays a vital role in acquiring the specific design (Kim et al., 2010). Cellulose is considered an excellent catalytic supporter for catalytic activity as this is the most abundant and renewable resource. Cellulose NFs developed by ES techniques enhance the hydrophilicity and mechanical properties to contact NPs. Thus, these properties allowed the adhesion and dispersion of NPs during the catalytic reactions. Gopiraman et al. (2018) used cellulose NFs with anionic charges to attain composite catalysts with Au, Ag, and Ni NPs. These mixed catalysts gained excellent adhesion and uniformly resulted in NP distribution. The cellulose NFs of Au, Ni, and Ag showed good efficiencies in removing 4 and 2 NPs compared to the literature available.

3.4 ELECTROSPUN NANOFIBERS IN CATALYTIC APPLICATIONS

Inorganic materials are used in electrospun catalysts to produce catalytic materials. The energy demand has increased daily due to the rapid increase in population and technology. Many researchers investigated sustainable and renewable technologies to replace conventional energy resources, reducing the associated costs and becoming more practical for on-board and off-board applications. Several renewable energy resources used the catalysts to reduce costs and enhance efficiency to meet energy demands. The catalyst designs are crucial for energy applications and should have high replicability and catalytic activities. Electrocatalytic oxidation of glucose (Ye et al., 2016), hydrogen sulfide, hydrazine, biomass, and water (Chen et al., 2017) is essential in the catalytic activities as they are related to conventional energy resource applications. Hydrogen gas production by oxidizing the water has been attracted by many researchers and considered an alternative approach to the wind, solar, etc. This method will not generate any pollutants and is highly stable for carrying energy. Even though conventional catalysts like oxides of iridium and ruthenium are highly active for this application, the durability and higher operation costs hinder their usage in industrial applications (Zou and Zhang, 2015). Chen et al. (2017) revealed that catalysts produced by the ES techniques are very efficient and stable for the oxidation of water to generate hydrogen. This method developed a catalyst adjusted with the proportions of the elements and changed the structure proportions. Further, using this method, they developed MOs and CNFs in a single approach, leading to the uniform distribution of MO particles on the carbon support surface area. Ye et al. (2016) even investigated the electrospun nanofibers as the catalysts for this reaction. Further, this group developed CNFs with platinum NPs and successfully performed non-enzymatic glucose oxidation.

Hydrazine oxidation by these methods is becoming more popular with the ongoing progress in the research of hydrazine fuel cells. In this regard, developing new catalytic anodes by changing the designs and modifications is required to attain high efficiency. Noble metals like platinum, palladium, and gold showed more reactivity, and nickel and copper oxides (non-noble metals) showed more excellent catalytic performances. During the reactions, the catalyst's performance is sensitive to the particles' size and structures.

ES techniques help attain the desired design during synthesizing different fibers from their precursors. Yoon et al. (2016) developed NFs with lanthanum manganite with an ES technique and used them as a supporting material for NPs of ruthenium (IV) oxide. Due to the homogenous distribution of the active sites, oxygen reduction by the catalytic activity was achieved and enhanced the electron transfer and kinetic reaction in the lithium–ion batteries. The electrospun NFs are even applied in hydrogen generation applications by the electrocatalytic process. Anis et al. (2017) developed nickel–tungsten trioxide NFs by ES as a catalyst for hydrogen generation. This preparation approach has many benefits by controlling the process parameters like the distance of ES, the voltage applied, and the flow rate of the solution. It is to be noted that this approach created new opportunities for the researchers to investigate the various compositions concerning the morphology and structure of fibers.

Furthermore, this approach enables high hydrogen evolution during the process. Niu et al. (2017) encapsulated molybdenum disulfide (MoS_2) in CNFs prepared using a catalyst's electrospun method. The catalyst prepared was used for artificial photosynthesis, which has become an emerging research area for environmental applications to generate hydrogen by splitting water. Qin et al. (2017) synthesized MoS_2@CdS/TiO_2 NFs by spinning techniques arbitrated by photodeposition to evaluate catalytic hydrogen. Characterization of synthesized materials showed that titanium oxide NFs acted as an excellent material for dispersion and enhanced the interactions with the substrates.

3.5 ELECTROSPUN NANOFIBERS FOR POLLUTANT REMOVAL

Technological advancements and population growth increased the environment's pollution levels and enhanced the stress on natural resources. Catalytic activity is becoming more prominent for degrading and disposing of pollutants. Guo et al. (2017) synthesized peroxyacetyl nitrate CNFs using electrospun methods in a single-step procedure. The synthesized catalyst and this method oxidized the nitrogen monoxide to nitrogen dioxide with low concentrations, simulating atmospheric conditions. Lee et al. (2017) conducted similar studies with polyvinyl chloride (PVC)-based zirconium oxide, silver dioxide, manganese oxide, and silver-manganese/zirconium dioxide NFs. They reported that the NFs degraded the soot and benzene with higher oxidation rates. Another significant oxidation reaction in the environment is catalytic oxidation of the volatile organic carbon (VOCs), as these substances adversely affect the environment and human health. Even though many researchers used different catalysts for the oxidations of VOCs, ES methods are not scientifically researched. In general, the MO catalysts are used for oxidation reactions in the energy field; however, the lower surface areas of the catalysts and faster aggregation lead to low-slung activity. Thus, ES techniques are studied to protect surface areas and reduce NP agglomeration.

Several metals have been used for the preparation of catalysts by different approaches. Moreno et al. (2015) developed silver/selenium oxide NFs by ES techniques using polyvinyl chloride (PVC) and reported the morphological features of NFs. The Taylor cone formation in the polymer solution is inhibited by maintaining high MO content and lower PVC concentrations. The viscosity and density of the keys used in reactions depend on the PVC content and the conductivity required. Furthermore, the higher viscosity of the solutions might result in bead formations on the NFs. The metal-to-polymer ratios lead to fiber-shaped catalyst-forming beads with high porosity. Organic pollutant degradation by catalytic/photocatalytic degradation is a new environmental and ecology protection topic. Synthesis of new active surface catalysts might be beneficial for degrading organic pollutants. Many researchers investigated several methods to immobilize the catalytic compounds.

Due to these reasons, developing new materials for the catalytic process has become more prominent in the current research. Zhou et al. (2020) synthesized zirconium dioxide NFs using ES techniques with magnetic and porous characteristics. The synthesized compounds showed high magnetic loads and iron ions for the Fenton reactions to degrade organic pollutants like methylene blue (MB) and phenols. The NFs developed by the electrospun method were subjected to heat treatment; further addition of magnetic NPs formed porous zirconium dioxide fibers. NFs of zirconium dioxide showed an average diameter and thinner walls with more defects and pores. Further, the hollow structures of NFs are not affected by the magnetic particle's impregnation. Ehsani and Aroujalian (2020) synthesized TiO2-doped polyether sulfone NFs to degrade the phenol with greater efficiencies from the aquatic environment. Later, Cheng et al. (2020) used the Mo_2N/MoO_2 composites to stain the MB and achieved a 99% efficiency. Table 3.1 shows a list of electrospun nanofibers used to remove the pollutants.

TABLE 3.1
Widely Used Electrospun Nanomaterial to Remove Environmental Pollutants

S. No.	Electrospun Nanofiber	Surface Area (m^2/g^{-1})	Pollutant Removed	% Removal	Reference
1.	Amidoximated polyacrylonitrile	-	Indigo carmine	90	Yazdi et al. (2018)
2.	Cellulose acetate-doped polyethylene oxide	24.7	Methylene blue	99	Zamel et al. (2019)
3.	Cellulose acetate-doped polydopamine	6.45	Methylene blue	99	Cheng et al. (2020)
4.	Graphene oxide-doped carboxymethyl cellulose	-	Iron	96	Yu et al. (2020)
5.	Graphene oxide-doped titanium dioxide/polyacrylonitrile	-	Malachite green	93	Du et al. (2020)
6.	Mo_2N/MoO_2 composites	-	Methylene blue	93	Cheng et al. (2020)
7.	Polyacrylonitrile-doped polyamide	6.45	Tetracycline	99	Pan et al. (2017)
8.	Polyamide and polyethylenimine	-	Bisphenol A	90	Maryšková et al. (2020)
9.	Polyaniline doped with p-toluene sulfonic acid	8.4	Methyl orange	90	Al-Qassar Bani et al. (2019)
10.	Poly-cyclodextrins	6.45	Methylene blue	99	Celebioglu et al. (2017)
11.	Poly(methyl methacrylate)-doped graphene oxide	16	Methylene blue	99	Mercante et al. (2017)
12.	Silver-doped polyacrylonitrile	-	Methyl orange	96	Chaúque et al. (2019)
13.	Titanium dioxide-doped carbon nanotubes	56.7	Methylene blue	90.6	Yu et al. (2020)
14.	Titanium dioxide-doped polyetherimide	24	Humic acid	80	Al-Ghafri et al. (2019)
15.	Titanium dioxide-doped polyethersulfone	18.4	Phenol	43	Ehsani and Aroujalian (2020)
16.	Titanium dioxide lignin	41	Methylene blue	87.2	Dai et al. (2020)
17.	ZIF-8-doped peroxyacetyl nitrate	9.7	Methylene green	90	Zhan et al. (2019)
18.	Zinc oxide nanoparticle-doped polyacrylonitrile	-	Reactive red 195	90	Phan et al. (2020)
19.	Zirconia	8.3	Allura red	90	Ávila-Martínez et al. (2020)
20.	ZnO nanofibers	24	Methylene blue	97	Elhousseini et al. (2020)
21.	ZnO nanofibers	24	Chromium	79	Elhousseini et al. (2020)

Ávila-Martínez et al. (2020) used zirconia nanofibers to remove the Allura red dye and achieved an efficiency of 90%. Al-Ghafri et al. (2019) used TiO2-doped polyetherimide to remove the humic acid from the aqueous medium and achieved 80% removal efficiencies. Yu et al. (2020) used TiO2-doped carbon nanotubes to fabricate the electrospun nanofibers, removed the MB from the water in their bench-scale studies, and achieved high removal efficiencies. Even though many researchers succeeded in good removal efficiencies in removing pollutants from different mediums, these are only confined to bench-scale studies (Nayl et al., 2022; Zhang et al., 2022; Schoeller et al., 2022; Mavukkandy et al., 2022; Sultana & Rahman, 2022; Panić et al., 2022). They are not moved to large-scale industrial applications, which might be attributed to the lack of funding for research in developing countries like India; a combined study in collaboration with academia and industries sustainably moves the country forward economically and reduces the stress of natural resources.

3.6   FUTURE CHALLENGES WITH ELECTROSPUN NANOFIBERS

In the early 2000s, significant concerns with ES were the fabrication of the nanomaterials using either synthetic or natural polymers for allied applications in science—later, many researchers by researching understood the controlling process parameters for the electrospun techniques for obtaining desired fibers. Electrospun techniques were later used for fiber generation using different polymers and composites. The incorporation of fibers in allied applications for various functions gained much prominence. Further to the available literature, establishing and using the ES units will help the industries find new electrospun nonwoven opportunities. These techniques will enhance their applicability in different applications of remediating environmental pollutants. Innovations might occur by introducing coaxial spinning techniques with other innate methods. Electrospun designs should be broadened to scale up and enhance nanofiber properties for future applications. The main disadvantages of these techniques are the establishment cost and requiring skilled labor for the operation. The initial establishment of these units in the industries requires more capital. The initiatives are not coming forward, especially in developing countries, which leads to increased stress on natural resources. The government has to provide subsidies for the industries to establish these units as they have various applications in different sectors. In addition, these techniques reduce environmental pollution compared to the natural methods used. Many researchers worldwide are still researching these techniques to understand and gain in-depth knowledge of these materials. Few are focusing on optimizing the process parameters to reduce the cost associated with the process. The paucity of literature on these new techniques is not open to many opportunities or applications in the allied fields; however, the availability of new characterization techniques and interest in this topic might lead to innovations in the research.

3.7   CONCLUSIONS

We can understand from the available literature that producing and manufacturing catalysts using ES techniques are reliable and cost-effective in developing countries like India. There are lacunae in the literature, and researchers worldwide should gain much focus and attention to produce a state-of-the-art technology for pollutant degradation. These materials are promising in degrading organic and inorganic pollutants compared to conventional catalysis materials. Adopting these technologies by developing countries is becoming complex as scaling-up studies are not conducted to date. There is a need for conducting collaborative research with academics to implement full-scale studies. Further adopting these technologies leads to the sustainability of the environment.

REFERENCES

Al-Ghafri, B., Lau, W. J., Al-Abri, M., Goh, P. S., & Ismail, A. F. (2019). Titanium dioxide-modified polyetherimide nanofiber membrane for water treatment. *Journal of Water Process Engineering*, 32, 100970.

Al-Qassar Bani Al-Marjeh, R., Atassi, Y., Mohammad, N., & Badour, Y. (2019). Adsorption of methyl orange onto electrospun nanofiber membranes of PLLA coated with pTSA-PANI. *Environmental Science and Pollution Research*, 26(36), 37282–37295.

Anis, S. F., Lalia, B. S., Mostafa, A. O., & Hashaikeh, R. (2017). Electrospun nickel–tungsten oxide composite fibers as active electrocatalysts for hydrogen evolution reaction. *Journal of Materials Science*, 52(12), 7269–7281.

Ávila-Martínez, A. K., Roque-Ruiz, J. H., Torres-Pérez, J., Medellín-Castillo, N. A., & Reyes-López, S. Y. (2020). Allura Red dye sorption onto electrospun zirconia nanofibers. *Environmental Technology & Innovation*, 18, 100760.

Bölgen, N., & Vaseashta, A. (2021). Electrospun nanomaterials: Applications in water contamination remediation. In Ashok Vaseashta, Carmen Maftei (Eds.), *Water Safety, Security and Sustainability* (pp. 197–213). Springer, Cham.

Celebioglu, A., Yildiz, Z. I., & Uyar, T. (2017). Electrospun crosslinked poly-cyclodextrin nanofibers: Highly efficient molecular filtration thru host-guest inclusion complexation. *Scientific Reports*, 7(1), 1–11.

Chaúque, E. F., Ngila, J. C., Ray, S. C., & Ndlwana, L. (2019). Degradation of methyl orange on Fe/Ag nanoparticles immobilized on polyacrylonitrile nanofibers using EDTA chelating agents. *Journal of Environmental Management*, 236, 481–489.

Chen, S., Du, H., Wei, Y., Peng, L., Li, Y., Gan, L., & Kang, F. (2017). Fine-tuning the cross-sectional architecture of antimony-doped tin oxide nanofibers as Pt catalyst support for enhanced oxygen reduction activity. *International Journal of Electrochemical Sciences*, 12, 6221–6231.

Cheng, J., Xing, Y., Wang, Z., Zhao, X., Zhong, X., & Pan, W. (2020). A novel efficient RhB absorbent of Mo_2N/MoO_2 composite nanofibers for wastewater treatment. *Journal of the American Ceramic Society*, 103(5), 2975–2978.

Choi, S. M., Kim, J. H., Jung, J. Y., Yoon, E. Y., & Kim, W. B. (2008). Pt nanowires prepared via a polymer template method: Its promise toward high Pt-loaded electrocatalysts for methanol oxidation. *Electrochimica Acta*, 53(19), 5804–5811.

Dai, Z., Ren, P., Cao, Q., Gao, X., He, W., Xiao, Y., … & Ren, F. (2020). Synthesis of TiO_2@ lignin based carbon nanofibers composite materials with highly efficient photocatalytic to methylene blue dye. *Journal of Polymer Research*, 27(5), 1–12.

Du, F., Sun, L., Huang, Z., Chen, Z., Xu, Z., Ruan, G., & Zhao, C. (2020). Electrospun reduced graphene oxide/TiO_2/poly (acrylonitrile-co-maleic acid) composite nanofibers for efficient adsorption and photocatalytic removal of malachite green and leucomalachite green. *Chemosphere*, 239, 124764.

Duru Kamaci, U., & Peksel, A. (2021). Enhanced catalytic activity of immobilized phytase into polyvinyl alcohol-sodium alginate based electrospun nanofibers. *Catalysis Letters*, 151(3), 821–831.

Ehsani, M., & Aroujalian, A. (2020). Fabrication of electrospun polyethersulfone/titanium dioxide (PES/TiO_2) composite nanofibers membrane and its application for photocatalytic degradation of phenol in aqueous solution. *Polymers for Advanced Technologies*, 31(4), 772–785.

Elhousseini, M. H., Isık, T., Kap, Ö., Verpoort, F., & Horzum, N. (2020). Dual remediation of waste waters from methylene blue and chromium (VI) using thermally induced ZnO nanofibers. *Applied Surface Science*, 514, 145939.

Filiz, B. C. (2020). Application of electrospun materials in catalysis. In Inamuddin Inamuddin, Rajender Boddula, Mohd Imran Ahamed, and Abdullah M. Asiri (Eds.), *Electrospun Materials and their Allied Applications* (pp. 113–130). Scrivener Publishing.

Formo, E., Lee, E., Campbell, D., & Xia, Y. (2008). Functionalization of electrospun TiO_2 nanofibers with Pt nanoparticles and nanowires for catalytic applications. *Nano Letters*, 8(2), 668–672.

Gopiraman, M., Deng, D., Saravanamoorthy, S., Chung, I. M., & Kim, I. S. (2018). Gold, silver and nickel nanoparticle anchored cellulose nanofiber composites as highly active catalysts for the rapid and selective reduction of nitrophenols in water. *RSC Advances*, 8(6), 3014–3023.

Guo, R., Jiao, T., Xing, R., Chen, Y., Guo, W., Zhou, J., ... & Peng, Q. (2017). Hierarchical AuNPs-loaded Fe_3O_4/polymers nanocomposites constructed by electrospinning with enhanced and magnetically recyclable catalytic capacities. *Nanomaterials*, 7(10), 317.

Hong, S. H., Lee, S., Nam, J. D., Lee, Y. K., Kim, T. S., & Won, S. (2008). Platinum-catalyzed and ion-selective polystyrene fibrous membrane by electrospinning andin-situ metallization techniques. *Macromolecular Research*, 16(3), 204–211.

Hosseini, S. R., Ghasemi, S., & Kamali-Rousta, M. (2017). Preparation of CuO/NiO composite nanofibers by electrospinning and their application for electro-catalytic oxidation of hydrazine. *Journal of Power Sources*, 343, 467–476.

Kang, H., Zhu, Y., Yang, X., Jing, Y., Lengalova, A., & Li, C. (2010). A novel catalyst based on electrospun silver-doped silica fibers with ribbon morphology. *Journal of Colloid and Interface Science*, 341(2), 303–310.

Karnena, M. K., Konni, M., & Saritha, V. (2020). Nano-catalysis process for treatment of industrial wastewater. In Gheorghe Duca and Ashok Vaseashta (Eds.), *Handbook of Research on Emerging Developments and Environmental Impacts of Ecological Chemistry* (pp. 229–251). IGI Global. https://www.igi-global.com/book/handbook-research-emerging-developments-environmental/231905

Kim, J. M., Joh, H. I., Jo, S. M., Ahn, D. J., Ha, H. Y., Hong, S. A., & Kim, S. K. (2010). Preparation and characterization of Pt nanowire by electrospinning method for methanol oxidation. *Electrochimica Acta*, 55(16), 4827–4835.

Konni, M., Dwarapureddi, B. K., Dash, S., Raj, A., & Karnena, M. K. (2022). Titanium dioxide electrospun nanofibers for dye removal: A review. *Journal of Applied and Natural Science*, 14(2), 450–458.

Lee, C., Shul, Y. G., & Einaga, H. (2017). Silver and manganese oxide catalysts supported on mesoporous ZrO_2 nanofiber mats for catalytic removal of benzene and diesel soot. *Catalysis Today*, 281, 460–466.

Liu, Y., Jiang, G., Li, L., Chen, H., Huang, Q., Du, X., & Tong, Z. (2017). Electrospun CeO_2/Ag@ carbon nanofiber hybrids for selective oxidation of alcohols. *Powder Technology*, 305, 597–601.

Maryšková, M., Schaabová, M., Tomankova, H., Novotný, V., & Rysová, M. (2020). Wastewater treatment by novel polyamide/polyethylenimine nanofibers with immobilized laccase. *Water*, 12(2), 588.

Mavukkandy, M. O., Ibrahim, Y., Almarzooqi, F., Naddeo, V., Karanikolos, G. N., Alhseinat, E., … & Hasan, S. W. (2022). Synthesis of polydopamine coated tungsten oxide@ poly (vinylidene fluoride-co-hexafluoropropylene) electrospun nanofibers as multifunctional membranes for water applications. *Chemical Engineering Journal*, 427, 131021.

Mercante, L. A., Facure, M. H., Locilento, D. A., Sanfelice, R. C., Migliorini, F. L., Mattoso, L. H., & Correa, D. S. (2017). Solution blow spun PMMA nanofibers wrapped with reduced graphene oxide as an efficient dye adsorbent. *New Journal of Chemistry*, 41(17), 9087–9094.

Moreno, I., Navascues, N., Irusta, S., & Santamaria, J. (2015). Electrospun Au/CeO_2 nanofibers: A highly accessible low-pressure drop catalyst for preferential CO oxidation. *Journal of Catalysis*, 329, 479–489.

Munnik, P., de Jongh, P. E., & de Jong, K. P. (2015). Recent developments in the synthesis of supported catalysts. *Chemical Reviews*, 115(14), 6687–6718.

Nayl, A. A., Abd-Elhamid, A. I., Awwad, N. S., Abdelgawad, M. A., Wu, J., Mo, X., … & Bräse, S. (2022). Review of the Recent Advances in Electrospun Nanofibers Applications in Water Purification. *Polymers*, 14(8), 1594.

Niu, F., Dong, C. L., Zhu, C., Huang, Y. C., Wang, M., Maier, J., … & Shen, S. (2017). A novel hybrid artificial photosynthesis system using MoS2 embedded in carbon nanofibers as electron relay and hydrogen evolution catalyst. *Journal of Catalysis*, 352, 35–41.

Ochanda, F., & Jones, W. E. (2005). Sub-micrometer-sized metal tubes from electrospun fiber templates. *Langmuir*, 21(23), 10791–10796.

Pan, S. F., Dong, Y., Zheng, Y. M., Zhong, L. B., & Yuan, Z. H. (2017). Self-sustained hydrophilic nanofiber thin film composite forward osmosis membranes: Preparation, characterization and application for simulated antibiotic wastewater treatment. *Journal of Membrane Science*, 523, 205–215.

Panić, J., Vraneš, M., Mirtič, J., Korošec, R. C., Zupančič, Š., Gadžurić, S., … & Bešter-Rogač, M. (2022). Preparation and characterization of innovative electrospun nanofibers loaded with pharmaceutically applicable ionic liquids. *International Journal of Pharmaceutics*, 615, 121510.

Phan, D. N., Rebia, R. A., Saito, Y., Kharaghani, D., Khatri, M., Tanaka, T., … & Kim, I. S. (2020). Zinc oxide nanoparticles attached to polyacrylonitrile nanofibers with hinokitiol as gluing agent for synergistic antibacterial activities and effective dye removal. *Journal of Industrial and Engineering Chemistry*, 85, 258–268.

Qin, N., Xiong, J., Liang, R., Liu, Y., Zhang, S., Li, Y., … & Wu, L. (2017). Highly efficient photocatalytic H_2 evolution over MoS_2/CdS-TiO_2 nanofibers prepared by an electrospinning mediated photodeposition method. *Applied Catalysis B: Environmental*, 202, 374–380.

Schoeller, J., Itel, F., Wuertz-Kozak, K., Fortunato, G., & Rossi, R. M. (2022). pH-responsive electrospun nanofibers and their applications. *Polymer Reviews*, 62(2), 351–399.

Shin, S. H., Purevdorj, O., Castano, O., Planell, J. A., & Kim, H. W. (2012). A short review: Recent advances in electrospinning for bone tissue regeneration. *Journal of Tissue Engineering*, 3(1), 2041731412443530.

Someswararao, M. V., Dubey, R. S., & Subbarao, P. S. V. (2021). Electrospun composite nanofibers prepared by varying concentrations of TiO_2/ZnO solutions for photocatalytic applications. *Journal of Photochemistry and Photobiology*, 6, 100016.

Soukup, K., Topka, P., Hejtmánek, V., Petráš, D., Valeš, V., & Šolcová, O. (2014). Noble metal catalysts supported on nanofibrous polymeric membranes for environmental applications. *Catalysis Today*, 236, 3–11.

Sultana, N., & Rahman, R. (2022). Electrospun nanofiber composite membranes based on cellulose acetate/nano-zeolite for the removal of oil from oily wastewater. *Emergent Materials*, 5(1), 145–153.

Thenmozhi, S., Dharmaraj, N., Kadirvelu, K., & Kim, H. Y. (2017). Electrospun nanofibers: New generation materials for advanced applications. *Materials Science and Engineering: B*, 217, 36–48.

Tucker, N., Stanger, J. J., Staiger, M. P., Razzaq, H., & Hofman, K. (2012). The history of the science and technology of electrospinning from 1600 to 1995. *Journal of Engineered Fibers and Fabrics*, 7(2_suppl), 155892501200702S10.

Wang, H., Sun, C., Cao, Y., Zhu, J., Chen, Y., Guo, J., … & Zou, G. (2017). Molybdenum carbide nanoparticles embedded in nitrogen-doped porous carbon nanofibers as a dual catalyst for hydrogen evolution and oxygen reduction reactions. *Carbon*, 114, 628–634.

Wen, S., Liang, M., Zou, R., Wang, Z., Yue, D., & Liu, L. (2015). Electrospinning of palladium/silica nanofibers for catalyst applications. *RSC Advances*, 5(52), 41513–41519.

Xu, P., Cen, C., Chen, N., Lin, H., Wang, Q., Xu, N.,... & Teng, Z. (2018). Facile fabrication of silver nanoparticles deposited cellulose microfiber nanocomposites for catalytic application. *Journal of Colloid and Interface Science*, 526, 194–200.

Xue, J., Xie, J., Liu, W., & Xia, Y. (2017). Electrospun nanofibers: New concepts, materials, and applications. *Accounts of Chemical Research*, 50(8), 1976–1987.

Yang, D., Yan, Z., Li, B., Higgins, D. C., Wang, J., Lv, H., … & Zhang, C. (2016). Highly active and durable Pt–Co nanowire networks catalyst for the oxygen reduction reaction in PEMFCs. *International Journal of Hydrogen Energy*, 41(41), 18592–18601.

Yazdi, M. G., Ivanic, M., Mohamed, A., & Uheida, A. (2018). Surface modified composite nanofibers for the removal of indigo carmine dye from polluted water. *RSC Advances*, 8(43), 24588–24598.

Ye, J. S., Liu, Z. T., Lai, C. C., Lo, C. T., & Lee, C. L. (2016). Diameter effect of electrospun carbon fiber support for the catalysis of Pt nanoparticles in glucose oxidation. *Chemical Engineering Journal*, 283, 304–312.

Yoon, K. R., Kim, D. S., Ryu, W. H., Song, S. H., Youn, D. Y., Jung, J. W., … & Kim, I. D. (2016). Tailored combination of low dimensional catalysts for efficient oxygen reduction and evolution in $Li–O_2$ batteries. *ChemSusChem*, 9(16), 2080–2088.

Yu, X., Lu, X., Qin, G., Li, H., Li, Y., Yang, L., … & Yan, Y. (2020). Large-scale synthesis of flexible TiO_2/N-doped carbon nanofibres: A highly efficient all-day-active photocatalyst with electron storage capacity. *Ceramics International*, 46(8), 12538–12547.

Zamel, D., Hassanin, A. H., Ellethy, R., Singer, G., & Abdelmoneim, A. (2019). Novel bacteria-immobilized cellulose acetate/poly (ethylene oxide) nanofibrous membrane for wastewater treatment. *Scientific Reports*, 9(1), 1–11.

Zhan, Y., Guan, X., Ren, E., Lin, S., & Lan, J. (2019). Fabrication of zeolitic imidazolate framework-8 functional polyacrylonitrile nanofibrous mats for dye removal. *Journal of Polymer Research*, 26(6), 1–11.

Zhang, Z., Wu, X., Kou, Z., Song, N., Nie, G., Wang, C., … & Mu, S. (2022). Rational design of electrospun nanofiber-typed electrocatalysts for water splitting: A review. *Chemical Engineering Journal*, 428, 131133.

Zhou, C., Liu, Z., Fang, L., Guo, Y., Feng, Y., & Yang, M. (2020). Kinetic and mechanistic study of rhodamine B degradation by H_2O_2 and $Cu/Al_2O_3/g-C_3N_4$ composite. *Catalysts*, 10(3), 317.

Zou, X., & Zhang, Y. (2015). Noble metal-free hydrogen evolution catalysts for water splitting. *Chemical Society Reviews*, 44(15), 5148–5180.

4 Nanoengineered Polymer Composites and Their Applications
A Systematic Assessment of Synthesis Methods

Amandeep Kaur and Shikha Madan

CONTENTS

DOI: 10.1201/9781003343912-4

4.1 INTRODUCTION

Researchers' fascination and interest in small things and vision that smaller things have greater potential promoted them to miniaturize things which resulted in the creation of laptops, palmtops in place of desktop computers, and microchips. Now, the world is still demanding smaller things, which will be more efficient than today's things. So, the scientists came up with a technology called "nanotechnology." [1] Nanotechnology is the science of tiny things. It is a controlled manipulation of size and shape at molecular level where at least one dimension of the material is 100 nm or less. Research in the field of materials with critical dimensions in the nanometer (nm) scale termed as "nanomaterials" has shown itself exciting areas in science for the past two decades. The huge interest in these materials is attributed to the fact that their (a) unique optical, chemical, physical, and electronic properties (b) material confinement in small structures, and (c) large surface-to-volume ratio. For example, aluminum (Al) can be perfectly safe; however, nano-sized aluminum finds application in producing bombs [2]. Also, nanophase ceramics attracted researchers due to their ductile nature at higher temperatures compared with coarse grain ceramics [3].

In the present decade, nanomaterials have gained a lot of recognition from scientists considering the remarkable improvement in the properties of polymers including high mechanical properties [4–5], thermal conductivity [6], and thermal stability [7]. Therefore, composites of polymers reinforced with nanomaterials find broad applications in interdisciplinary fields from biomedical to marine technology. Both 1D and 2D nanomaterials can be reinforced with polymers depending on their applications. Properties and applications of these nanoengineered polymer composites are briefly explained in the next section.

Carbon nanotubes (CNTs) fabricated by Iijima in 1991 [8] have received a lot of attention in microelectronics, photovoltaics, and biosensors because of their exceptional optical and electrical properties [9]. CNTs are 1D nanomaterials and are commonly employed as nanocarriers to deliver drugs, genes, and other therapeutic agents since they have a high aspect ratio [10]. In the present work, the usage of CNTs for tissue engineering applications has been discussed. CNTs incorporated into polymers exhibit an improvement in the induction of angiogenesis, lessen thrombosis, and effect gene expression for tissue repair [11].

Graphene was the first material that provided enormous motivation to the scientists for 2D nanomaterials. The remarkable advancement of graphene-based materials in the present year results in the emergence of other 2D nanomaterials namely boron nitride nanosheet (BNNS), molybdenum disulfide (MoS_2), niobium diselenide, etc. Two-dimensional nanomaterials acquire a larger surface area in contrast with the nanoparticles (0D) and nanotubes (1D). The large surface-to-volume ratio permits strong interfacial properties with polymers among all lower-dimension materials. Therefore, 2D nanomaterials have become an ideal candidate for improving reinforcing efficiency of polymer composites [12–15].

This present chapter includes the synthesis, properties, and applications of nanoengineered polymer composites. We will broadly emphasize 1D and 2D nanomaterials, for instance, CNTs, graphene and its derivatives, MoS_2 nanosheet, and BNNS. In the middle section, the methods used for the synthesis of nanopolymer composites will be discussed. Subsequently, the applications of nanomaterials reinforced with polymers including tissue engineering, strain sensors, thermal dissipation, three-dimensional (3D) printing techniques, wearable textiles, and tribological applications will be explained.

4.2 STRUCTURE AND PROPERTIES OF NANOMATERIALS

4.2.1 CARBON NANOTUBES

CNTs are thin, long cylinders in the form of nanotubes composed of carbon atoms which are joined in a series of benzene rings [4]. CNTs are classified into single-walled carbon nanotubes (SWCNTs) and multi-walled carbon nanotubes (MWCNTs) as shown in Figure 4.1.

SWCNTs are single long wrapped graphene sheets with an sp^2 hybrid bond. MWCNTs consist of multi-layers of CNTs, concentrically surrounding a central CNT with increasing diameter. SWCNTs have an internal diameter of 1–2 nm, whereas MWCNTs have an internal diameter of 5–20 nm. CNTs also show an excellent electrical conductivity of 105 S/cm^2 and exhibit a thermal conductivity of approximately 5002 W/(m K) [16].

SWCNT structure consists of tips and sidewalls. Three different structures can be obtained from a graphene sheet rolled into cylinders namely armchair, zigzag, and chiral. The electrical conductivity of SWCNTs is fully dependent on their chirality. The zigzag- and chiral-type CNTs have got properties comparable to semiconductors, while the armchair-type CNTs have identical electrical properties to metals.

4.2.2 GRAPHITE

Graphite is a naturally occurring allotrope that is made up of hexagonal carbon sheets bundled on top of each other.

4.2.2.1 Graphene

Graphene is a monolayer of carbon sheet with a hexagonal honeycomb lattice structure. It is popularly known as a "super" material because of its distinctive properties. Novoselov et al. in 2004 obtained the few-layered graphene by the scotch tape transferring technique and characterized its excellent electrical property [17]. Its properties include electrical conductivity (100 S/cm) with Young's modulus of ~1.2 TPa, optical transmittance (~98%), and large surface area (~2598 m^2/g). Despite excellent properties, the growth of pristine graphene faces challenges on high scales. Due to this crucial point, the origin of graphene led to its derivatives including GO (graphene oxide) and rGO (reduced graphene oxide). The synthesis of these derivatives is comparatively easier.

4.2.2.2 Graphene Oxide

The discovery of GO was held before the discovery of graphene. In 1859, GO was first synthesized by oxidation and exfoliation of graphite. Until the discovery of graphene, GO remained somewhat negligible. It was after the discovery of graphene that GO also gained attention as a possible way to obtain graphene. From that point, research on GO picked up pace and it was found to be an excellent candidate for a variety of applications [18].

4.2.2.3 Reduced Graphene Oxide

The approach of synthesis of graphene by reducing GO has fascinated many researchers. In order to obtain rGO structures, chemical, thermal, or photothermal reduction methods are adopted although it is difficult to obtain a pristine graphene structure. Despite best efforts and acute reduction processes, rGO still carries structural defects and residual oxygen. This process affects various parameters such as mechanical and structural properties, reactivity, and dispersibility of GO. It restores the sp^2 structure by reducing the oxygen-related compounds. Reduction of GO enhances electrical conductivity by ~6200 S/cm and excellent mobility of ~3200 cm^2/(V s). In contrast to GO, rGO

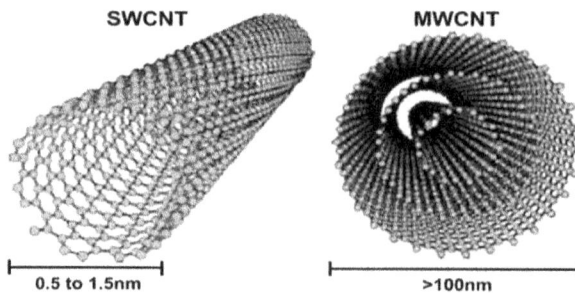

(a) Carbon nanotube, Single walled, Multi walled

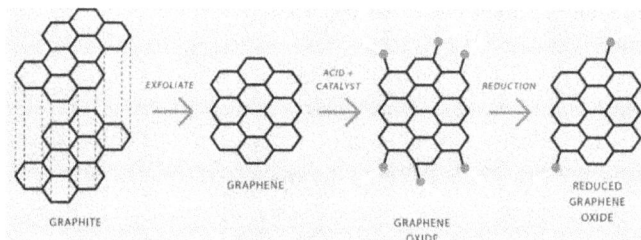

(b) Graphite, Graphene, Graphene oxide and Reduced graphene oxide

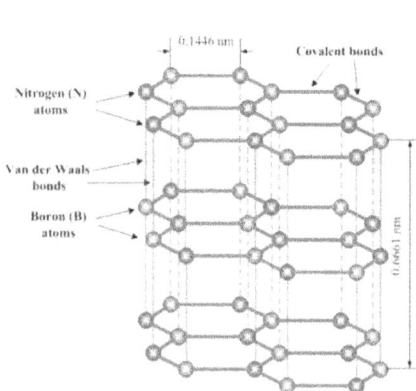

(c) Structural View of BNNS   **(d) MoS2 Nanosheets**

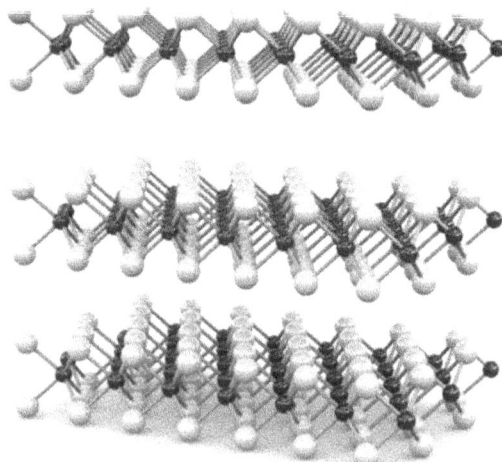

FIGURE 4.1 (a) Structure of nanotubes: SWCNT and MWNT, (b) graphite, graphene, graphene Oxide, and reduced graphene oxide, (c) BNNS, and (d) MoS₂ nanosheets. (Photograph by the Author.)

attains a hydrophobic behavior because of the increased carbon-to-oxygen ratio. Eventually, this reduction lowers the dispersibility of rGO. Although graphene is not completely retrieved by the reduction of GO, this structure holds valuable properties such as controllable functionality, high electrical and thermal conductance, and cheap and scalable preparation process [19]. However, the dispersibility of rGO lowers after reduction. Even though the graphene structure is not completely retrieved by the reduction of GO, this structure holds valuable properties such as enhanced electrical and thermal parameters and requires a cheap and scalable preparation process [19].

4.2.3 BORON NITRIDE NANOSHEET (BNNS)

Boron nitride (BN) is one of the lightest compounds among the three and four groups in the periodic table. BN consists of the same numbers of B and N atoms arranged in a hexagonal manner, similar to carbon atoms in graphene. Figure 4.1 shows the monolayer of BN and multilayer which is termed as BNNS due to the van der Waals interaction between the adjacent layers which holds the BN layers together. The distance between consecutive sheets is ~0.3 nm [20]. The thermal conductivity varies from ~350 to 2500 W/(m K), which is comparable to that of graphene. The hardness of graphene and BNNS is 335 and 267 TPa, respectively. Hence, BNNSs are considered to be a suitable candidate for reinforcement in composites. Researchers reported that modulus and tensile strength for nanosheets are in the range of 210–5120 and 9–17 TPa [20].

4.2.4 2D MOLYBDENUM DISULFIDE (MoS2)

MoS_2 has a structure comparable to graphene. Two closely packed sulfur atom layers are sandwiched between one layer of molybdenum atoms and all layers are bonded by van der Waals forces as shown in Figure 4.1. The stiffness of monolayer MoS_2 is ~200±50 N/m, which indicates an effective Young's modulus of ~220±100 GPa, which is comparable to that of steel. This material is extremely crystalline in nature and has zero defects. The thin structure of MoS_2 shows excellent mechanical, thermal, electrical, and optical properties. It is therefore used as a reinforcing material in the composites for numerous applications such as flexible electronics, photonics, energy storage, and sensors [21,22].

4.3 SYNTHESIS TECHNIQUES

4.3.1 SOLUTION-BASED PROCESSING TECHNIQUE

For the synthesis of nanotubes reinforced with polymer composites, the most preferable technique is the solution processing method. In this technique, the solution is prepared by mixing nanomaterials and polymers using the same/different solvents. Subsequently, the solution is mixed well using mechanical stirring or ultra-sonication. A thin film of a such composite solution is formed by drop casting on the substrate. Following this, a nanoengineered polymer composite film is formed by evaporating the solvent as shown in Figure 4.2 [23,24].

4.3.2 MELT PROCESSING TECHNIQUE

Melt processing is a method specifically employed for insoluble polymers, for example, thermoplastics. Thermoplastics become soft when they are subjected to heat. Nanomaterials are added to the melted polymer and blended using shear mixing [23,24]. Figure 4.3 shows the flowchart for the bulk samples using melt processing.

4.3.3 IN SITU POLYMERIZATION

In situ polymerization is a promising technique that permits nanofillers to be dispersed homogeneously in the conducting as well as insulating polymer matrix, thereby providing a robust interaction between the polymer and the nanofiller. This process is carried out by mixing the monomers/prepolymers with nanofillers in a proper solvent to form a uniform solution. Eventually, polymerization takes place by adjusting the temperature and time as shown in Figure 4.4. Various researchers have reported that it is an outstanding method to obtain a uniform dispersion of CNTs and GO. This method is a popular method for the preparation of composites where polymers are insoluble in solvents or for thermal instability [25].

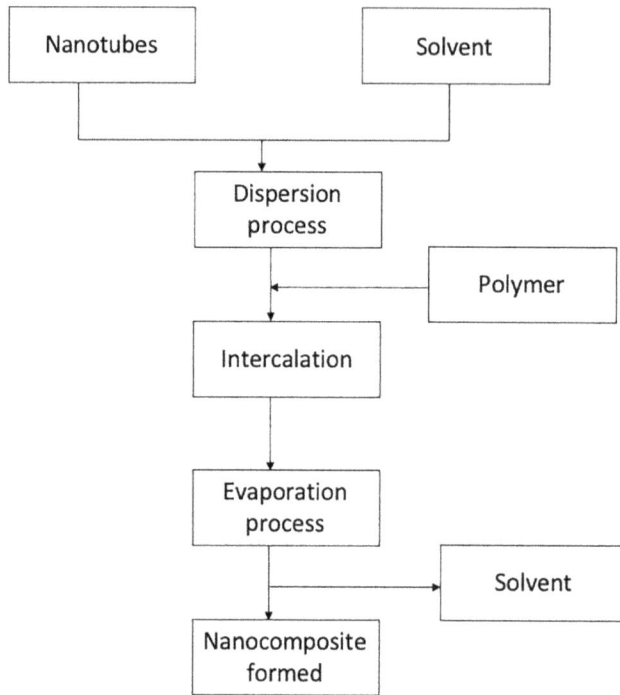

FIGURE 4.2 Flowchart presenting solution-based processing technique. (Source: Author.)

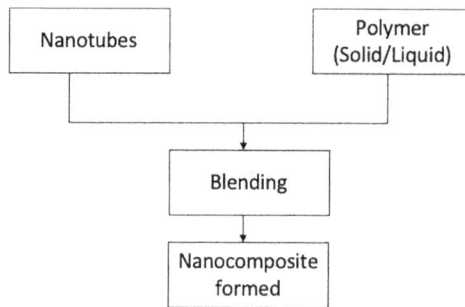

FIGURE 4.3 Flowchart presenting melt processing technique. (Source: Author.)

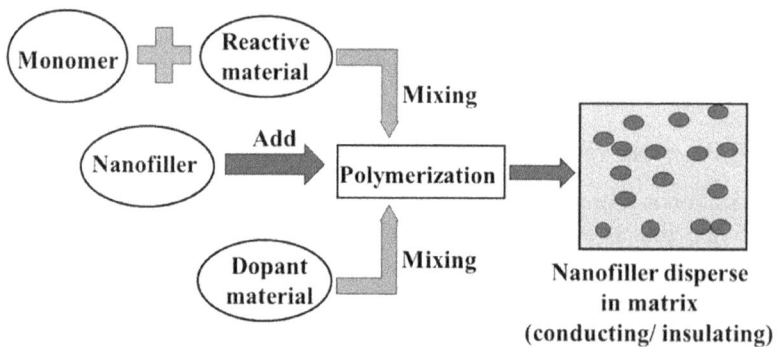

FIGURE 4.4 In situ polymerization. (Source: Author.)

4.4 SALIENT FACTORS INFLUENCING THE PERFORMANCE OF COMPOSITES

As we have discussed various ways to obtain nanocomposites in the previous section, these nano-engineered polymer composites have found applications in various fields. However, it has been observed that various parameters affect the performance of these composites. These influencing factors are discussed as follows.

4.4.1 GEOMETRY

The geometry of nano-sized materials, for example, thickness, lateral dimension, and aspect ratio affect the overall performance of the nanocomposite. The important factors that influence the mechanical properties of polymer composites are the lateral dimension and thickness [16].

4.4.2 VOLUME CONTENT

The volume of nanomaterials is a prominent parameter for deciding the mechanics, coefficient of friction, and wear rate of the composite. A small amount of nanomaterial can show remarkable growth in various properties of composites. It may be noted that an increase in nanomaterial quantity in the nanocomposites increases the strength and modulus. However, if the amount of nanomaterials is more than required, it results in self-agglomerates of the nanomaterial, affecting the overall performance of the composite [26–28].

4.4.3 DISPERSION

The fabrication of nanocomposites has been dominantly influenced by the extent of dispersion of nanomaterials. Nanomaterials have a tendency to form agglomerates due to the strong van der Waals interaction between them which leads to a big challenge for scientists. Researchers have obtained a homogeneous dispersion of nanomaterials by modifying or functionalizing them [29].

4.4.4 ORIENTATION

Current studies have shown that the orientation of nanomaterials reinforced with polymers impacts their mechanical and tribological properties. An orientation of nanomaterials in composites in a proper distribution enhances the interactions between the nanofiller and the matrices, which in turn results in efficient stress transfer whereas random distribution of nanomaterials results in modest load transfer ability [30].

4.4.5 INTERFACE BONDING

Interface bonding plays an important role in deciding the performance of the composite. If the interface bonding between the polymer and the nanomaterial is good, the final composite has a high Young's modulus in comparison to that of the pristine polymer [16,19].

4.4.6 PROCESSING PARAMETERS

In the previous section, different processing parameters including pressure, temperature, and time for the synthesis process of nanocomposites have been described. These parameters directly influence the quality of the composite. For obtaining an efficient nanoengineered composite, the above-mentioned parameters should be optimized.

4.5 APPLICATIONS OF NANOENGINEERED POLYMER COMPOSITE

4.5.1 APPLICATIONS BASED ON SURFACE PROPERTIES

4.5.1.1 Tribology

Tribology deals with the field of lubrication, friction control, and prevention of wear of surfaces with motion under loading conditions. Figure 4.5 shows the application of tribology in various sectors [31,32].

Currently, nanoengineered polymer composites with excellent potential have gained tremendous popularity in tribological applications. Two-dimensional nanomaterials such as BNNS, MoS_2, and graphene act as excellent lubricants as they offer less friction and high thermal conductivity.

Min et al. reported that the incorporation of a small amount of BNNS results in a significant reduction in the coefficient of friction and wear rate [33]. These 2D nanosheets are compatible with polymers [34]. Apart from this, 2D nanosheets are low in density, which results in excellent dispersion in nanocomposites. Therefore, at the interface, thermal conductivity, mechanical properties, and stress transfer between the polymer and nanosheets show a significant improvement.

Due to agglomeration in 2D nanosheets, which reveals weak van der Waals interactions between subsequent layers of ultra-thin shape with respect to one-dimensional nanorods, sliding between consequent layers of nanomaterials happens smoothly as it effortlessly enters the contact area of the sliding surfaces and exhibits the self-lubricant property [35,36].

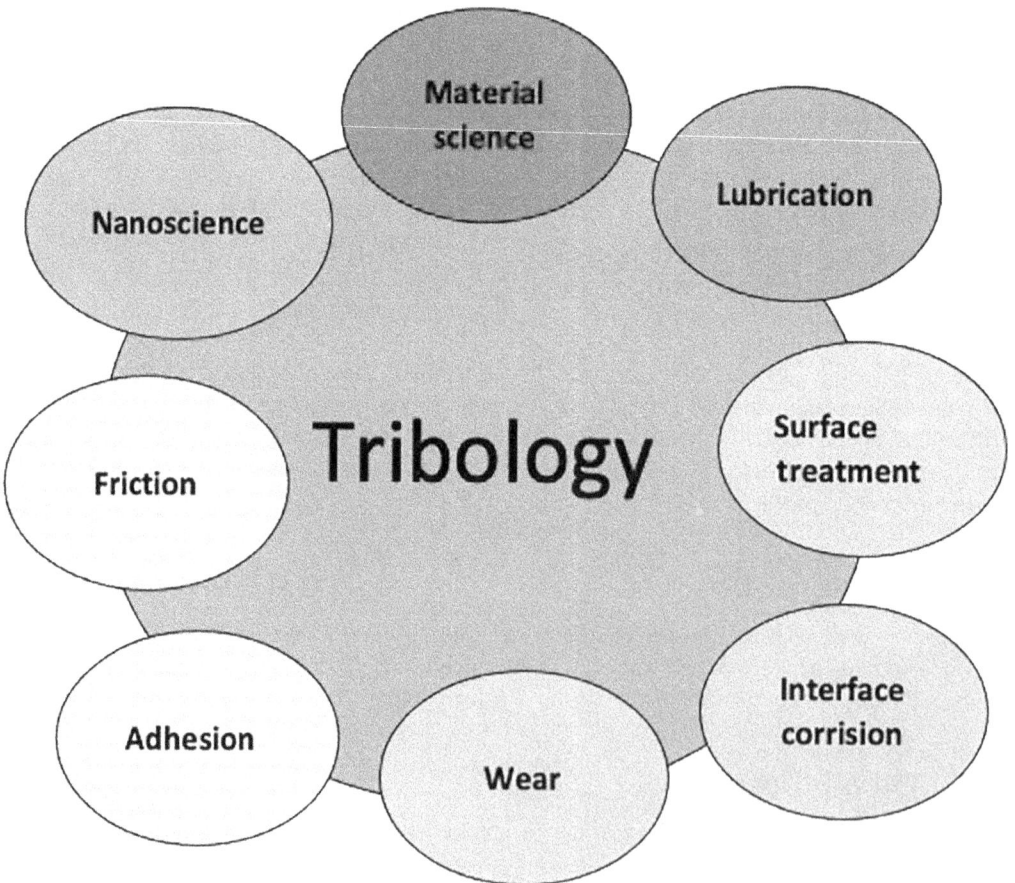

FIGURE 4.5 Applications of tribology in various sectors [31,32]. (Photograph by the Author.)

Jialin Liu et al. [12] stated that 2D nanomaterials improve structural defects and tribological performance by controlling the friction heat between adjacent layers. Besides this, the increase in humidity reduces the friction in graphene and BNNS due to the fact that water molecules dissociate in hydrogen and oxygen. These hydrogen ions repair the damaged graphene by reacting with the carbon-dangling bonds. As a result, graphene maintains and increases its lubricant nature. Table 4.1 shows the comparison of various nanoengineered polymer composites for tribological applications [36–39].

Qiu et al. [40] also investigated the melting property of composites made up of MoS_2. It is observed that the addition of MoS_2 nanoparticles reduces the friction as well as the melting wear of the composites. Therefore, 2D nanomaterials have proved to be promising fillers that eventually enhance the tribology in composites [41,42].

S. Li reported [43] that when the polytetrafluoroethylene (PTFE) composites rub with their metal counterparts due to the easy shear of PTFE lamellae. Figure 4.6 shows the friction and transfer film mechanism, respectively.

It is investigated that transfer film formation involves parameters such as chemical reaction, mechanical compression, tension, and shear. During interlayer sliding, chain breakdown of the carbon–carbon bond of PTFE can occur. Afterward, carboxylic acid end groups chelate to the metal that affects the wear rate of the material significantly.

4.5.1.2 3D Printing

Three-dimensional printing technique which is also known as digital fabrication technology is an encouraging method for fabricating composite structures due to its low cost and easy fabrication. This technology is extensively used in the healthcare industry, automation, agriculture, and locomotive industry for customization and production [44–47].

When a polymer and specific nanomaterial are mixed to form a composite, a low-viscosity material is formed because of the strong interaction between the polymers and nanomaterials. Fei et al. investigated that incorporation of graphene into partially hydrolyzed polyacrylamide shows an increase in shear rate resulting in low viscosity [48]. At a high shear rate and low viscosity, the nanoengineered polymer composite persuades the printing process smoothly. Besides this, an improvement in the mechanical and thermal conductivity of 3D-printed composite polymers is also observed [12].

TABLE 4.1
Tribological Properties of 2D Nanoengineered Polymer Composites

S. No.	Polymer Matrix	Weight % Nanomaterial	Coefficient of Friction	Wear Rate
1	PTFE	~5.1 wt.% graphene	0.2	~7×10^{-6}
2	EP	~0.45% BNNS	0.52	~17×10^{-5}
3	Polyurethane	~3.2% MoS_2	0.15	~9.2×10^{-5}

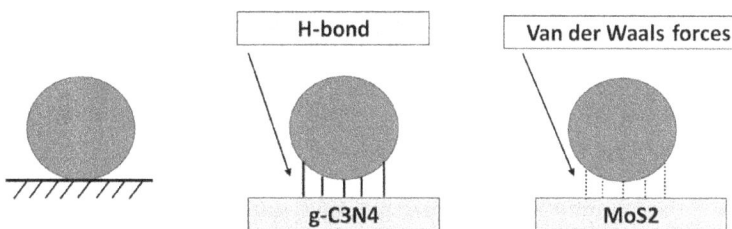

FIGURE 4.6 Schematic representation showing friction mechanism in composites.

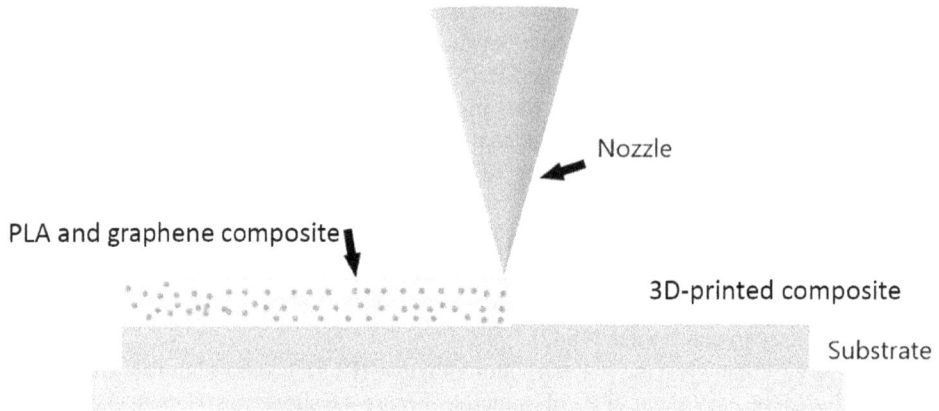

FIGURE 4.7   Schematic diagram of 3D-printed composites. (Photograph by the Author.)

Nanoengineered polymer composites can be fabricated with oriented 2D nanosheets in the polymer matrix [49,50]. Figure 4.7 shows the steps to be followed for making a 3D printing film. Firstly, the GO membrane is formed by spraying GO on the glass substrate. Afterward, polylactic acid (PLA) is projected on the membrane. After that, PLA/GO sheet is cast off from the glass substrate. Finally, for getting the 3D printing, the casted-off PLA/GO sheet is put onto the polymer as shown in Figure 4.7.

4.5.1.3   Tissue Engineering

Tissue engineering is a multidisciplinary area that is related to the knowledge of engineering and life sciences to restore, repair, and replace damaged or unhealthy tissues. It is achieved by incorporating living cells or any biological alternatives such as biomolecules and synthesis or natural materials which are biocompatible and degradable that can help maintain and improve the architecture of tissues/organs [51].

Polymers act as suitable materials used in tissue engineering applications because of their biodegradable as well as biocompatible properties. In spite of that fact, some polymers having inert nature restricted their uses in tissue engineering. Thus, the composite of polymers with other bioactive materials is commonly used for such an application. The composite of CNT with a polymer improves the flexibility, biocompatibility, and structural, physical, and chemical properties and also lowers thrombosis and frames up gene expression for tissue engineering. CNTs limit themselves in terms of solubility as they are insoluble in most organic solvents even in water, which makes them hard to use for tissue engineering applications. Also, during manufacturing, defects such as structure and randomness lead to genotoxicity. In order to enhance the solubility rates and minimize the toxicity of CNTs, they are generally combined with polymers [16].

In the field of tissue engineering, researchers mostly employ CNTs as a reinforcing material in polymers. They aid in the fabrication of scaffolds for neural, cardiac, and bone tissue engineering. Incorporation of nanotubes promotes the conductivity of nerve-related scaffolds and improves nerve cell responses. Besides this, the elastic strength and biological response are also enhanced [16,51].

Several researchers have reported that CNTs can be employed as a substrate for tissue engineering as they have the ability to carry neuron attachment, provide a form of extended neurites and also enhance cell differentiation. However, the toxicity of nanotubes is a big challenge. Bare polymer scaffolds are biocompatible but they lack electrical conductivity and appropriate tensile strength. However, this nanoengineered CNT polymer composite solves various problems and offers high electrical conductivity and extraordinary mechanical properties.

The electrospinning method is used for the fabrication of a 2D film for neural tissue engineering. MWCNT-based composites exhibit low electrical resistance with high Young's modulus and excellent biocompatibility [16,51].

4.5.2 MECHANICAL PROPERTIES

4.5.2.1 Young's Modulus and Tensile Strength

Young's modulus predicts the elasticity possessed by the material, which gives the estimation of how much deformation in the material can occur as well as the power needed to deform it. However, tensile strength is the degree of the maximum stress that a material can bear. Both these factors are important in the case of nanoengineered polymer composites as the composite developed for engineering applications may undergo static and dynamic stress. It is desired that the nanocomposite remains intact under various conditions.

Researchers observed that Young's modulus and tensile strength of pristine polymers are limited [52]. When 2D nanosheets such as graphene, BNNS, and MoS_2 are incorporated into the polymers, Young's modulus and tensile strength of the composites have enhanced drastically as shown in Table 4.2. Also, it has been reported that the incorporation of graphene into the PVA matrix shows excellent reinforcing efficiency in Young's modulus [52].

The tensile strength of nanoengineered polymer composites is accredited to the proper stress transfer at the interface. Precisely, nanosheets adsorb chains of polymer at the interface which limits the movement of polymer chains. Eventually, it leads to an improvement in the strength of the composite [52,53].

In the case of MoS_2 nanosheets reinforced with polymers, an improvement in tensile strength is observed. These nanoengineered polymer composites exhibit the following advantages. Firstly, it improves reinforcing efficiency. Secondly, it decreases the number of voids in the composite. Voids in the polymer matrix lead to defects that reduce the strength of the polymer matrix. H. Peng et al. reported that voids of microscale are certainly visible in the case of pristine polyethylene oxide (PEO), although PEO reinforced with the MoS_2 nanosheet composite is void free under the same magnification [54].

Reinforcement of nanomaterials also improves the mechanical properties of composites. The nanoengineered polymer composite shows better durability and protection in various cases such as the crash of automobiles, smashing of ships, and birds striking the plane. As a result, this nanoengineered composite shows excellent fracture toughness by improving the impact resistance [61].

TABLE 4.2
Improvement in Young's Modulus and Tensile Strength of Different Nanocomposites

Nanoengineered Composites	Improvement in Young's Modulus	Improvement in Tensile Strength	Reference
Epoxy:Graphene	~55%	~125%	[55]
UHMWPE:Graphene	~40%	~72%	[56]
PMMA:BNNS	~20%	~10%	[57]
PVA:Graphene	~70%	~14%	[58]
PVA:Graphene	~15%	~16%	[59]
PVA: MoS_2	~12%	~15%	[60]

4.5.3   Applications Based on Thermal Properties

4.5.3.1   Thermal Dissipation

Thermal management systems are essential in electronic devices as they generate heat continuously. All electronic gadgets such as battery units in electric vehicles and smartphones produce heat. This heat affects the reliability and life span of components and leads to premature device failure. As the size of electronic devices reduces, the heat produced inside the device needs to be sank out. Therefore, thermal management materials are used in the development of gadgets so the heat generated within the gadgets can be dispersed away frequently. Heat sinks are generally interfaced with electronic devices for effective heat dissipation [62].

A survey shows that pristine polymers are used as a heat sink in electric automobiles and gadgets. However, pristine polymers offer low thermal conductivity [63]. It was reported by various researchers that by adding a small amount of nanomaterials in polymers, an efficient heat-conducting path is formed, leading to improved thermal conductivity [64–67].

Graphene exhibits amazing thermal and electrical conductivity. Therefore, graphene-based nanoengineered composites are suitable for developing electronic gadgets and vehicles. If the proportion of graphene is increased up to a certain level, the thermal conductivity of nanoengineered composites will improve further [68].

It has been reported that composites incorporating BNNS show superior electrical insulation and thermal stability. Hence, it has found applications in electronic gadgets and electric vehicles. Thermal conductivity of BNNS- and poly(diallyldimethylammonium chloride)-based composites shows a significant improvement when the portion of the above nanomaterials is increased from 0% to 90% [69].

Dispersion and alignment of nanomaterials are two factors that impact the thermal property of the composite. When BNNSs are dispersed linearly in a polymer, the resulting composite transports the heat through the thermal conduction routes. If the nanomaterial in a polymer is oriented in one direction, then heat transfer occurs more efficiently [70].

Besides various advantages, these nanoengineered composites offer some limitations also. It includes defects, nonuniformity at the polymer nanomaterial interface, and anisotropic thermal conductivity. However, these limitations can be resolved through the functionalization of nanomaterials. Functionalized nanomaterials lead to better composite characteristics.

4.5.4   Strain Sensor Applications

A strain sensor is a sensor that detects and responds to mechanical motion. This change is converted into an electrical signal leading to a change in the resistance of the sensor. Polymer-based strain sensors exhibit low sensitivity [71,72]. Gauge factor of the strain sensor indicates its sensitivity. It has been observed that the addition of nanomaterials such as MoS_2, graphene, and CNTs improves the sensitivity of polymer-based strain sensors. Graphene-based strain sensors show a better performance than nanotubes [73,74].

4.5.4.1   Nanoengineered Strain Sensor

Nanoengineered strain sensors have been developed using MoS_2 and polyethylene terephthalate (PET) by Rana et al. as shown in Figure 4.8. MoS_2 nanosheets are encapsulated in the polymer and contacts are taken from the gold electrodes. The structure is fabricated on the PET substrate. As the composite is stretched, the tunneling gap in the MoS_2 grain increases from d_0 to $d_0 + \Delta d_0$, as shown in Figure 4.8. As the tunneling gap is increased, carriers experience difficulty in reaching the other end, leading to a decrease in current in comparison to its initial value.

The resistance, voltage, and current of the sensor can be noted at different stress levels. The developed nanoengineered strain sensor exhibits outstanding piezoresistive properties, offers better

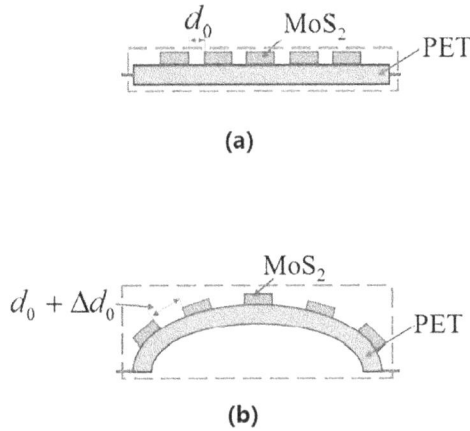

FIGURE 4.8 Schematic representation of strain sensor at different states. (a) MoS$_2$/PET film and (b) stressed MoS$_2$/PET film. (Photograph by the Author.)

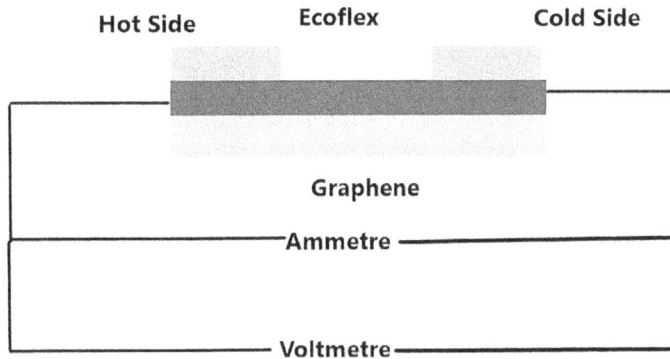

FIGURE 4.9 Structure of thermoelectric device. (Photograph by the Author.)

durability, and can bear pressure up to 14 MPa [75]. In place of MoS$_2$, graphene and BNNS-based polymer composites are also reported for the development of strain sensors [76–78].

4.5.4.2 Nanoengineered Thermoelectric Device

Zhang et al. developed self-driven strain sensors using graphene. It works on the principle of temperature differences between the skin and the ambient environment. The developed sensor gives good reproducibility and has a high strain resolution with a low response time. Figure 4.9 shows the schematic representation of the developed device. When a temperature gradient is applied on the hot and cold sides, the thermal gradient ($\Delta T = T_1 - T_2$) across the composite film is achieved.

The developed nanoengineered composite generates voltage and current due to the temperature gradient. When the temperature gradient is around 14 K, nano-ampere current starts flowing with a voltage of ~0.5 mV. This device produces voltage irrespective of the strain applied. The current generated depends on the temperature difference. This device produces a large current–voltage signal due to its amazing thermoelectric properties. It can be easily used as a self-powered strain sensor. Such sensors have found application in structural health monitoring and numerous other fields. Huge structures like bridges and buildings undergo invisible damage due to external load. The developed nanoengineered sensors are efficient in sensing this damage and suitable arrangements can be done to repair them on time [79–83].

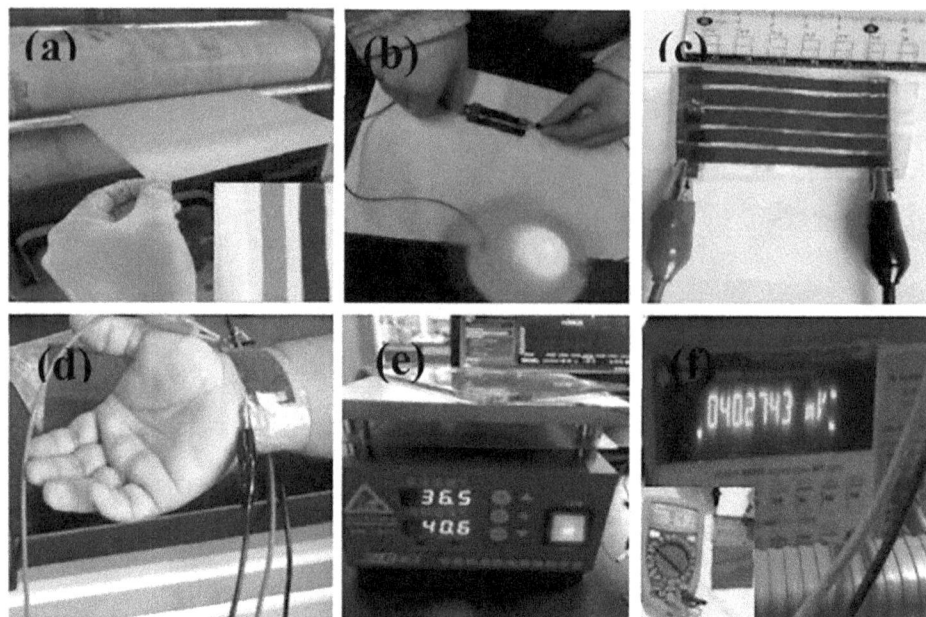

FIGURE 4.10 Fabrication of wearable textiles: (a) GO coating process on fabric, (b) sheet resistance of composite, (c) stretching and bending of composite, (d) triboelectric nanogenerator connected to human wrist, (e) Seebeck measurement setup with controlled thermal gradient, and (f) output open circuit (mV) [19].

4.5.5 Nanoengineered Wearable Textiles for Energy Harvesting

The demand for self-driven wearable textile devices has increased over the past few years. Graphene-based wearable devices have been recently designed for energy harvesting and storage. Khoso et al. developed a wearable thermoelectric generator (TEG) by making a composite of rGO as an n-type material and poly(3,4-ethylenedioxythiophene) polystyrene sulfonate as a p-type material. This composite is used as a coating for fabric and the final fabric when worn by humans will be used for energy harvesting from the heat generated by the human body as shown in Figure 4.10.

The textile shows an improvement in the Seebeck coefficient up to 150 mV/K and a power factor up to 60 mW/(m K). It can translate human body heat into electrical energy. This TEG is capable of generating an open circuit output voltage of around 13–120 mV at an ambient fixed temperature of 20°C [19].

4.6 CONCLUSION

The synthesis and techniques for the development of polymers and nanomaterials have been discussed briefly. There are various parameters that affect the performance of these composites such as the volume of nanomaterials, geometry, orientation, and dispersion of nanomaterials in the polymer matrix. Nanoengineered polymer composites have found application in tribology. It is because of the reduction in the coefficient of friction and wear rate. A low-viscosity material is formed when polyamide and BNNS are mixed. At a high shear rate and low viscosity, the nanoengineered polymer composite persuades a smooth printing process. Nanoengineered polymer composites are suitable for tissue engineering as they provide better nerve cell responses. They also promote the conductivity, elastic strength, and biological responses of the cardiac-related scaffolds. Reinforcement of nanomaterials such as graphene, BNNS, and MoS_2 to the polymer matrix results in higher Young's modulus and tensile strength whereas the incorporation of BNNS exhibits superior electrical insulation and thermal stability. It has found applications in electronic gadgets and electric vehicles.

The thermal conductivity of the above-mentioned composite is enhanced drastically by the addition of BNNS. It is possible to develop strain sensors using the PET substrate and MoS_2 nanosheets. The developed nanoengineered polymer composite is highly durable and can bear high pressure. This self-powered strain sensor generates a large current–voltage signal when the temperature gradient is applied to it owing to its superior thermoelectric properties. When the temperature difference is around 14 K, 125 nA current begins to flow and a voltage of ~0.5 mV is generated. It can efficiently translate human heat into electrical energy.

4.7 FUTURE SCOPE

With the advancement of these nanoengineered composites, their applications in various sectors such as aviation and aerospace will be explored. Apart from that, it is proposed that nanoengineered composites will be employed to repair tissue defects, promote bone tissue regeneration, and fabricate those electronic devices which can be available for medical applications. In the near future, various fabrication techniques will be adopted for manufacturing hierarchical and multi-material scaffolds, nearly reflecting the natural environment for cell proliferation and differentiation.

REFERENCES

1. Sharon M, Sharon M. 2007. *Nano Forms of Carbon and Its Applications*. Mumbai: MONAD Nanotech Pvt. Ltd.
2. Liu YD, Zhao P, Chan YS, et al. 2021. Effects of nano-sized aluminum on detonation characteristics and metal acceleration for RDX-based aluminized explosive. *Def Technol* 17(2):327–337.
3. Palmero P. 2015. Structural ceramic nanocomposites: A review of properties and powders'synthesis methods. *Nanomaterials* 5(2), 656–696.
4. Domun N, Hadavinia H, Zhang T et al. 2015. Improving the fracture toughness and the strength of epoxy using nanomaterials-a review of the current status. *Nanoscale* 7(23):10294–10329.
5. Lau D, Jian W, Yu Z et al. 2018. Nano-engineering of construction materials using molecular dynamics simulations: Prospects and challenges. *Compos Pt B-Eng* 43:282–291.
6. Huang C, Qian X, Yang R. 2018. Thermal conductivity of polymers and polymer nanocomposites. *Mat Sci Eng R* 32:1–22.
7. Hu K, Kulkarni DD, Choi I, Tsukruk VV. 2014 Graphene-polymer nanocomposites for structural and functional applications. *Prog Polym Sci* 39(11):1934–1972.
8. Iijima S. 1991. Helical microtubules of graphitic carbon. *Nature* 354:56–58.
9. Venkataraman A, Amadi VE, Chen Y, Papadopoulos C. 2019. Carbon nanotube assembly and integration for applications. *Nanoscale Res Lett* 14:220–267.
10. Caoduro C, Hervouet E, Girard-Thernier C, et al. 2017. Carbon nanotubes as gene carriers: Focus on internalization pathways related to functionalization and properties. *Acta Biomater* 49:36–44.
11. Zheng S, TiaN X, Ouyang J., Shen Y, et al. 2022. Carbon nanomaterials for drug delivery and tissue engineering. *Front Chem*. DOI: 10.3389/fchem.2022.990362.
12. Liu J, Hui D, Lau D. 2022. Two-dimensional nanomaterial-based polymer composites: Fundamentals and applications. *Nanotechnol Rev*. 11:770–792.
13. Gowda A, Esler D, Tonapi S, Nagarkar K, Srihari K. 2004. Voids in thermal interface material layers and their effect on thermal performance. *Proceedings of 6th Electronics Packaging Technology Conference IEEE, Singapore*. Cat. No.04EX971: 41–46.
14. Morishita T, Okamoto H. 2016. Facile exfoliation and non covalent superacid functionalization of boron nitride nanosheets and their use for highly thermally conductive and electrically insulating polymer nanocomposites. *ACS Appl Mater Interfaces* 8(40):27064–27073.
15. Veca LM, Lu F, Meziani MJ, Cao L, Zhang P, Qi G, et al. 2009. Polymer functionalization and solubilization of carbon nanosheets. *Chem Commun* (Cambridge, UK) 18:2565–2567.
16. Huang B. 2020. Carbon nanotubes and their polymeric composites: The applications in tissue engineering. *Biomanuf Rev* 5(3):1–26.
17. Novoselov SK, Geim KA, Morozov VS, et al. 2004. Electric field effect in atomically thin carbon films. *Science*, 306(5696):666–669.

18. Nanografi Nano Technology. 2012. Graphene Oxide. In nanografi.com/. Retrieved 2023, from https://nanografi.com/graphene/graphene-oxide/

19. Khoso AN, Jiao X, GuangYu X. 2021. Enhanced thermoelectric performance of graphene based nanocomposite coated self-powered wearable e-textiles for energy harvesting from human body heat. *Royal Soc Chem Adv* 11:16675. DOI: 10.1039/d0ra10783b.

20. Sarmazdeh RS, Morteza S, Dizaji Z, Kang KA. *Two-Dimensional Nanomaterials*. DOI: 10.5772/intechopen.85263.

21. Yi M, Zhang C. 2018. The synthesis of two-dimensional MoS_2 nanosheets with enhanced tribological properties as oil additives. *RSC Adv* 8(17):9564–9573.

22. Bertolazzi S, Brivio J, Kis A. 2011. Stretching and breaking of ultrathin MoS_2. *ACS Nano* 5(12):9703–9709.

23. Coleman NJ, Khan U, Blau WJ, Gun'ko KY, 2006. Small but strong: A review of the mechanical properties of carbon nanotube–polymer composites. *Carbon* 44:1624–1652.

24. Du HJ, Bai J, Cheng MH. 2007. The present status and key problems of carbon nanotube based polymer composites. *Express Polym Lett* 1:253–273.

25. Shukla V. 2010. Review on the electromagnetic interference shielding materials fabricated by iron ingredient. *Nanoscale Adv* 5:1–34.

26. Bhargava S, Koratkar N, Blanchet TA. 2015. Effect of platelet thickness on wear of graphene–polytetrafluoroethylene (PTFE) composites. *Tribol Lett* 59(1):17.

27. Polschikov SV, Nedorezova PM, Klyamkina AN, Kovalchuk AA, Aladyshev AM, Shchegolikhin AN, Shchegolikhin VG, Muradyan VE. 2013. Composite materials of graphene nanoplatelets and polypropylene, prepared by in situ polymerization. *J Appl Polym Sci* 127(2):904–911.

28. Costa P, Nunes-Pereira J, Oliveira J, Silva J, Moreira JA, Carabineiro SAC, Buijnsters JG, LancerosMendez S. 2017. High-performance graphene-based carbon nanofiller/polymer composites for piezoresistive sensor applications. *Compos Sci Technol* 153:241–252.

29. Saboori A, Moheimani SK, Dadkhah M, Pavese M, Badini C, Fino P. 2018. An overview of key challenges in the fabrication of metal matrix nanocomposites reinforced by graphene nanoplatelets. *Metals* 8(3):172.

30. Huang T, Lu RG, Su C, Wang HN, Guo Z, Liu P, Huang ZY, Chen HM, Li TS. 2012. Chemically modified graphene/polyimide composite films based on utilization of covalent bonding and oriented distribution. *ACS Appl Mater Interfaces* 4(5):2699–2708.

31. Bhusan B. 2002. *Introduction to Tribology*. New York: John Wiley & Sons, Inc.

32. Basu KS. 2012. *Fundamentals of Tribology*. New Delhi: PHI Learning Private Ltd.

33. Min YJ, Kang KH, Kim DE. 2018. Development of polyimide films reinforced with boron nitride and boron nitride nanosheets for transparent flexible device applications. *Nano Res* 11(5):2366–2378.

34. Wang H, Xie G, Zhu Z, Ying Z, Zeng Y. 2014. Enhanced tribologicalperformance of the multi-layer graphene filled poly(vinylchloride) composites. *Compos Part A-Appl Sci Manuf*.67:268–273.

35. Affdl JCH, Kardos J. 1976. The Halpin-Tsai equations: A review. *Polym Eng Sci* 16(5):344–352.

36. Zhengjia JI, Lin Z, Guoxin X, et al. 2020. Mechanical and tribological properties of nanocomposites incorporated with two-dimensional materials. *Friction* 8(5):813–846.

37. Li Z, Yang WJ, Wu YP, Wu SB, Cai ZB. 2017. Role of humidity in reducing the friction of graphene layers on textured surfaces. *Appl Surf Sci* 403:362–370.

38. Chen Z, He X, Xiao C, Kim SH. 2018. Effect of humidity on friction and wear-A critical review. *Lubricants* 6(3):74.

39. Chen ZY, Yan HX, Guo LL, Li L, Yang PF, Liu B. 2018. A novel polyamide-type cyclophosphazene functionalized rGO/WS_2 nanosheets for bismaleimide resin with enhanced mechanical and tribological properties. *Compos Part A: Appl Sci Manuf* 121:18–27.

40. Qiu SL, Hu YX, Shi YQ, Hou YB, Kan YC, Chu F K, Sheng H.B, Yuen RK.K, Xing W.Y. 2018. In Situ growth of polyphosphazene particles on molybdenum disulfide nanosheets for flame retardant and friction application. *Compos Part A: Appl Sci Manuf* 114:407–417.

41. Xu Z, Lou WJ, Zhao GQ, Zhao Q, Xu N, Hao JY, Wang XB. 2019. Preparation of WS_2 nanocomposites via mussel-inspired chemistry and their enhanced dispersion stability and tribological performance in polyalkylene glycol. *J Dispersion Sci Technol* 40(5):737–744.

42. Lv Y, Wang W, Xie GX, Luo JB. 2018. Self-lubricating PTFE-based composites with black phosphorus nanosheets. *Tribol Lett* 66(2):61–66.

43. Li S, Duan CJ, Li X, Shao MC, Qu CH, Zhang D, Wang QH, Wang TM, Zhang XR. 2020. The effect of different layered materials on the tribological properties of PTFE composites. *Friction* 8(3):542–552.

44. Nieto A, Bisht A, Lahiri D, Zhang C, Agarwal A. 2017. Graphene reinforced metal and ceramic matrix composites: A review. *Int Mater Rev* 62(5):241–302.

45. Birenboim M, Nadiv R, Alatawna A, et al. 2019. Reinforcement and workability aspects of graphene-oxide-reinforced cement nano-composites. *Compos Part B: Eng* 161:68–76.

46. Li YJ, Ge BZ, Wu ZH, Xiao GQ, Shi ZQ, Jin ZH. 2017. Effects of h-BN on mechanical properties of reaction bonded β-SiAlON/h-BN composites. *J Alloys Compd* 703:180–187.

47. Li Q, Cai DL, Yang ZH, Duan XM, Li DX, Sun YS, Wang S J, Jia D C, Joachim B, Zhou Y. 2019. Effects of BN on the microstructural evolution and mechanical properties of BAS-BN composites. *Ceram Int* 45(2):1627–1633.

48. Fei M.M, Lin R.Z, Lu YW, Zhang XL, Bian RJ, Cheng JG, Luo PF, Xu CX, Cai DY. 2017. MXene- reinforced alumina ceramic composites. *Ceram Int* 43(18):17206–17210.

49. Belmonte M, Ramírez C, González-Julián J, Schneider J, Miranzo P, Osendi MI. 2013. The beneficial effect of graphene nanofillers on the tribological performance of ceramics. *Carbon* 61:431–435.

50. Porwal H, Tatarko P, Saggar R, Grasso S, Kumar Mani M, Dlouhý I, Dusza J, Reece MJ. 2014. Tribological properties of silica–graphene nano-platelet composites. *Ceram Int* 40(8):12067–12074.

51. Tanaka M, Sato Y, Haniu H, Nomura H, Kobayashi S, Takanashi S, Okamoto M, Takizawa T, Aoki K, Usui Y. 2017. A three-dimensional block structure consisting exclusively of carbon nanotubes serving as bone regeneration scaffold and as bone defect filler. *PLoS One* 12(2):e0172601.

52. Lau D, Broderick K, Buehler MJ, Büyüköztürk O. 2014. A robust nanoscale experimental quantification of fracture energy in a bilayer material system. *Proc Natl Acad Sci USA* 111(33):11990.

53. Wang X, Xing W, Feng X, Song L, Hu Y. 2017. MoS$_2$/polymer nanocomposites: Preparation, properties, and applications. *Polym Rev (Philadelphia, PA, US)* 57(3):440–466.

54. Peng H, Wang D, Zhang L, Li M, Liu M, Wang C, et al. 2020. Amorphous cobalt borate nanosheets grown on MoS$_2$ nanosheet for simultaneously improving the flame retardancy and mechanical properties of polyacrylonitrile composite fiber. *Compos Pt B-Eng* 201:108298.

55. Vadukumpully S, Paul J, Mahanta N, Valiyaveettil S. 2011. Flexible conductive graphene/poly(vinyl chloride) composite thin films with high mechanical strength and thermal stability. *Carbon* 49(1):198–205.

56. Bhattacharyya A, Chen S, Zhu M. 2014. Graphene reinforced ultra high molecular weight polyethylene with improved tensile strength and creep resistance properties. *Express Polym Lett* 8(2):74–84.

57. Zhi C, Bando Y, Tang C, Kuwahara H, Golberg D. 2009. Large-scale fabrication of boron nitride nanosheets and their utilization in polymeric composites with improved thermal and mechanical properties. *Adv Mater* 21(28):2889–2893.

58. Bao C, Guo Y, Song L, Hu Y. 2011. Poly(vinyl alcohol) nanocomposites based on graphene and graphite oxide: A comparative investigation of property and mechanism. *J Mater Chem* 21(36):13942–13950.

59. Zhang J, Lei W, Chen J, Liu D, Tang B, Li J, et al. 2018. Enhancing the thermal and mechanical properties of polyvinyl alcohol (PVA) with boron nitride nanosheets and cellulose nanocrystals. *Polymer* 148:101–108.

60. O'Neill A, Khan U, Coleman JN. 2012. Preparation of high concentration dispersions of exfoliated MoS$_2$ with increased flake size. *Chem Mater.* 24(12):2414–2421.

61. Ávila AF, Neto AS, Nascimento Junior H. 2011. Hybrid nanocomposites for mid-range ballistic protection. *Int J Impact Eng* 38(8):669–676.

62. Cai Q, Scullion D, Gan W, et al. 2019. High thermal conductivity of high-quality monolayer boron nitride and its thermal expansion. *Sci Adv* 5(6):eaav0129.

63. Han Z, Fina A. 2011. Thermal conductivity of carbon nanotubes and their polymer nanocomposites: A review. *Prog Polym Sci* 36(7):914–944.

64. Hong H, Jung YH, Lee JS, et al. 2019. Anisotropic thermal conductive composite by the guided assembly of boron nitride nanosheets for flexible and stretchable electronics. *Adv Funct Mater* 29(37):1902575.

65. Liu X, Zhang G, Zhang Y-W. 2014. Thermal conduction across graphene cross-linkers. *J Phys Chem C.* 118(23):12541–12547.

66. Wang Y, Xu L, Yang Z, et al. 2018. High temperature thermal management with boron nitride nanosheets. *Nanoscale* 10(1):167–73.

67. Han S, Meng Q, Qiu Z, Osman A, Cai R, Yu Y, et al. 2019. Mechanical, toughness and thermal properties of 2D material-reinforced epoxy composites. *Polymer* 184:121884.

68. Shen X, Wang Z, Wu Y, Liu X, He YB, Kim JK. 2016. Multilayer graphene enables higher efficiency in improving thermal conductivities of graphene/epoxy composites. *Nano Lett* 16(6):3585–3593.

69. Wu Y, Xue Y, Qin S, Liu D, Wang X, Hu X, et al. 2017. BN nanosheet/ polymer films with highly anisotropic thermal conductivity for thermal management applications. *ACS Appl Mater Interfaces* 9(49):43163–43170.

70. Lu H, Liang F, Gou J. 2011. Nanopaper enabled shape-memory nanocomposite with vertically aligned nickel nanostrand: Controlled synthesis and electrical actuation. *Soft Matter* 7(16):7416–7423.

71. Laukhina E, Pfattner R, Ferreras LR, Galli S, Mas-Torrent M, Masciocchi N, et al. 2010. Ultrasensitive piezoresistive all-organic flexible thin films. *Adv Mater* 22(9):977–981.
72. Bae SH, Lee Y, Sharma BK, Lee HJ, Kim JH, Ahn JH. 2013. Graphene-based transparent strain sensor. *Carbon* 51:236–242.
73. Tan C, Dong Z, Li Y, Zhao H, et al. 2020. A high performance wearable strain sensor with advanced thermal management for motion monitoring. *Nat Commun* 11(1):3530.
74. Tran L, Kim J. 2018. A comparative study of the thermoplastic polyurethane/carbon nanotube and natural rubber/carbon nanotube composites according to their mechanical and electrical properties. *Fibers Polym* 19(9):1948–1955.
75. Rana V, Gangwar P, Meena JS, Ramesh AK, Bhat KN, Das, et al. 2020. A highly sensitive wearable flexible strain sensor based on polycrystalline MoS$_2$ thin film. *Nanotechnology* 31(38):385501.
76. Boland Conor S, Khan U, Ryan G, Barwich S, Charifou R, Harvey A, et al. 2016. Sensitive electromechanical sensors using viscoelastic graphene-polymer nanocomposites. *Science* 354(6317):1257–1260.
77. Lu L, Yang B, Liu J. 2020. Flexible multifunctional graphite nanosheet/electrospun-polyamide 66 nanocomposite sensor for ECG, strain, temperature and gas measurements. *Chem Eng J*. 400:125928.
78. Zhang D, Zhang K, Wang Y, Wang Y, Yang Y. 2019. Thermoelectric effect induced electricity in stretchable graphene-polymer nanocomposites for ultrasensitive self-powered strain sensor system. *Nano Energy* 56:25–32.
79. Zhu M, Du X, Liu S, Li J, Wang Z, Ono T. 2021. A review of strain sensors based on two-dimensional molybdenum disulfide. *J Mater Chem C* 9:9083–9101.
80. Tang X, Cheng D, Ran J, Li D, He C, Bi S, et al. 2021. Recent advances on the fabrication methods of nanocomposite yarn-based strain sensor. *Nanotechnol Rev* 10(1):221–236.
81. Nurazzi NM, Abdullah N, Demon SZN, Halim NA, Azmi AFM, Knight VF, et al. 2021. The frontiers of functionalized graphene based nanocomposites as chemical sensors. *Nanotechnol Rev* 10(1):330–369.
82. Sagadevan S, Shahid MM, Yiqiang Z, Oh W-C, Soga T, Anita Lett J, et al. 2021. Functionalized graphene-based nanocomposites for smart optoelectronic applications. *Nanotechnol Rev* 10(1):605–635.
83. Naghib SM, Behzad F, Rahmanian M, Zare Y, Rhee KY. 2020. A highly sensitive biosensor based on methacrylated graphene oxide-grafted polyaniline for ascorbic acid determination. *Nanotechnol Rev* 9(1):760–767.

5 Polymer Nanocomposites for Energy Harvesting

Harun Güçlü and Murat Yazıcı

CONTENTS

5.1 INTRODUCTION

Energy is one of the most vital requirements of the present. Every day, the need for sustainable and clean energy is increasing. In this area, scientific research is conducted on the quest for new and more eco-friendly energy sources and the development of systems that create energy from new sources. Electrochemical batteries meet the power needs of small-scale electronic devices used today. Chemical-containing external power supplies have a limited lifespan and must be replaced after a certain period. Replacing the batteries of micro-/nanosized electronic devices is challenging and harms the environment as chemical waste [1]. Due to these difficulties, studies on energy-harvesting technologies have increased intensively in the last decade to operate micro-/nanoscale devices without needing an external power source [2]. It is possible to harvest energy from solar, wind, and mechanical vibrations in our environment. Piezoelectric energy harvesting is one of the techniques available for energy harvesting [3]. Materials with piezoelectric properties can convert mechanical energy into electrical energy, depending on the dipole moments in their structures [4]. They can also convert electrical energy into mechanical energy as a converse piezoelectric effect. With the help of piezoelectric nanogenerators, mechanical vibrations are converted into useful energy. Energy-harvesting systems consist of four main groups.

The first component is the energy source, the second component is the piezoelectric nanogenerator, the third component is the converter, and the fourth component is the storage part. Figure 5.1 shows the energy-harvesting system. Piezoelectric nanogenerators produce the electrical signal

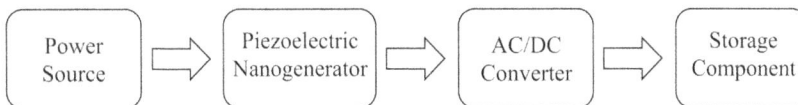

FIGURE 5.1 Piezoelectric energy-harvesting system.

DOI: 10.1201/9781003343912-5

output as a form of alternating current. These signals are stored in a storage element such as a capacitor using the AC-DC converter circuit. In Section 5.2, the materials used in piezo-nanogenerators are mentioned in detail. Then, the most commonly used spin-coating and electrospinning methods are explained. In Section 5.4, the mechanical vibration energy-harvesting performance of the electrospun BaTiO$_3$/PVDF piezo-nanogenerator is investigated.

5.2 MATERIALS

The piezoelectric effect was first discovered in single crystals by the Curie brothers in 1880 [5]. From the mid-1950s, piezoceramics with high electrical-mechanical coupling factors such as PZT (lead zirconate titanate) [6] began to take their place in the literature. In the following years, studies have focused on lead-free piezoceramics (BaTiO$_3$ (barium titanate), ZnO (zinc oxide), KNaNbO$_3$ (potassium sodium niobate)) due to the damage of lead to human health and the environment [7]. Due to the brittle nature of piezo-crystals and piezoceramics, their use under mechanical vibrations is valid for limited applications. On the other hand, piezo polymers were carried out well in the condition of large strain and vibrations, as they have a more flexible and higher fatigue life structure. Electroactive polymers are the types of polymers that attract the most attention as smart materials. The most widely known and high piezo-property polymer is polyvinylidene fluoride (PVDF). Kawai (1969) first found the piezoelectric properties of stretched and polarized PVDF [8]. PVDF and copolymers are widely used because of intense dipole moments among monomers [9]. α, β, γ, δ, and ε are the crystalline phases of PVDF in the semicrystalline form [10]. The β phase shows the strongest dipole moment in TTT (plane zigzag) formation [11]. PVDF and its copolymers PVDF-TrFE are widely used as nanogenerators in energy-harvesting applications due to their high piezoelectric properties, flexibility, and low production costs. Recent studies on piezo-polymer nanocomposites produced by adding piezoceramic additives or nanoadditives to obtain more efficient energy outputs from piezo-polymer structures have attracted significant attention. The 3-1 directional energy output of the PVDF/BaTiO$_3$ piezo-polymer nanocomposite structure was determined to be approximately 26% higher than the pure PVDF thin film [12]. The β phase transformation was directly obtained in PVDF thin films cooled by melting with multiwalled carbon nanotubes according to FTIR and X-ray experiments [13]. The d$_{33}$ piezoelectric coefficient which reaches up to 32 pC/N in the 30 wt.% BaTiO$_3$/P (VDF-TrFE) composite structure produced using the compression casting technique was determined via measurements [14]. Mendes et al. [15] investigated the effects of filler types and ratios in the PVDF/BaTiO$_3$ nanocomposite structure on the β phase and dielectric properties and determined that smaller ceramic fillers improved the piezoelectric output of the nanocomposite structure. In the study conducted to achieve a more β phase ratio in PVDF thin films, it was seen that the structure containing 0.2% by weight of MWCNT had a β phase over 84% according to FTIR measurements. In addition, d33 coefficient values reached as high as 25 pC/N [16]. Yang et al. [17] produced TiO$_2$/MWCNT/PVDF nanocomposite via solution cast method and mechanical rolling technique. Due to the oriented TiO$_2$/MWCNT, the d$_{33}$ coefficient was measured as 41 pC/N, which is approximately twice that of pure PDVF. PVDF/KNN/ZnO hybrid piezoelectric nanogenerator structure was produced using the electrospinning method. The maximum voltage, current, and power density values were determined as 8.3 V, 5 μA, and 10.38 μW/cm^{-2} under the bending technique using a sewing machine [18]. Piezoelectric outputs of electrospun hybrid piezoelectric nanogenerator (PVDF/KNN/CNT) with different KNN and CNT additive ratios were examined. Composition of 3 wt.% KNN and 0.1 wt.% CNT hybrid nanogenerator showed optimum results with 23.24 V and 9.9 μA according to tests [19]. Vibration energy-harvesting performance of aligned PVDF/PZT nanogenerator was tested with different PZT additive ratios. The highest piezoelectric output power was measured as 6.35 μW under 1 Mohm load resistance at 10 wt.% PZT in PVDF/PZT nanogenerator [20]. Studies show that piezo nanoceramic additives increase the piezoelectric potential of the structure. On the other hand, nanoconductive additives positively affect charge transfer in the dielectric structure [21,22].

5.3 FABRICATION PROCESSES

There are different techniques for producing PVDF-based thin-film piezoelectric structures in the β phase. The most widely known of these techniques are spin-coating and electrospinning methods. Other techniques include stretching under uniaxial temperature [23], hot pressing [24], solvent casting [25], and melt quenching [26]. The following two sections give detailed information about spin-coating and electrospinning techniques.

5.3.1 SPIN-COATING TECHNIQUE

The widely used production technique in the production of piezoelectric polymer thin-film sensors is the spin-coating technique. The polymer solution is dropped into the middle of a thin substrate. The substrate is covered with a specific layer thickness with the effect of centrifugal force. After the solvent evaporates, the thin film is removed from the substrate. With this coating technique, thin films of desired thickness can be produced in a controlled manner. Figure 5.2 shows the spin-coating technique.

Numerous publications on spin-coated PVDF polymer nanocomposites have appeared in the scientific literature. Cardoso et al. [27] produced PVDF thin films by spin-coating technique and investigated the effects of polymer-solvent mixing ratio, spin-coating speed, and crystallization temperature on β phase ratio. It was determined that higher β phase ratios in PVDF thin films were obtained with the temperature-controlled spin-coating device [28]. d_{33} coefficient was measured as 20 pC/N in PVDF thin films ranging from 300 nm to 25 μm produced by producing thickness-controlled thin films with spin-coating [29]. Moreover, it has been revealed that phase transformations and microstructure change significantly due to annealing thin films after spin-coating [30]. Spin-coating technique stands out with its low cost and easy production operation. However, it is more suitable for laboratory-scale studies due to the limitation of the part sizes and is not suitable for mass production.

5.3.2 ELECTROSPINNING TECHNIQUE

The electrospinning method, used to obtain nanostructures with piezoelectric properties, is a more effective production technique than the spin-coating technique. Nanofiber membrane thin films are obtained using the electrospinning technique. The most important advantage of this method is that the mechanical stretching and polarization processes are combined in a single process. Briefly, the polymer solution is fed into the tip of the needle with the help of a syringe pump. A high electric field is applied between the needle tip and the collector at a certain distance. The polymer solution is scattered from the needle tip and collected to the collector in fiber form under the electric field. During this procedure, the polymer chains are subjected to a stretching force while concurrently undergoing a polarization process. Figure 5.3 demonstrates the electrospinning process.

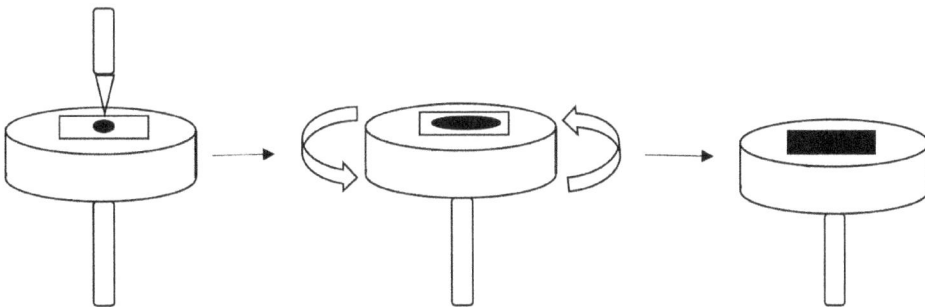

FIGURE 5.2   Spin-coating thin-film production technique.

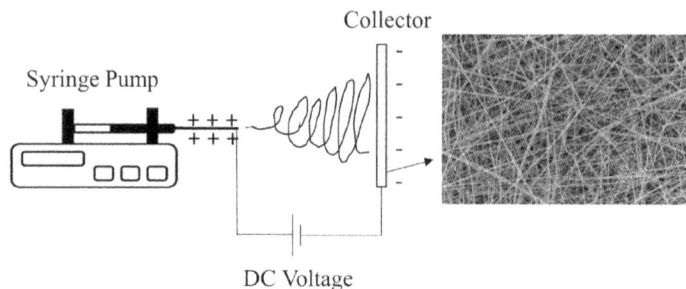

FIGURE 5.3 Electrospinning technique.

The polymer chains of the PVDF membrane structure produced by this method are formed in the β phase, and the piezoelectric effect is gained to the structure by directing the dipole moments in the structure. Electrospinning is determined primarily by the solvent ratio, polymer concentration, syringe feeding rate, applied voltage, and needle tip-to-collector distance. Numerous studies have examined the effect of these parameters, which are used in the electrospinning method, on the piezoelectric properties of the nanofiber mats. The energy-harvesting performances of PVDF, P(VDF-TrFE), and P(VDF-TrFE)/BaTiO$_3$ polymer composites produced by the electrospinning method in 2013 were investigated. Samples were produced under different electrospinning parameters (applied electrical voltage, feeding speed, needle tip dimensions), and their morphologies were investigated by the SEM method. According to SEM measurements, the feeding rate was the most critical parameter affecting the fiber diameters. The piezoelectric power outputs of the PVDF and P(VDF-TrFE)/BaTiO$_3$ nanocomposite structures were measured as 0.02 and 0.01 µW, respectively. It has been interpreted that this is because, although BaTiO$_3$ crystals have a high electromechanical coupling coefficient, the ceramic additives in the nanofiber structure produced by the electrospinning method increase the damping rate of the structure. In addition, one of the notable results of the study was that the output power values of the pure PDVF structure and the nanocomposite structure did not change between 100 Hz and 1 kHz [31]. Abolhasani et al. [32] investigated the morphological and electrical properties of PVDF/graphene nanocomposite structures produced by the electrospinning method. In nanocomposites with graphene additive ratios of 0.1 wt.%, 1 wt.%, 3 wt.%, and 5 wt.%, β phase ratios were determined as 76%, 74%, and 75%, respectively. While the crystallization degree of PVDF without additives is 50%, the β phase rate is 77%. In this study, the open-circuit voltage of the additive-free PVDF was 3.8 V, when the 0.1 wt.% graphene-added nanocomposite was measured to be around 7.9 V. In 2019, using P(VDF-TrFE) produced by electrospinning method and coaxially arrayed MWCNT (multiwalled carbon nanotube) composite membrane structure as a wearable nanogenerator was investigated. The highest rate of crystallization and β phase was determined by the characterization tests performed in 3 wt.% MWCNTs. 18.23 V output voltage and 2.14 A current were measured at nanocomposite containing 3 wt.% MWCNTs in tests performed under 20% strain and 1 Hz frequency. As the MWCNT ratio increases from 0% to 3%, the output voltage has increased, but the voltage values have decreased between 3% and 9%. The highest power output was obtained as 19.6 microwatts under 20% strain, 1 Hz frequency, and 10 Mohm loads [33]. The open-circuit voltage was measured to be 70 V in the energy-harvesting experiment. Also, power density was determined as 66 µW/cm^2 at a 106-ohm load [34]. Zeyrek Ongun et al. [35] investigated the piezo-output values of rGO/PVDF flexible thin-film structures obtained by the electrospinning method. 4.38 V was measured as the highest voltage value under 5 Hz in 0.8 wt.% rGO/PVDF nanogenerator. The use of bismuth chloride (BiCl$_3$)/PVDF fiber structure produced by the electrospinning method as a piezoelectric nanogenerator was investigated by Zhang et al. [36]. The output value of 2 wt.% BiCl$_3$/PVDF fiber structure was measured as a 1.1 V, and it was stated that this value was 4.76 times higher than pure PVDF. Furthermore, the greatest current value was discovered to be 2 µA, and the power density was found to be 0.2 µW/cm^2. The authors also mentioned that the positive effect on the piezoelectric output value is due to the decrease in nanogenerator thickness. The piezoelectric properties of the structures produced due to the electrospinning technique are directly dependent on the electrospinning parameters. These parameters affect each other. Therefore, studies have been carried out to determine the values that will make the β-phase ratio optimal by using appropriate optimization methods for the parameters of solution concentration, solvent ratio, needle tip, needle tip and collector distance, applied voltage, and production time. In the study by Abolhasani et al. [37], an optimization model was developed using virtual neural networks to predict fiber diameters, β-phase ratio, output voltage, and crystallization degrees of P(VDF-TrFE) piezoelectric fiber structure produced using the electrospinning method. The proposed optimization model using artificial neural networks was validated with two experimental groups as four different inputs (Concentration, applied voltage, feed rate, membrane thickness) and four different outputs (fiber diameter, output voltage, β-phase ratio, crystallization degree). According to the optimization results, it was found that the error rate on the fiber diameter was 2%, for the β phase 0.6%, for the voltage 2%, and the crystallization degree 3%. 20% as concentration, 25 kV as applied voltage, 1 mL/h as feeding rate, and 2 hours as membrane thickness are optimum inputs parameters for the model. In addition, the sensitivity analysis results showed that the most significant effect on the output was the polymer concentration ratio, followed by the applied voltage and feeding rate, respectively. In another optimization study using the Taguchi method, the electrospinning parameters were optimized for the highest β-phase ratio as the output function. It has been shown that the parameters that affect the β-phase ratio the most are DMF/acetone mixing ratio, feed rate, the distance between needle and collector, and an applied voltage, respectively. According to optimization results, achieving an 80% β-phase ratio, the optimum parameters are given below: 60% DMF/40% acetone mixture by volume, the flow rate of 0.8 mL/s, the distance between the needle and the collector of 16 cm, and applied voltage values of 14 kV [38].

5.4 POLYMER NANOCOMPOSITE ENERGY-HARVESTING APPLICATION

Studies conducted in recent years show that increasing the β phase and inducing more dipole moments in the structure are the two main methods of increasing the piezoelectric effect in the polymer composite structure. The electrospinning method stands out as a suitable production method because it combines the mechanical stretching and polarization processes in a single step. Electrospun PVDF and BaTiO$_3$/PVDF piezo-nanocomposite energy generator have been developed for the application of piezoelectric energy harvesting from vibration. Morphological analyses of electrospun nanofiber composite thin film were performed using SEM and FTIR. Piezoelectric output values were measured using an oscilloscope in the dynamic test set-up.

5.4.1 ELECTROSPUN PIEZO-POLYMER NANOGENERATOR

Poly(vinylidene fluoride) powder (PVDF, M_w~534,000 g/mole) and barium titanate (BaTiO$_3$ powder, purity > 99.5%) were used as an electro active polymer and piezo-ceramic additive. N,N-dimethylformamide (DMF, M_w = 73.09 g/mole) and acetone (M_w = 58.08 g/mole, 95% pure) were used as a solvent. 15 wt.% PVDF and 6:4 vol/vol acetone/DMF ratio were selected as solution parameters for electrospun PVDF and BaTiO$_3$/PVDF nanogenerators. 5 wt.% BaTiO$_3$ ceramic additives and acetone/DMF mixture determined by measuring using a precision balance were sonicated in an ultrasonic bath for 2 hours. The primary purpose of this process is to provide a homogeneous mixture of ceramic additives in the solvent. As a second step, the polymer containing 15 wt.% PVDF was added to the BaTiO$_3$/acetone/DMF mixture. The polymer was dissolved in the solvent using a magnetic stirrer until a transparent and homogeneous solution was obtained. As a result of the processes, two 10 mL homogeneous PVDF and BaTiO$_3$/PVDF mixtures were obtained. Each polymer solution was transferred into 10 mL plastic syringes. Plastic syringes are connected to the syringe

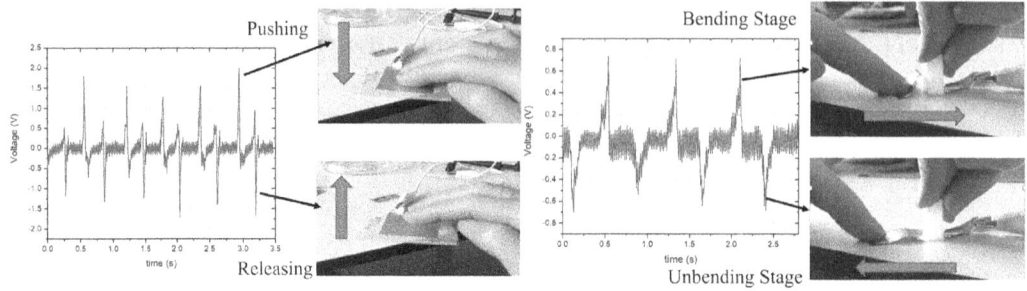

FIGURE 5.4   Open-circuit voltage values of impact and bending conditions.

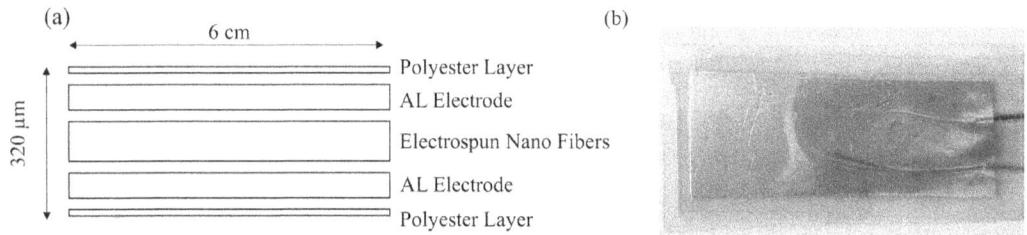

FIGURE 5.5   Electrospun PVDF and $BaTiO_3$/PVDF piezoelectric nanogenerator: (a) schematic view and (b) reel image.

pump purchased from New Era Pump System Inc. A flat collector type was chosen. The polymer solution feed rate was determined as 0.8 mL/h. The distance between the needle tip and the collector was set as 15 cm. The inner diameter of the needle used was 0.4 mm, and the electrospinning time was selected as 4 hours. As a result of electrospinning, approximately 50 microns thick PVDF and $BaTiO_3$/PVDF piezo-polymer nanocomposite fiber mats were produced. A preliminary study was carried out to demonstrate the piezoelectric properties of the fabricated structures. Electrospun mats cut in 3×3 cm dimensions were laminated between two aluminum electrodes. The response of the electrospun PVDF structure to manual impact and bending movements is shown in Figure 5.4. Open-circuit voltage values were collected via TEKTRONIX MDO3034 oscilloscope.

Electrospun fiber mats were cut in 2×6 cm dimensions and placed between two 12-micron aluminum electrodes to prepare the piezo-polymer nanogenerator. The outer parts were wrapped with Kapton tape. In the last stage, the whole structure was hot-pressed with a polyester layer to give flexibility to the structure. Copper cables were connected to the lower and upper electrodes to collect the piezoelectric output. Nanogenerators with a total thickness of 320 microns and a size of 12 cm^2 were prepared for energy harvesting from vibration. The piezo-polymer nanogenerator is shown in Figure 5.5.

5.4.2   MATERIAL CHARACTERIZATION

Microstructural examination of electrospun PVDF and $BaTiO_3$/PVDF nanofiber structures was performed using SEM (scanning electron microscope-ZEISS Evo). FTIR (Fourier-transform infrared spectroscopy-Perkin Elmer) analyses were performed to determine the β-phase ratios in both fiber structures. Image-J program was used for the determination of fiber diameters. SEM images and fiber diameters of electrospun PVDF and $BaTiO_3$/PVDF are given in Figure 5.6.

Nanofiber formation was observed in both structures. The main reason for nanofiber formation is that the electric field force exceeds the surface tension in the Taylor cone at the tip of the syringe. As a result of the measurements performed in Image-J software, the average fiber diameters were

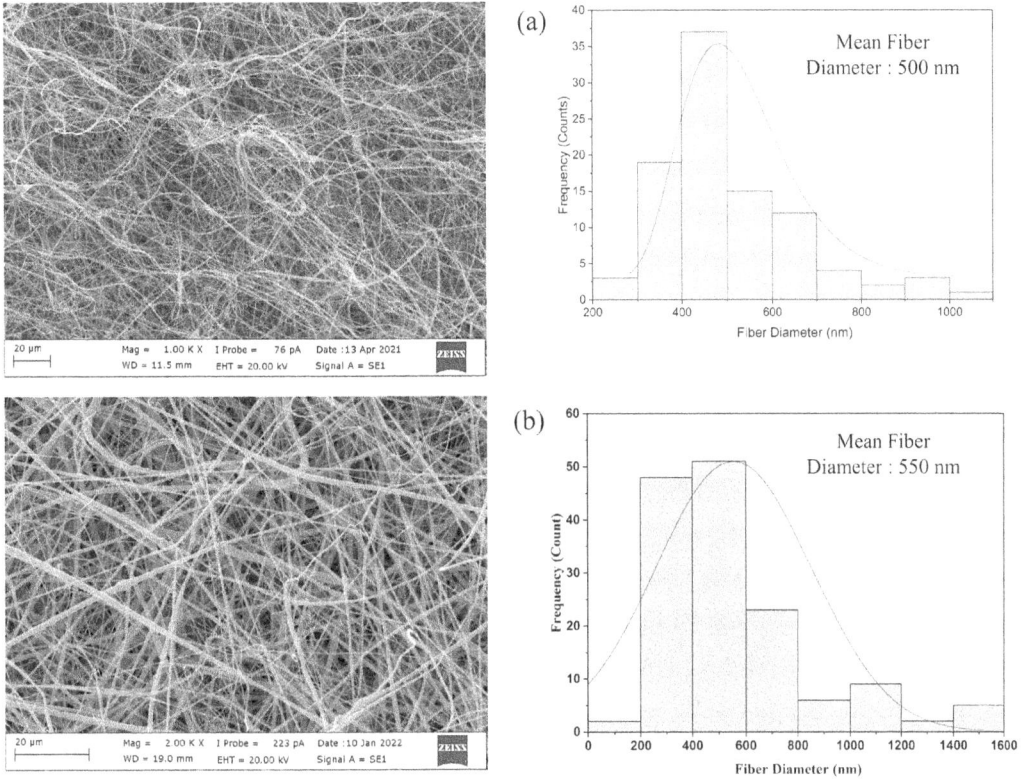

FIGURE 5.6   SEM images and fiber diameters of electrospun: (a) PVDF and (b) BaTiO₃/PVDF.

determined as 500 nm in the PVDF structure without additives and 550 nm in the BaTiO$_3$ doped structure. The reason for higher fiber diameters in the BaTiO$_3$/PVDF structure is that the ceramic additives create interruptions in the jet flow at the needle tip during the electrospinning process. In addition, BaTiO$_3$ ceramic grain sizes affect increasing the fiber diameter. FTIR results and β-phase ratios of PVDF and BaTiO$_3$/PVDF nanofiber structures are shown in Figure 5.7. The specific absorption bands of the α and β phases are 760 and 840 cm^{-1}, respectively. β phase ratios are calculated according to Equation 5.1.

$$F(\beta) = \frac{A_\beta}{1.26A_\alpha + A_\beta} \tag{5.1}$$

Expressions A_α and A_β in Equation (5.1) represent the α- and β-phase absorption peaks. The β-phase ratio of the pure PVDF nanofiber structure was calculated as 80%, and the β phase ratio of the BaTiO$_3$/PVDF nanocomposite structure was calculated as 83%. Barium titanate ceramic particles in the structure induced more dipole moments in nanofibers.

5.4.3   VIBRATION ENERGY HARVESTING

Electrospun piezo-polymer nanogenerators were connected as a uni-morph type into a dynamical system for piezoelectric energy harvesting from vibration. The dynamic test system consists of a vibration generator, DC power supply, function generator, and power amplifier, and the piezoelectric output values were measured with an oscilloscope under 3 Mohm load resistance. The dynamic test system is shown in Figure 5.8.

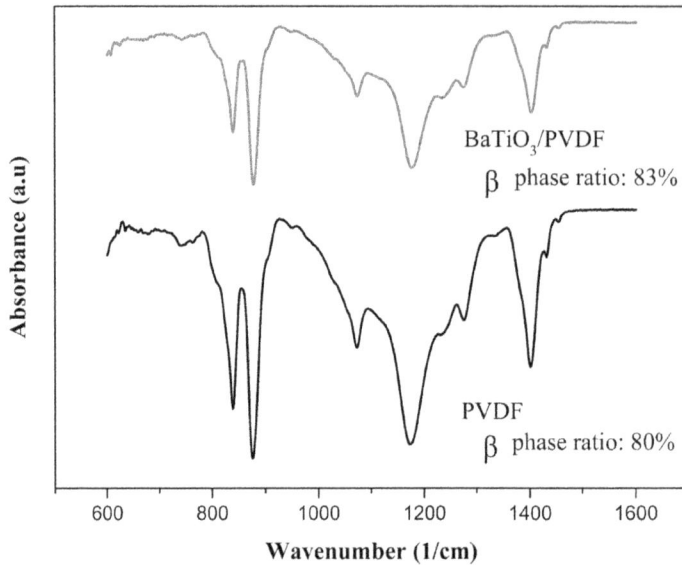

FIGURE 5.7   FTIR results and β phase ratios of PVDF and BaTiO₃/PVDF nanofiber.

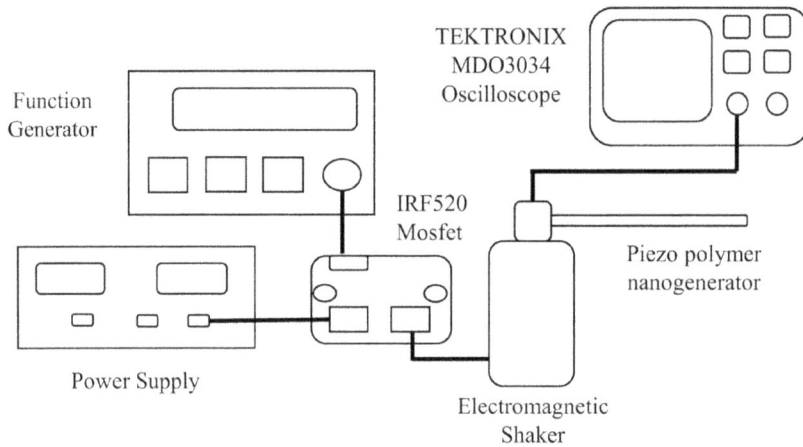

FIGURE 5.8   The dynamic test system for piezoelectric vibration energy harvesting.

The vibration test was carried out at 17 Hz, which is the resonance frequency of the piezo-polymer nanocomposite structure. As a result of the measurements, the peak-to-peak open-circuit voltage of approximately 1.5 V was measured in the piezo-nanogenerator made of pure PVDF. On the other hand, the open-circuit voltage of BaTiO₃/PVDF was determined as about 2 V. The power values and densities at 3 Mohm load for PVDF and BaTiO₃/PVDF were calculated as 0.135 µW, 0.012 µW/cm², and 0.170 µW, 0.015 µW/cm², respectively. Figure 5.9 shows the open-circuit voltage and power values. According to the tests, it was determined that the BaTiO₃-doped piezo-nanocomposite structure reached higher power values. There are two main reasons why the ceramic-doped nanocomposite structure produces a better piezo-output. The nanoceramic additives have started the nucleation process in the fiber by inducing more dipole moments in the PVDF. As a result, more β phases were formed in the nanofiber structure. Moreover, BaTiO₃ ceramic particles also contributed to the load transfer throughout the structure during vibration, as they showed piezo properties.

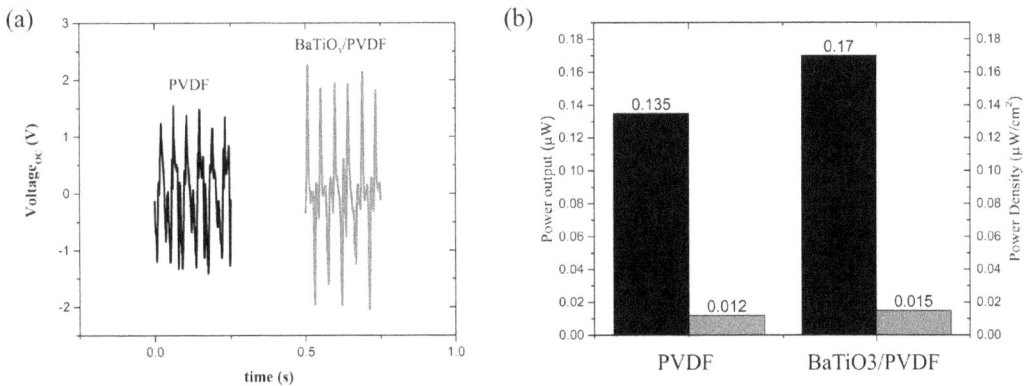

FIGURE 5.9   (a) The open-circuit voltage and (b) power values of PVDF and BaTiO$_3$/PVDF nanogenerator.

5.5   CONCLUSION

This study introduced piezoelectric materials, and a literature review about piezo-polymer nanocomposites was included. Second, the most widely used methods for the fabrication of piezo-polymer nanocomposite structures were described. The studies on the electrospinning method and the piezo-polymer nanogenerator produced were explained in detail. Energy harvesting from vibration was performed with the BaTiO$_3$/PVDF nanogenerator produced using electrospun PVDF-based piezo-polymer nanocomposite. Piezoceramic nano additives positively affected piezoelectric power outputs according to dynamic test measurements. Nanogenerators produced from piezo-polymer nanocomposites are a suitable candidate for micro-/nanoelectronic devices as a self-powering source.

REFERENCES

1. Sezer N, Koç M. A comprehensive review on the state-of-the-art of piezoelectric energy harvesting. *Nano Energy* 2021; 80: 105567.
2. Lu L, Ding W, Liu J, et al. Flexible PVDF based piezoelectric nanogenerators. *Nano Energy* 2020; 78: 105251.
3. Shepelin NA, Glushenkov AM, Lussini VC, et al. New developments in composites, copolymer technologies and processing techniques for flexible fluoropolymer piezoelectric generators for efficient energy harvesting. *Energy Environ Sci* 2019; 12: 1143–1176.
4. Mishra S, Unnikrishnan L, Nayak SK, et al. Advances in piezoelectric polymer composites for energy harvesting applications: A systematic review. *Macromol Mater Eng* 2019; 304: 1800463.
5. Ramadan KS, Sameoto D, Evoy S. A review of piezoelectric polymers as functional materials for electromechanical transducers. *Smart Mater Struct* 2014; 23: 033001.
6. Yan J, Liu M, Jeong YG, et al. Performance enhancements in poly(vinylidene fluoride)-based piezoelectric nanogenerators for efficient energy harvesting. *Nano Energy* 2019; 56: 662–692.
7. Surmenev RA, Orlova T, Chernozem RV, et al. Hybrid lead-free polymer-based nanocomposites with improved piezoelectric response for biomedical energy-harvesting applications: A review. *Nano Energy* 2019; 62: 475–506.
8. Kawai H. The piezoelectricity of poly (vinylidene Fluoride). *Jpn J Appl Phys* 1969; 8: 975–976.
9. Salimi A, Yousefi AA. FTIR studies of β-phase crystal formation in stretched PVDF Films. *Polym Test* 2003; 22: 699–704.
10. Broadhurst MG, Davis GT, McKinney JE, et al. Piezoelectricity and pyroelectricity in polyvinylidene fluoride—A model. *J Appl Phys* 1978; 49: 4992–4997.
11. Giannetti E. Semi-crystalline fluorinated polymers. *Polym Int* 2001; 50: 10–26.
12. Kakimoto K, Fukata K, Ogawa H. Fabrication of fibrous BaTiO$_3$-reinforced PVDF composite sheet for transducer application. *Sensors Actuators A Phys* 2013; 200: 21–25.
13. Ke K, Pötschke P, Jehnichen D, et al. Achieving β-phase poly(vinylidene fluoride) from melt cooling: Effect of surface functionalized carbon nanotubes. *Polymer (Guildf)* 2014; 55: 611–619.

14. Vacche SD, Oliveira F, Leterrier Y, et al. The effect of processing conditions on the morphology, thermomechanical, dielectric, and piezoelectric properties of P(VDF-TrFE)/BaTiO$_3$ composites. *J Mater Sci* 2012; 47: 4763–4774.

15. Mendes SF, Costa CM, Caparros C, et al. effect of filler size and concentration on the structure and properties of poly(vinylidene fluoride)/BaTiO$_3$ nanocomposites. *J Mater Sci* 2012; 47: 1378–1388.

16. Kim GH, Hong SM, Seo Y. Piezoelectric properties of poly(vinylidene fluoride) and carbon nanotube blends: β-phase development. *Phys Chem Chem Phys* 2009; 11: 10506.

17. Yang L, Ji H, Zhu K, et al. Dramatically improved piezoelectric properties of poly(vinylidene fluoride) composites by incorporating aligned TiO$_2$@MWCNTs. *Compos Sci Technol* 2016; 123: 259–267.

18. Bairagi S, Ali SW. A hybrid piezoelectric nanogenerator comprising of KNN/ZnO nanorods incorporated PVDF electrospun nanocomposite webs. *Int J Energy Res* 2020; 44: 5545–5563.

19. Bairagi S, Ali SW. Investigating the role of carbon nanotubes (CNTs) in the piezoelectric performance of a PVDF/KNN-based electrospun nanogenerator. *Soft Matter* 2020; 16: 4876–4886.

20. Koç M, Paralı L, Şan O. Fabrication and vibrational energy harvesting characterization of flexible piezoelectric nanogenerator (PEN) based on PVDF/PZT. *Polym Test* 2020; 90: 106695.

21. Dubey U, Kesarwani S, Verma RK. Incorporation of graphene nanoplatelets/hydroxyapatite in PMMA bone cement for characterization andenhanced mechanical properties of biopolymer composites. *J Thermo Compos Mater* 2023; 36(5): 1978–2008.

22. Kesarwani S, Verma RK. A critical review on synthesis, characterization and multifunctional applications of reduced graphene oxide (rGO)/composites. *Nano;* 2021 16(09): 2130008.

23. Gomes J, Serrado Nunes J, Sencadas V, et al. Influence of the β-phase content and degree of crystallinity on the piezo- and ferroelectric properties of poly(vinylidene fluoride). *Smart Mater Struct* 2010; 19: 065010.

24. Fu J, Hou Y, Zheng M, et al. Flexible piezoelectric energy harvester with extremely high power generation capability by sandwich structure design strategy. *ACS Appl Mater Interfaces* 2020; 12: 9766–9774.

25. Cai X, Lei T, Sun D, et al. A critical analysis of the α, β and γ phases in poly(vinylidene fluoride) using FTIR. *RSC Adv* 2017; 7: 15382–15389.

26. Ramlee NA, Tominaga Y. Structural and physicochemical properties of melt-quenched poly(ethylene carbonate)/poly(lactic acid) blends. *Polym Degrad Stab* 2019; 163: 35–42.

27. Cardoso VF, Minas G, Costa CM, et al. Micro and nanofilms of poly(vinylidene fluoride) with controlled thickness, morphology and electroactive crystalline phase for sensor and actuator applications. *Smart Mater Struct* 2011; 20: 087002.

28. Ramasundaram S, Yoon S, Kim KJ, et al. Direct preparation of nanoscale thin films of poly(vinylidene fluoride) containing β-crystalline phase by heat-controlled spin coating. *Macromol Chem Phys* 2008; 209: 2516–2526.

29. Cardoso VF, Minas G, Lanceros-Méndez S. Multilayer spin-coating deposition of poly(vinylidene fluoride) films for controlling thickness and piezoelectric response. *Sensors Actuators A Phys* 2013; 192: 76–80.

30. Cardoso VF, Costa CM, Minas G, et al. Improving the optical and electroactive response of poly(vinylidene fluoride–trifluoroethylene) spin-coated films for sensor and actuator applications. *Smart Mater Struct* 2012; 21: 085020.

31. Nunes-Pereira J, Sencadas V, Correia V, et al. Energy harvesting performance of piezoelectric electrospun polymer fibers and polymer/ceramic composites. *Sensors Actuators A Phys* 2013; 196: 55–62.

32. Abolhasani MM, Shirvanimoghaddam K, Naebe M. PVDF/graphene composite nanofibers with enhanced piezoelectric performance for development of robust nanogenerators. *Compos Sci Technol* 2017; 138: 49–56.

33. Zhao C, Niu J, Zhang Y, et al. Coaxially aligned MWCNTs improve performance of electrospun P(VDF-TrFE)-based fibrous membrane applied in wearable piezoelectric nanogenerator. *Compos Part B Eng* 2019; 178: 107447.

34. Tiwari S, Gaur A, Kumar C, et al. Enhanced piezoelectric response in nanoclay induced electrospun PVDF nanofibers for energy harvesting. *Energy* 2019; 171: 485–492.

35. Zeyrek Ongun M, Oguzlar S, Doluel EC, et al. Enhancement of piezoelectric energy-harvesting capacity of electrospun β-PVDF nanogenerators by adding GO and rGO. *J Mater Sci Mater Electron* 2020; 31: 1960–1968.

36. Zhang D, Zhang X, Li X, et al. Enhanced piezoelectric performance of PVDF/BiCl$_3$/ZnO nanofiber-based piezoelectric nanogenerator. *Eur Polym J* 2022; 166: 110956.
37. Abolhasani MM, Shirvanimoghaddam K, Khayyam H, et al. Towards predicting the piezoelectricity and physiochemical properties of the electrospun P(VDF-TrFE) nanogenrators using an artificial neural network. *Polym Test* 2018; 66: 178–188.
38. Gee S, Johnson B, Smith AL. Optimizing electrospinning parameters for piezoelectric PVDF nanofiber membranes. *J Memb Sci* 2018; 563: 804–812.

6 Nanophytomedicine and Their Applications
A Brief Overview[1]

Karthika Paul, B. H. Jaswanth Gowda, Fouad Damiri,
Hemalatha Y. R., Chandan R. S., and Mohammed Berrada

CONTENTS

6.1 INTRODUCTION

Nowadays, plant-based medicines are regaining their potential to treat various wide range of diseases, just like synthetic chemicals, due to the emergence of extraction and isolation of specific phytochemicals from plants that possess potent activity [1,2]. The World Health Organization (WHO) defined herbal medicine as "the practice of medicine involving herbs, herbal ingredients, herbal formulations, and fabricated herbal extracts, which primarily consists of therapeutically active phytoconstituents isolated from plant parts or plant extracts, or even their combinations" [3,4]. Phytoconstituents are isolated from the parts of natural plants or the by-product of the plant materials. Even today, herbal medicines are the priority of 80% of the world's population, specifically underdeveloped nations, due to their lower cost and negligible side effects [3,5,6]. Unfortunately, phytochemicals face bioavailability issues due to their poor solubility, permeability, short half-life, lack of targeting ability, and many more [7–9]. Therefore, immediate calls need to be taken to improve the bioavailability of phytoconstituents to improve their therapeutic efficacy.

Nanotechnology is a combination of science and engineering at the nanometer range. Nanomaterials have gained significant interest due to their unique physicochemical properties. The materials engineered at the nano-level generally possess a size of 1–1000 nm [10–12]. Nanoparticles (NPs) can be developed from a wide range of materials and are broadly classified into three types, i.e., lipid-based NPs, inorganic NPs, and polymer-based NPs. Some examples of lipid-based NPs include solid

---

[1] Dr Karthika Paul and Dr B. H. Jaswanth Gowda have equally contributed to this chapter.

DOI: 10.1201/9781003343912-6

lipid nanoparticles, nanostructured lipid carriers, liposomes, niosomes, transferosomes, ethosomes, phytosomes, etc. [13–16]. Examples of inorganic NPs are mesoporous silica NPs, gold NPs, zinc oxide NPs, cerium oxide NPs, silver NPs, and so on [17–19]. Lastly, examples of polymer-based NPs include micelles, dendrimers, and matrix/capsule-based NPs (Figure 6.1). Among all other types of NPs, polymer-based NPs have gained tremendous interest due to their simple preparation technique, biocompatibility, long-term storage stability, and precise drug-delivering ability [20–22].

So far, polymeric NPs have been used to improve many synthetic drugs' solubility, permeability, and bioavailability. It is also witnessed that the functionalized polymeric NPs specifically deliver the drugs to the target site by protecting them from various environmental and physiological conditions. Nevertheless, these NPs can achieve controlled drug delivery for long-term treatment efficacy [23,24]. In this context, the present review describes the different types of polymer-based NPs for the delivery of various phytochemicals. Further, we have also canvassed the application of polymer-based nanophytomedicine in diverse biomedical applications such as cancer, neurological diseases, cardiovascular diseases, wound healing, antimicrobial activity, and diabetes mellitus.

6.2 POLYMER-BASED NANOPARTICLES

Polymer-based NPs have received considerable attention due to their ability to improve the therapeutic efficacy of a wide range of phytomedicine. Among many, polymer-based NPs are most widely preferred because of their simple preparation techniques, biocompatibility, precise drug-delivering ability, and long-term storage stability [25–27]. These NPs are fabricated using polymers, whether of synthetic or of natural origin. Polymeric NPs load the phytoconstituents via adsorption, conjugation, encapsulation, or dispersion based on the need. For instance, some phytochemicals that can readily undergo degradation in certain physiological fluids, pH, light, and temperature are preferred to encapsulate using polymeric NPs to protect therapeutic activity. In contrast, to improve the solubility and permeability profile of certain phytochemicals, adsorption or dispersion of these agents in a polymeric matrix is followed. In other cases, where the phytoconstituents need to be delivered sustainably for a prolonged period, the conjugation, encapsulation, or dispersion methods

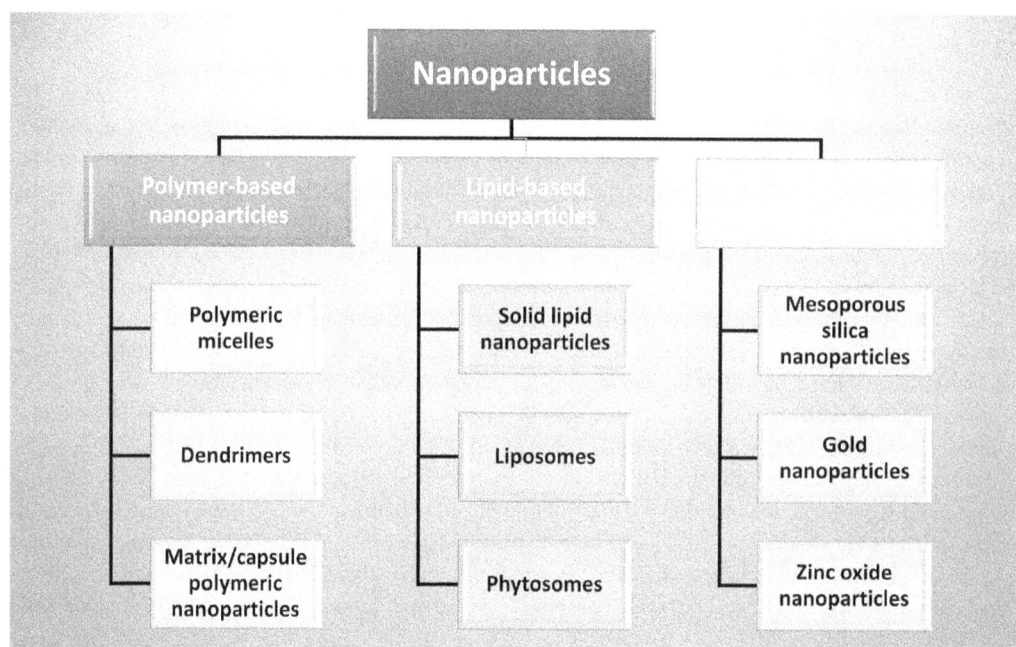

FIGURE 6.1   Different types of nanoparticles.

of loading are favored [20,28]. These NPs are further categorized into different types, including polymeric micelles, dendrimers, and polymeric nanoparticles.

6.2.1 Polymeric Micelles

Polymeric micelles (PMs) are self-assembled nanosized core-shell structures with a hydrophilic head and hydrophobic tail. They are usually made of amphiphilic polymers. Unlike lipid vesicles, PMs are renowned for their ability to hold hydrophobic drugs in their core specifically. The size of PMs generally ranges from 5 to 500 nm [29–31]. They can substantially improve the bioavailability of phytochemicals by enhancing their solubility with the help of a hydrophilic surface. Due to their hydrophilicity, they can easily escape from being taken up by the reticuloendothelial system. Additionally, they have gained tremendous interest due to their ability to penetrate tissues, controlled drug release, and stability [30]. Further, the PMs can also be conjugated with several ligands to achieve active targeted drug delivery. Some commonly loaded phytoconstituents into PMs are curcumin, resveratrol, berberine, etc., for various applications. For instance, Zhang et al. developed curcumin-loaded PMs via a radical polymerization technique for controlled release [32]. Another study by Qiqi Li et al. prepared naringenin-loaded PMs to treat dry eye disease via topical application [33].

6.2.2 Dendrimers

Dendrimers are unique tree-like with excellent drug-carrier properties. They are hyperbranched macromolecules that are developed using the step-by-step addition of repetitive monomers. A dendrimer is derived from the Greek words "Dendron" and "meros," which stand for "tree" and "part," respectively [34]. A single dendrimer comprises three regions, i.e., core, repeating units, and terminal groups. Due to their branched structure, dendrimers possess an excellent surface-to-volume ratio with the ability to load the drugs inside the voids. Further, their terminal functional groups support the conjugation of a variety of drug molecules and ligands for site-specific/targeted delivery [35]. Basically, the dendrimers are developed using three techniques, i.e., convergent method, divergent method, and double exponential mixed method. Several types of dendrimers include polyamidoamine (PAMAM) dendrimers, polypropylene imine/polypropylene amine (PPI/POPAM) dendrimers, tecto dendrimers, peptide dendrimers, and so on, categorized based on the materials used to synthesize them [36]. These dendrimers are exorbitantly utilized to improve the therapeutic efficacy of phytochemicals. For example, Karami et al. synthesized curcumin- and doxorubicin-conjugated pH-sensitive glycodendrimer for tumor-targeted drug delivery [37]. Another study by Gallien et al. fabricated curcumin-attached dendrimers for the potential application of glioblastoma treatment [38].

6.2.3 Polymeric Nanoparticles

The term polymeric NPs and polymer-based NPs are quite confusing and could lead to many misunderstandings. The polymer-based NPs represent any type of NPs made of different synthetic and natural polymers via unique preparation techniques. For example, dendrimers, micelles, and polymersomes [25]. However, polymeric NPs are specific sub-type of polymer-based NPs synthesized using synthetic or natural polymers like poly lactic-co-glycolic acid (PLGA), polyethylene glycol (PEG), chitosan, alginate, and albumin, to yield either matrix or capsule-typed NPs. The matrix type of NPs involves homogeneous drug dispersion throughout the polymer matrix to enable controlled release. However, the capsule type covers the drug molecule inside the polymeric shell to protect the drug from physiological conditions until it reaches its target site [39–41]. Recently, Kohli et al. prepared berberine-loaded chitosan-alginate NPs via ionic gelation technique to enhance the oral bioavailability [42]. Another study by Jiang and team fabricated curcumin-encapsulated zein-alginate NPs to protect the curcumin from gastric acid, thereby encouraging intestinal delivery [43].

Moon et al. encapsulated quercetin in soybean-chitosan NPs to improve quercetin's stability and bioavailability [44].

6.3 APPLICATIONS OF POLYMER-BASED NANOPHYTOMEDICINE

Phytomedicine-loaded polymeric nanoparticles are widely utilized in many biomedical applications to enhance the therapeutic efficacy of phytoconstituents [45]. Some common applications include cancer therapy, treatment of Alzheimer's disease, Parkinson's disease, ischemic stroke, atherosclerosis, chronic wounds, and metabolic syndromes like diabetes mellitus and obesity (Table 6.1) (Figure 6.2).

6.3.1 CANCER THERAPY

Cancer is a significant health problem in the world and the second leading cause of death. It has been anticipated that around 1.9 million new cases will occur by the end of 2021. Regarding treatment approaches, immunotherapy and targeted are facing issues due to their unaffordable cost. In contrast, chemotherapy is economical but associated with several side effects [46]. Therefore, plant-based therapeutic agents have evolved due to their potential anticancer properties without adverse effects [47]. However, their drawbacks, such as poor solubility and permeability, instability

TABLE 6.1
List of Polymer-Based Nanophytomedicine in Various Applications

Sl. No.	Type of NPs	Phytoconstituent	Particle Size	Application	Reference
1	PLGA-based micelles	Curcumin	154 nm	To overcome chemoresistance effectively by enhancing the oral bioavailability of curcumin	[50]
2	Chitosan NPs	Gallic acid/Quercetin	214 nm	For targeted treatment of colon cancer	[51]
3	Chitosan-PLGA NPs	Curcumin	200 nm	To treat Alzheimer's disease via intranasal route	[58]
4	PLGA NPs	Rosmarinic acid/ Curcumin	<90 nm	To treat β-amyloid-induced neuronal toxicity in the brain	[59]
5	PLGA NPs	Bioperine/Curcumin	293 nm	To treat atherosclerosis	[62]
6	Chitosan NPs	Curcumin	40 nm	To overcome the cardiotoxicity induced by hydroxyapatite NPs	[63]
7	Cellulose acetate butyrate NPs	Resveratrol	248 nm	To accelerate wound healing	[73]
8	Polyvinyl alcohol-sodium alginate-based NPs	Berberine	300 nm	To treat diabetic wound	[74]
9	PLGA-dextran-based micelles	Curcumin	498 nm	To enhance the antibacterial efficacy against two different strains of *Pseudomonas* spp.	[77]
10	Lecithin-chitosan NPs	Curcumin	236–333 nm	To enhance the antibacterial and antifungal efficacy against Gram-negative bacteria, Gram-positive bacteria, and *Candida* fungi	[78]
11	Chitosan-alginate NPs	Quercetin	91 nm	To treat diabetes mellitus	[82]
12	Chitosan NPs	*Stevia rebaudiana* leaf extract	327 nm	To treat diabetes mellitus	[83]

in environmental and physiological conditions, and lack of tumor targeting and controlled release ability, have hindered their ability to treat the cancer effectively. To overcome these issues, researchers have come up with nanotechnology-based approaches to alleviate all the above-mentioned limitations [48,49]. Eskandari et al. developed curcumin-loaded PLGA-levan micelles to overcome the chemoresistance of tumors by improving the bioavailability of curcumin via oral administration [50]. Previously, curcumin has been reported to inhibit NF-κB, a chief factor in chemoresistance development. The developed micelles were stable for up to three months and enhanced curcumin's solubility 10,000-fold more than pure curcumin. Further, in vitro and in vivo studies showed reduced NF-κB levels caused by the chemotherapeutic drug gemcitabine. Based on this result, the developed PLGA-levan micelles are considered promising candidates for curcumin delivery. Another study by Patil and Killendar developed gallic acid- and quercetin-loaded chitosan NPs for sustained release in colon cancer [51]. The developed NPs delivered the loaded phytoconstituents in the target site, i.e., colon exhibited a better anticancer effect than free drugs. With this evidence, it can be concluded that polymer-based NPs play a significant role in phytomedicine delivery.

6.3.2 Neurological Disorders Therapy

Neurological disorders (NDs) are recognized as a major threat to public health. Some common types of NDs include ischemic stroke, Parkinson's, Huntington's, Alzheimer's, dementia, and prion diseases. According to WHO statistics of 2019, many conditions related to the central nervous system (CNS) have exhibited the highest mortality rate globally. Ischemic stroke holds second place, followed by Alzheimer's and dementia [52]. Although the pathophysiology of these diseases differs, the administered therapeutic agents need to reach the brain to exhibit their action. The NDs are noncurable but can be controlled with medications necessitating long-term treatment. However, synthetic drugs are associated with several side effects with prolonged use [53]. Therefore, plant-based therapeutic agents have made their way to treat NDs without side effects [54,55]. But the major drawbacks of these phytochemicals are their inability to cross the blood-brain barrier (BBB) to exhibit their therapeutic efficacy in the brain. The BBB is a chief gateway between blood and the brain, which only allows essential nutrients and certain low molecular weight lipophilic drugs to the brain. Therefore, NPs have come into the limelight to perform as a carrier system to deliver drugs into the target site in brain by permeating through BBB [56,57]. Recently, Dhas and Mehta developed curcumin-loaded chitosan-coated PLGA core-shell NPs to treat Alzheimer's disease via the intranasal route [58]. It was found that the NPs were uptaken by caveolae-mediated endocytosis to reach the brain. Further, in vivo model exhibited maximum accumulation of NPs in the brain after administering intranasally. These results confirmed that the chitosan-coated PLGA core-shell NPs could potentially treat Alzheimer's disease by reducing the oxidative stress in brain. Another interesting investigation by Kuo and Tsai developed rosmarinic acid and curcumin-encapsulated polyacrylamide-cardiolipin-PLGA NPs (R-C@PCP NPs) functionalized with 83-14 monoclonal antibody for targeted treatment of β-amyloid induced neuronal toxicity [59]. The results from cell line studies showed positive results by encouraging the developed R-C@PCP NPs to permeate through BBB to exhibit neuroprotection potentially.

6.3.3 Cardiovascular Diseases Therapy

Cardiovascular diseases (CVDs) are one of the top causes of the highest number of deaths per WHO 2019 report [52]. As per World Heart Federation, CVDs are responsible for approx. 17.3 million deaths every year. It is also speculated that annual deaths will increase to 23 million by the end of 2030. Some common CVDs include coronary heart disease, stroke, peripheral arterial disease, aortic disease, atherosclerosis, myocardial infarction, etc. Recently, many phytoconstituents such as resveratrol, curcumin, berberine, naringenin, etc., have exhibited good effects on various CVDs [60,61]. However, they possess poor solubility, half-life, and targetability, which results in declined

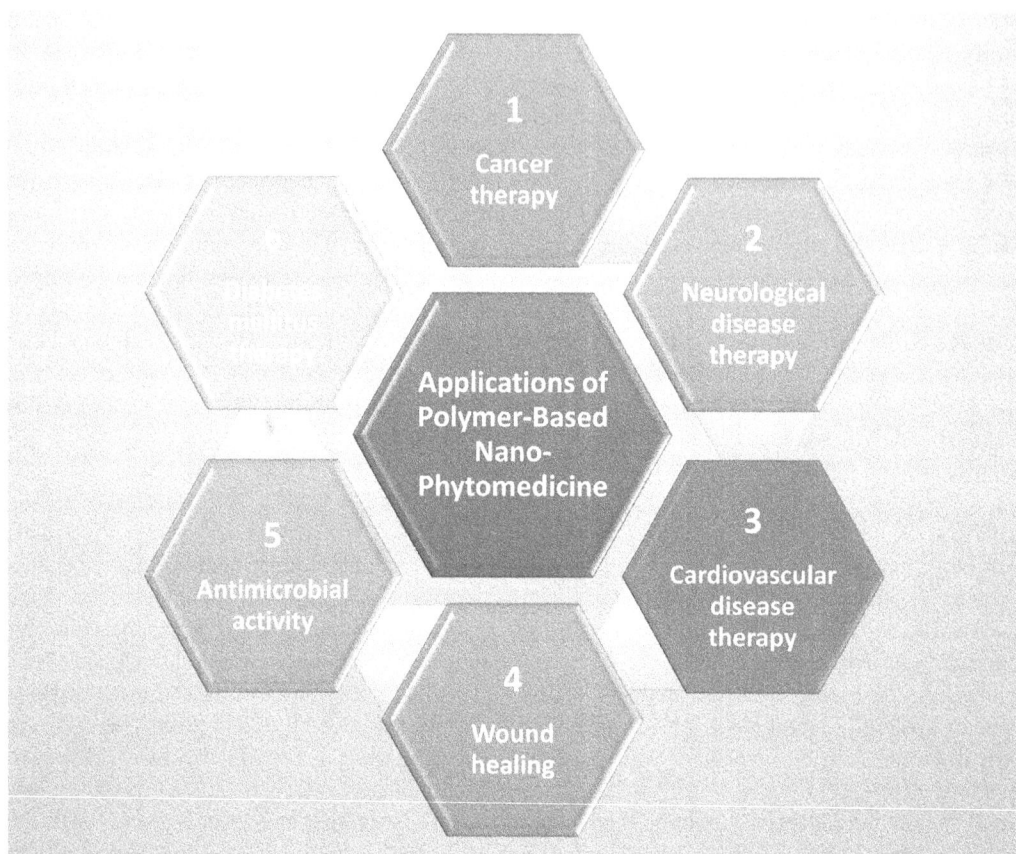

FIGURE 6.2 Applications of nanophytomedicine in diverse diseases.

therapeutic efficacy. In this regard, NPs have been utilized to overcome these issues. A recent study by Pillai and team fabricated bioperine- and curcumin-loaded PLGA NPs to target atherosclerosis plaques and prevent their foam formation [62]. The developed NPs helped upregulation of SCAR-B1 expression, followed by the downregulation of signal transducer and activator of transcription 3 (STAT-3), cluster of differentiation 36 (CD-36), monocyte chemoattractant protein-1 (CCL2/MCP-1), and nuclear factor kappa B (NF-κB) activity resulting in improved antiatherosclerotic activity. Recently, Mosa and colleagues developed curcumin-loaded chitosan NPs for cardioprotective activity [63]. Studies have shown that oral delivery of hydroxyapatite NPs exhibited potential risks of a heart attack in rats, and this study was particularly developed to overcome this cardiotoxicity issue. Results from animal studies showed that curcumin-loaded chitosan NPs significantly reduced the risk of heart attack and thus can be a potential candidate in future studies.

6.3.4 Wound Healing

Wound healing is a complex process consisting of four phases, i.e., hemostasis, inflammation, proliferation, and remodeling [64]. Generally, wounds are categorized into two types such as acute and chronic wounds based on the time taken to completely heal. The acute wound heals in a normal time period by correctly following the above-mentioned phases. However, if there is any interruption in any of the wound healing phases, it could transform the acute wound into a chronic one [65,66]. The principal reason behind prolonged wound healing is a bacterial infection, which interferes in the inflammatory phase. Other than this, disease conditions like diabetes slow down the fast wound recovery. A common treatment for accelerating the wound healing process includes synthetic antibiotics (kills microbes), growth factors (promotes angiogenesis), reactive oxygen species (ROS) scavengers (reduce free radicals), and anti-inflammatory agents [67]. However, some medications cause side effects, and others are expensive. In this concern, phytochemicals like quercetin, arnebin-1, picroliv, and curcumin come into the picture. Certain phytochemicals possess antioxidant and anti-inflammatory properties; some possess angiogenic properties, and some exhibit significant antibacterial activity [68,69]. But, these plant-based chemicals are unstable in environmental and physiological conditions along with rapid bodily elimination. Therefore, nanotechnology-based systems have been widely adopted to enhance the therapeutic efficacy of phytochemicals in wound healing [70–72]. Amanat and team prepared resveratrol-loaded cellulose acetate butyrate NPs (RES@CAB NPs) and further incorporated it into carboxymethyl cellulose wafer (R-NPs@wafer) to accelerate wound healing [73]. The studies in rats bearing incisional wound exhibited fast healing upon treatment with R-NPs@wafer than free resveratrol and plain wafer. Another interesting study by Zhang et al. investigated berberine-loaded polyvinyl alcohol and sodium alginate-based NPs (B@PS NPs) for treating diabetic wounds [74]. The developed B@PS NPs inhibited the TNF-α, NF-κB, and IL-6 expressions followed by increasing the SMA, VEGF, and CD 31 expressions, leading to the activation of Sirt 1 that predominantly enhanced the wound healing in diabetic rats. All these results strongly suggest that NPs-based systems are the potential candidates for accelerating the wound healing process.

6.3.5 ANTIBACTERIAL AND ANTIFUNGAL ACTIVITY

Bacterial and fungal infections are the major causes of many chronic infectious diseases leading to mortality. These infections are commonly treated with synthetic antibiotics due to the fastest recovery and cost-effectiveness. Unfortunately, the increased use of antibiotics in either mild or severe conditions is making way for the emergence of antibiotic-resistant bacterial strains [75]. Studies have also found that drug-resistance bacteria possess a resistance gene called NDM-1. In this regard, phytoconstituents belonging to a wide range of groups, such as alkaloids, terpenoids, lectins and polypeptides, polyphenols, and essential oils, have exhibited excellent antimicrobial activity against many drug-resistant bacterial strains [76]. However, to enhance the solubility, permeability, cellular uptake, and targeting ability, along with controlled release ability, these phytoconstituents are delivered with the aid of NPs. Barros and team synthesized curcumin-loaded PLGA-dextran-based micelles for exhibiting antibacterial activity [77]. The results revealed that the developed NPs significantly killed two different strains of *Pseudomonas* spp. compared to free curcumin showing NPs are the suitable candidate to exhibit increased antibacterial activity. Another interesting study by Valencia et al. developed curcumin-loaded lecithin-chitosan NPs for enhanced antibacterial and antifungal activity [78]. It was found that these NPs significantly inhibited several Gram-negative and Gram-positive bacteria along with various strains of *Candida* fungi than free curcumin. These results depicted the NPs as the noteworthy approach for enhanced antibacterial and antifungal activity.

6.3.6 DIABETES MELLITUS THERAPY

Diabetes mellitus is a noncurable metabolic disease that substantially affects patients' quality of life. As per WHO, approx. 1.5 million patients in the world die annually. It is estimated that nearly 366 million people worldwide will be affected by this disease by 2030. Despite constant noncurable treatment using synthetic drugs and insulin, several plant-based drugs have exhibited a certain level of cure [79,80]. For instance, momordicines I and II have been witnessed to stimulate significant insulin secretion in MIN6 β-cells, arrubiin has increased the insulin level, glucose transporter-2 gene expressions in INS-1 cells, and scrophuside showed inhibitory activity against α-glucosidase [81]. However, these phytochemicals need to be protected in the physiological system until they reach their target to exhibit therapeutic action along with the prolonged release. Thus, researchers have developed different types of NPs to necessitate the targeted delivery and controlled release of phytochemicals for efficient diabetes mellitus treatment. Mukhopadhyay and colleagues developed quercetin-loaded chitosan-alginate-based core-shell NPs for the treatment of diabetes mellitus [82]. The developed NPs exhibited a significant hypoglycemic effect in a diabetic rat after administering them intraperitonially compared to free quercetin administered orally. Another study by Venkatachalam and team developed *Stevia rebaudiana* leaf extract (SRLE)-loaded chitosan NPs for treating diabetes mellitus [83]. These NPs significantly reduced the serum glucose level in diabetic rats. It was found that the serum levels of various enzymes such as alkaline phosphatases, superoxide dismutase, reduced glutathione, serum glutamic oxaloacetic transaminase, etc., were normal after treating the rats with SRLE-chitosan NPs. By considering these results, the phytomedicine-based NPs can be considered promising candidates in the treatment of diabetes mellitus.

6.4 CONCLUSION

- Phytochemicals have been proven to treat a wide range of diseases, such as cancer, stroke, diabetes mellitus, chronic wounds, bacterial infections, etc.
- Due to the poor bioavailability of phytochemicals, their therapeutic efficacy is not as significant as synthetic chemicals.
- Nanotechnology has been used to upgrade the therapeutic efficacy of phytoconstituents.
- Among many nanomaterials, polymer-based NPs are considered the best to deliver phytochemicals due to their simple preparation process, long-term stability, and precise drug delivery.
- Polymeric micelles, dendrimers, and polymeric NPs are the outstanding polymer-based NPs that have been utilized to deliver both hydrophilic and hydrophobic phytoconstituents.
- The polymer-based nanophytomedicine has exhibited a significant therapeutic effect on diverse disease conditions compared to plain phytoconstituents, encouraging nanotechnology in phytomedicine delivery.
- Polymer-based NPs have already been present in the market for the delivery of synthetic drugs such as Abraxane® (albumin-bound NPs for paclitaxel delivery). Similarly, in the near future, it is expected to witness phytoconstituents-loaded polymer-based NPs in the market, which greatly improves the therapeutic efficacy of phytoconstituents.

REFERENCES

1. S.K. Dubey, S. Parab, V.P.K. Achalla, A. Narwaria, S. Sharma, B.H. Jaswanth Gowda, P. Kesharwani, Microparticulate and nanotechnology mediated drug delivery system for the delivery of herbal extracts, *J Biomater Sci Polym Ed*. 33 (2022) 1531–1554. https://doi.org/10.1080/09205063.2022.2065408.
2. M. Ekor, The growing use of herbal medicines: Issues relating to adverse reactions and challenges in monitoring safety, *Front Neurol*. 4 (2014) 177. https://doi.org/10.3389/FPHAR.2013.00177/BIBTEX.
3. WHO Global Centre for Traditional Medicine, (n.d.). https://www.who.int/initiatives/who-global-centre-for-traditional-medicine (accessed October 10, 2022).
4. Traditional, Complementary and Integrative Medicine, (n.d.). https://www.who.int/health-topics/traditional-complementary-and-integrative-medicine#tab=tab_1 (accessed October 10, 2022).
5. C.C. Falzon, A. Balabanova, Phytotherapy: An introduction to herbal medicine, *Prim Care*. 44 (2017) 217–227. https://doi.org/10.1016/j.pop.2017.02.001.
6. L. Ang, H.W. Lee, A. Kim, M.S. Lee, Herbal medicine for the management of COVID-19 during the medical observation period: A review of guidelines, *Integr Med Res*. 9 (2020) 100465. https://doi.org/10.1016/J.IMR.2020.100465.

7. N. Sayed, A. Khurana, C. Godugu, Pharmaceutical perspective on the translational hurdles of phytoconstituents and strategies to overcome, *J Drug Deliv Sci Technol.* 53 (2019) 101201. https://doi.org/10.1016/J.JDDST.2019.101201.

8. N. Jamil, M. Naqiuddin, M. Zairi, N. A'in, M. Nasim, F. Pa'ee, Influences of environmental conditions to phytoconstituents in clitoria ternatea (butterfly pea flower) – A review, *J.Sci Technol.* 10 (2018) 208–228. https://doi.org/10.30880/jst.2018.10.02.029.

9. M.A. Barkat, Goyal, H.A. Barkat, M. Salauddin, F.H. Pottoo, E.T. Anwer, Herbal medicine: Clinical perspective and regulatory status, *Comb Chem High Throughput Screen.* 24 (2020) 1573–1582. https://doi.org/10.2174/1386207323999201110192942.

10. B.H.J. Gowda, M.G. Ahmed, S. Chinnam, K. Paul, M. Ashrafuzzaman, M. Chavali, R. Gahtori, S. Pandit, K.K. Kesari, P.K. Gupta, Current trends in bio-waste mediated metal/metal oxide nanoparticles for drug delivery, *J Drug Deliv Sci Technol.* 71 (2022) 103305. https://doi.org/10.1016/J.JDDST.2022.103305.

11. M. Ashrafuzzaman, *An Introduction to Drug Carriers.* (2021) pp. 1–214. https://doi.org/10.52305/MZZS5968.

12. U. Hani, R.A.M. Osmani, S. Yasmin, B.H.J. Gowda, H. Ather, M.Y. Ansari, A. Siddiqua, M. Ghazwani, A. al Fatease, A.H. Alamri, M. Rahamathulla, M.Y. Begum, S. Wahab, Novel drug delivery systems as an emerging platform for stomach cancer therapy, *Pharmaceutics.* 14 (2022), 1576. https://doi.org/10.3390/PHARMACEUTICS14081576.

13. B. García-Pinel, C. Porras-Alcalá, A. Ortega-Rodríguez, F. Sarabia, J. Prados, C. Melguizo, J.M. López-Romero, Lipid-based nanoparticles: Application and recent advances in cancer treatment, *Nanomaterials.* 9 (2019). https://doi.org/10.3390/NANO9040638.

14. R. Kumar, Lipid-based nanoparticles for drug-delivery systems. In *Nanocarriers for Drug Delivery: Nanoscience and Nanotechnology in Drug Delivery.* (2019) pp. 249–284. https://doi.org/10.1016/B978-0-12-814033-8.00008-4.

15. S. A, M.G. Ahmed, B.H.J. Gowda, S. Surya, Formulation and characteristic evaluation of tacrolimus cubosomal gel for vitiligo (2022) 1–10. https://doi.org/10.1080/01932691.2022.2139716.

16. A. Sanjana, M.G. Ahmed, J.B.H. Gowda, Development and evaluation of dexamethasone loaded cubosomes for the treatment of vitiligo, *Mater Today Proc.* 50 (2022) 197–205. https://doi.org/10.1016/J.MATPR.2021.04.120.

17. J.J. Giner-Casares, M. Henriksen-Lacey, M. Coronado-Puchau, L.M. Liz-Marzán, Inorganic nanoparticles for biomedicine: Where materials scientists meet medical research, *Materials Today.* 19 (2016) 19–28. https://doi.org/10.1016/J.MATTOD.2015.07.004.

18. J. Meena, A. Gupta, R. Ahuja, M. Singh, S. Bhaskar, A.K. Panda, Inorganic nanoparticles for natural product delivery: A review, *Environ Chem Lett.* 18(6) (2020) 2107–2118. https://doi.org/10.1007/S10311-020-01061-2.

19. S. Kesarwani, R.K. Verma, A critical review on synthesis, characterization and multifunctional applications of reduced graphene oxide (rGO)/composites. *Nano.* 16 (2021). https://doi.org/10.1142/S1793292021300085.

20. S. Pramanik, V. Sali, Connecting the dots in drug delivery: A tour d'horizon of chitosan-based nanocarriers system, *Int J Biol Macromol.* 169 (2021) 103–121. https://doi.org/10.1016/J.IJBIOMAC.2020.12.083.

21. A. Zielinska, F. Carreiró, A.M. Oliveira, A. Neves, B. Pires, D. Nagasamy Venkatesh, A. Durazzo, M. Lucarini, P. Eder, A.M. Silva, A. Santini, E.B. Souto, Polymeric nanoparticles: Production, characterization, toxicology and ecotoxicology, *Molecules.* 25 (2020). https://doi.org/10.3390/MOLECULES25163731.

22. K.C. de Castro, J.M. Costa, M.G.N. Campos, Drug-loaded polymeric nanoparticles: A review. 71 (2020) 1–13. https://doi.org/10.1080/00914037.2020.1798436.

23. M. Taib, F. Damiri, Y. Bachra, M. Berrada, L. Bouyazza, Recent advances in micro- and nanoencapsulation of bioactive compounds and their food applications. In *Nanotechnology in Intelligent Food Packaging.* (2022) pp. 271–289. https://doi.org/10.1002/9781119819011.CH11.

24. B. Begines, T. Ortiz, M. Pérez-Aranda, G. Martínez, M. Merinero, F. Argüelles-Arias, A. Alcudia, Polymeric nanoparticles for drug delivery: Recent developments and future prospects, *Nanomaterials.* 10 (2020) 1403. https://doi.org/10.3390/NANO10071403.

25. M.J. Mitchell, M.M. Billingsley, R.M. Haley, M.E. Wechsler, N.A. Peppas, R. Langer, Engineering precision nanoparticles for drug delivery, *Nat Rev Drug Discov* 20(2) (2020) 101–124. https://doi.org/10.1038/s41573-020-0090-8.

26. C. Vauthier, K. Bouchemal, Methods for the preparation and manufacture of polymeric nanoparticles, *Pharm Res.* 26 (2008) 1025–1058. https://doi.org/10.1007/S11095-008-9800-3.

27. U. Dubey, S. Kesarwani, R.K. Verma, Incorporation of graphene nanoplatelets/hydroxyapatite in PMMA bone cement for characterization and enhanced mechanical properties of biopolymer composites. (2022). https://doi.org/10.1177/08927057221086833.

28. B.L. Banik, P. Fattahi, J.L. Brown, Polymeric nanoparticles: The future of nanomedicine, *Wiley Interdiscip Rev Nanomed Nanobiotechnol.* 8 (2016) 271–299. https://doi.org/10.1002/WNAN.1364.

29. S. Perumal, R. Atchudan, W. Lee, A review of polymeric micelles and their applications, Polymers (Basel). 14 (2022). https://doi.org/10.3390/POLYM14122510.

30. Z. Ahmad, A. Shah, M. Siddiq, H.B. Kraatz, Polymeric micelles as drug delivery vehicles, RSC Adv. 4 (2014) 17028–17038. https://doi.org/10.1039/C3RA47370H.

31. M. Ghezzi, S. Pescina, C. Padula, P. Santi, E. del Favero, L. Cantù, S. Nicoli, Polymeric micelles in drug delivery: An insight of the techniques for their characterization and assessment in biorelevant conditions, *J Control Release.* 332 (2021) 312–336. https://doi.org/10.1016/J.JCONREL.2021.02.031.

32. S.F. Zhang, W. Hu, X. Yan, D. Wang, W. Yang, J. Zhang, Z. Liu, Chondroitin sulfate-curcumin micelle with good stability and reduction sensitivity for anti-cancer drug carrier, Mater Lett. 304 (2021) 130667. https://doi.org/10.1016/J.MATLET.2021.130667.

33. Q. Li, X. Wu, S. Xin, X. Wu, J. Lan, Preparation and characterization of a naringenin solubilizing glycyrrhizin nanomicelle ophthalmic solution for experimental dry eye disease, *Eur J Pharm Sci.* 167 (2021) 106020. https://doi.org/10.1016/J.EJPS.2021.106020.

34. V. Patel, P. Patel, J.V. Patel, P.M. Patel, Dendrimer as a versatile platform for biomedical application: A review, *J Ind Chem Soc.* 99 (2022) 100516. https://doi.org/10.1016/J.JICS.2022.100516.

35. M.A. Mintzer, M.W. Grinstaff, Biomedical applications of dendrimers: A tutorial, *Chem Soc Rev.* 40 (2010) 173–190. https://doi.org/10.1039/B901839P.

36. P. Patel, V. Patel, P.M. Patel, Synthetic strategy of dendrimers: A review, *J Ind Chem Soc.* 99 (2022) 100514. https://doi.org/10.1016/J.JICS.2022.100514.

37. S. Karimi, H. Namazi, Synthesis of folic acid-conjugated glycodendrimer with magnetic β-cyclodextrin core as a pH-responsive system for tumor-targeted co-delivery of doxorubicin and curcumin, *Colloids Surf A Physicochem Eng Asp.* 627 (2021) 127205. https://doi.org/10.1016/J.COLSURFA.2021.127205.

38. J. Gallien, B. Srinageshwar, K. Gallo, G. Holtgrefe, S. Koneru, P.S. Otero, C.A. Bueno, J. Mosher, A. Roh, D.S. Kohtz, D. Swanson, A. Sharma, G. Dunbar, J. Rossignol, Curcumin loaded dendrimers specifically reduce viability of glioblastoma cell lines, Molecules. 26 (2021) 6050. https://doi.org/10.3390/MOLECULES26196050.

39. S. Sharma, A. Parmar, S. Kori, R. Sandhir, PLGA-based nanoparticles: A new paradigm in biomedical applications, *TrAC Trends Anal Chem.* 80 (2016) 30–40. https://doi.org/10.1016/J.TRAC.2015.06.014.

40. A.G. Niculescu, A.M. Grumezescu, Applications of chitosan-alginate-based nanoparticles—An up-to-date review, *Nanomaterials.* 12 (2022) 186. https://doi.org/10.3390/NANO12020186.

41. A. Dodero, S. Alberti, G. Gaggero, M. Ferretti, R. Botter, S. Vicini, M. Castellano, An up-to-date review on alginate nanoparticles and nanofibers for biomedical and pharmaceutical applications, *Adv Mater Interfaces.* 8 (2021) 2100809. https://doi.org/10.1002/ADMI.202100809.

42. K. Kohli, A. Mujtaba, R. Malik, S. Amin, M.S. Alam, A. Ali, M.A. Barkat, M.J. Ansari, Development of natural polysaccharide-based nanoparticles of berberine to enhance oral bioavailability: Formulation, optimization, ex vivo, and in vivo assessment, *Polymers (Basel).* 13 (2021). https://doi.org/10.3390/POLYM13213833.

43. F. Jiang, L. Yang, S. Wang, X. Ying, J. Ling, X. kun Ouyang, Fabrication and characterization of zein-alginate oligosaccharide complex nanoparticles as delivery vehicles of curcumin, *J Mol Liq.* 342 (2021) 116937. https://doi.org/10.1016/J.MOLLIQ.2021.116937.

44. H. Moon, P. Lertpatipanpong, Y. Hong, C.T. Kim, S.J. Baek, Nano-encapsulated quercetin by soluble soybean polysaccharide/chitosan enhances anti-cancer, anti-inflammation, and anti-oxidant activities, *J Funct Foods.* 87 (2021) 104756. https://doi.org/10.1016/J.JFF.2021.104756.

45. S. Kumari, A. Goyal, M. Garg, Phytoconstituents based novel nano-formulations: An approach, ECS Trans. 107 (2022) 7365–7379. https://doi.org/10.1149/10701.7365ECST/XML.

46. S. Gavas, S. Quazi, T.M. Karpiński, Nanoparticles for cancer therapy: Current progress and challenges, *Nanoscale Res Lett.* 16 (2021) 173. https://doi.org/10.1186/S11671-021-03628-6.

47. A.S. Choudhari, P.C. Mandave, M. Deshpande, P. Ranjekar, O. Prakash, Phytochemicals in cancer treatment: From preclinical studies to clinical practice, Front Pharmacol. 10 (2020) 1614. https://doi.org/10.3389/FPHAR.2019.01614/BIBTEX.

48. C. Melim, M. Magalhães, A.C. Santos, E.J. Campos, C. Cabral, Nanoparticles as phytochemical carriers for cancer treatment: News of the last decade, Expert Opin Drug Deliv. 19 (2022) 179–197. https://doi.org/10.1080/17425247.2022.2041599.

49. T. Khan, P. Gurav, PhytoNanotechnology: Enhancing delivery of plant based anti-cancer drugs, Front Pharmacol. 8 (2018) 1002. https://doi.org/10.3389/FPHAR.2017.01002/BIBTEX.

50. Z. Eskandari, F. Bahadori, V.B. Yenigun, M. Demiray, M.S. Eroğlu, A. Kocyigit, E.T. Oner, Levan enhanced the NF-κB suppression activity of an oral nano PLGA-curcumin formulation in breast cancer treatment, *Int J Biol Macromol.* 189 (2021) 223–231. https://doi.org/10.1016/J.IJBIOMAC.2021.08.115.

51. P. Patil, S. Killedar, Formulation and characterization of gallic acid and quercetin chitosan nanoparticles for sustained release in treating colorectal cancer, *J Drug Deliv Sci Technol.* 63 (2021) 102523. https://doi.org/10.1016/J.JDDST.2021.102523.

52. World Health Organization, The top 10 causes of death, (2020). https://www.who.int/news-room/fact-sheets/detail/the-top-10-causes-of-death (accessed October 10, 2022).

53. M.M. Mehndiratta, V. Aggarwal, Neurological disorders in India: Past, present, and next steps, Lancet Glob Health. 9 (2021) e1043–e1044. https://doi.org/10.1016/S2214-109X(21)00214-X.

54. G.P. Kumar, F. Khanum, Neuroprotective potential of phytochemicals, *Pharmacogn Rev.* 6 (2012) 81. https://doi.org/10.4103/0973-7847.99898.

55. J. Kim, H.J. Lee, K.W. Lee, Naturally occurring phytochemicals for the prevention of Alzheimer's disease, *J Neurochem.* 112 (2010) 1415–1430. https://doi.org/10.1111/J.1471-4159.2009.06562.X.

56. T. Bhattacharya, G.A.B.E. Soares, H. Chopra, M.M. Rahman, Z. Hasan, S.S. Swain, S. Cavalu, Applications of phyto-nanotechnology for the treatment of neurodegenerative disorders, *Materials.* 15 (2022). https://doi.org/10.3390/MA15030804.

57. M. Ovais, N. Zia, I. Ahmad, A.T. Khalil, A. Raza, M. Ayaz, A. Sadiq, F. Ullah, Z.K. Shinwari, Phyto-therapeutic and nanomedicinal approaches to cure Alzheimer's disease: Present status and future opportunities, Front Aging Neurosci. 10 (2018) 284. https://doi.org/10.3389/FNAGI.2018.00284/BIBTEX.

58. N. Dhas, T. Mehta, Intranasal delivery of chitosan decorated PLGA core /shell nanoparticles containing flavonoid to reduce oxidative stress in the treatment of Alzheimer's disease, *J Drug Deliv Sci Technol.* 61 (2021) 102242. https://doi.org/10.1016/J.JDDST.2020.102242.

59. Y.C. Kuo, H.C. Tsai, Rosmarinic acid- and curcumin-loaded polyacrylamide-cardiolipin-poly(lactide-co-glycolide) nanoparticles with conjugated 83-14 monoclonal antibody to protect β-amyloid-insulted neurons, *Mater Sci Eng C Mater Biol Appl.* 91 (2018) 445–457. https://doi.org/10.1016/J.MSEC.2018.05.062.

60. R.M. Pop, A. Popolo, A.P. Trifa, L.A. Stanciu, Phytochemicals in cardiovascular and respiratory diseases: Evidence in oxidative stress and inflammation, *Oxid Med Cell Longev.* 2018 (2018). https://doi.org/10.1155/2018/1603872.

61. B. Pagliaro, C. Santolamazza, F. Simonelli, S. Rubattu, Phytochemical compounds and protection from cardiovascular diseases: A state of the art, *Biomed Res Int.* 2015 (2015). https://doi.org/10.1155/2015/918069.

62. S.C. Pillai, A. Borah, M.N.T. Le, H. Kawano, K. Hasegawa, D. Sakthi Kumar, Co-delivery of curcumin and bioperine via PLGA nanoparticles to prevent atherosclerotic foam cell formation, Pharmaceutics. 13 (2021). https://doi.org/10.3390/PHARMACEUTICS13091420.

63. I.F. Mosa, H.H. Abd, A. Abuzreda, A.B. Yousif, N. Assaf, Chitosan and curcumin nanoformulations against potential cardiac risks associated with hydroxyapatite nanoparticles in wistar male rats, *Int J Biomater.* 2021 (2021). https://doi.org/10.1155/2021/3394348.

64. A.C.D.O. Gonzalez, Z.D.A. Andrade, T.F. Costa, A.R.A.P. Medrado, Wound healing - A literature review, An Bras Dermatol. 91 (2016) 614. https://doi.org/10.1590/ABD1806-4841.20164741.

65. K. Raziyeva, Y. Kim, Z. Zharkinbekov, K. Kassymbek, S. Jimi, A. Saparov, Immunology of acute and chronic wound healing, Biomolecules. 11 (2021). https://doi.org/10.3390/BIOM11050700.

66. T.N. Demidova-Rice, M.R. Hamblin, I.M. Herman, Acute and impaired wound healing: Pathophysiology and current methods for drug delivery, Part 1: Normal and chronic wounds: Biology, causes, and approaches to care, *Adv Skin Wound Care.* 25 (2012) 304. https://doi.org/10.1097/01.ASW.0000416006.55218.D0.

67. G. Han, R. Ceilley, Chronic wound healing: A review of current management and treatments, Adv Ther. 34 (2017) 599. https://doi.org/10.1007/S12325-017-0478-Y.

68. R.L. Thangapazham, S. Sharad, R.K. Maheshwari, Phytochemicals in wound healing, *Adv Wound Care (New Rochelle).* 5 (2016) 230. https://doi.org/10.1089/WOUND.2013.0505.

69. A. Shah, S. Amini-Nik, The role of phytochemicals in the inflammatory phase of wound healing, *Int J Mol Sci.* 18 (2017). https://doi.org/10.3390/IJMS18051068.

70. A. Qadir, S. Jahan, M. Aqil, M.H. Warsi, N.A. Alhakamy, M.A. Alfaleh, N. Khan, A. Ali, Phytochemical-based nano-pharmacotherapeutics for management of burn wound healing, Gels. 7 (2021) 209. https://doi.org/10.3390/GELS7040209.

71. B. Gorain, M. Pandey, N.H. Leng, C.W. Yan, K.W. Nie, S.J. Kaur, V. Marshall, S.P. Sisinthy, J. Panneerselvam, N. Molugulu, P. Kesharwani, H. Choudhury, Advanced drug delivery systems containing herbal components for wound healing, *Int J Pharm.* 617 (2022) 121617. https://doi.org/10.1016/J.IJPHARM.2022.121617.

72. V.V.S.R. Karri, G. Kuppusamy, S.V. Talluri, S.S. Mannemala, R. Kollipara, A.D. Wadhwani, S. Mulukutla, K.R.S. Raju, R. Malayandi, Curcumin loaded chitosan nanoparticles impregnated into collagen-alginate scaffolds for diabetic wound healing, *Int J Biol Macromol.* 93 (2016) 1519–1529. https://doi.org/10.1016/J.IJBIOMAC.2016.05.038.

73. S. Amanat, S. Taymouri, J. Varshosaz, M. Minaiyan, A. Talebi, Carboxymethyl cellulose-based wafer enriched with resveratrol-loaded nanoparticles for enhanced wound healing, Drug Deliv Transl Res. 10 (2020) 1241–1254. https://doi.org/10.1007/S13346-020-00711-W.

74. P. Zhang, L. He, J. Zhang, X. Mei, Y. Zhang, H. Tian, Z. Chen, Preparation of novel berberine nanocolloids for improving wound healing of diabetic rats by acting Sirt1/NF-κB pathway, *Colloids Surf B Biointerfaces.* 187 (2020) 110647. https://doi.org/10.1016/J.COLSURFB.2019.110647.

75. S. bin Zaman, M.A. Hussain, R. Nye, V. Mehta, K.T. Mamun, N. Hossain, A review on antibiotic resistance: Alarm bells are ringing, *Cureus.* 9 (2017). https://doi.org/10.7759/CUREUS.1403.

76. T. Khare, U. Anand, A. Dey, Y.G. Assaraf, Z.-S. Chen, Z. Liu, V. Kumar, Exploring phytochemicals for combating antibiotic resistance in microbial pathogens, *Front Pharmacol.* 12 (2021) 1726. https://doi.org/10.3389/FPHAR.2021.720726.

77. C.H.N. Barros, D.W. Hiebner, S. Fulaz, S. Vitale, L. Quinn, E. Casey, Synthesis and self-assembly of curcumin-modified amphiphilic polymeric micelles with antibacterial activity, *J Nanobiotechnol.* 19 (2021) 1–15. https://doi.org/10.1186/S12951-021-00851-2/FIGURES/8.

78. M.S. Valencia, M.F. da Silva Júnior, F.H. Xavier-Júnior, B. de O. Veras, P.B.S. de Albuquerque, E.F. de O. Borba, T.G. da Silva, V.L. Xavier, M.P. de Souza, M. das G. Carneiro-da-Cunha, Characterization of curcumin-loaded lecithin-chitosan bioactive nanoparticles, *Carbohydr Polym Technol Appl.* 2 (2021) 100119. https://doi.org/10.1016/J.CARPTA.2021.100119.

79. B. Silver, K. Ramaiya, S.B. Andrew, O. Fredrick, S. Bajaj, S. Kalra, B.M. Charlotte, K. Claudine, A. Makhoba, EADSG guidelines: Insulin therapy in diabetes, *Diabetes Ther.* 9 (2018) 449–492. https://doi.org/10.1007/S13300-018-0384-6.

80. S. Alam, M.M.R. Sarker, T.N. Sultana, M.N.R. Chowdhury, M.A. Rashid, N.I. Chaity, C. Zhao, J. Xiao, E.E. Hafez, S.A. Khan, I.N. Mohamed, Antidiabetic phytochemicals from medicinal plants: Prospective candidates for new drug discovery and development, Front Endocrinol (Lausanne). 13 (2022) 1. https://doi.org/10.3389/FENDO.2022.800714.

81. S.M. Firdous, Phytochemicals for treatment of diabetes, EXCLI J. 13 (2014) 451. /pmc/articles/PMC4464495/ (accessed October 10, 2022).

82. P. Mukhopadhyay, S. Maity, S. Mandal, A.S. Chakraborti, A.K. Prajapati, P.P. Kundu, Preparation, characterization and in vivo evaluation of pH sensitive, safe quercetin-succinylated chitosan-alginate core-shell-corona nanoparticle for diabetes treatment, Carbohydr Polym. 182 (2018) 42–51. https://doi.org/10.1016/J.CARBPOL.2017.10.098.

83. V. Perumal, T. Manickam, K.S. Bang, P. Velmurugan, B.T. Oh, Antidiabetic potential of bioactive molecules coated chitosan nanoparticles in experimental rats, *Int J Biol Macromol.* 92 (2016) 63–69. https://doi.org/10.1016/J.IJBIOMAC.2016.07.006.

7 Polymer Nanocomposite Coatings

Characterisation Techniques and Applications

*Hemalata Jena and Sudesna Roy**

CONTENTS

7.1 INTRODUCTION

Polymer nanocomposites (PNCs) are an emerging class of materials that have found wide applications in many technical, research and industrial areas. Among these, PNC coatings are specifically used as coatings of a few millimetres thickness on hard substrates. Their uses are multiple and will be discussed in detail in Section 7.3.

Here the primary component or the matrix is a polymer material. The polymer may be thermoplastic based or thermosetting based. The secondary component or reinforcement material may be any class of material or compound. The principal constraint is to limit the size of the reinforcement, at least one dimension, in the nanometre range (< 100 nm). The transition from micro- to nanosized reinforcement increases the surface-to-volume ratio of the reinforcement. This in turn enhances the dominance of the surface properties and affects all the aggregate properties of the composite drastically. The mechanical properties like strength, hardness and toughness of the composite increase drastically due to the reduction in grain size component [1]. The increase in the grain boundary area, due to decreased grain size below 100 nm, increases the grain boundary contribution. Beyond a threshold grain size value, the grain boundary area increases drastically and contributes mostly to the enhanced properties of the nanomaterials.

Nanocomposite coatings, specifically, are limited to the thin hard coatings of the composite of dimension a few microns to millimetres in thickness. They may be single-layered or multi-layered depending on their applications. The fabrication technique involves a coating process like the

DOI: 10.1201/9781003343912-7

95

sol–gel method, vapour deposition method, cold spraying, thermal spraying, in-situ polymerisation and electrochemical deposition, to name a few [2]. In each of these methods, the polymer matrix is mixed with the nanodimensional reinforcement and applied to the substrate. In some cases, like PVD, CVD and in-situ polymerisation methods, the reinforcement is formed in-situ during fabrication. This is advantageous over the mix-in blending methods since the interfacial adherence or bonding between the reinforcement and the matrix is improved. Most failures for composites occur at the interfacial sites due to de-bonding between reinforcement and matrix. Moreover, for nanosized reinforcement due to the increase in surface area, the interfacial bonding becomes even more important. The preparation and fabrication are unique, and details of such can be found elsewhere [2].

This chapter focuses on the characterisation techniques that are usually relied on nanocomposite coatings. Basic and advanced characterisation methods are employed to gauge the viability of the performance of the coatings in different applications. Here, the characterisation methods have been categorised into different sections depending on their properties, viz. macro- and micro-structure, thermal properties, rheological properties, biological properties and other advanced specific techniques. The advanced techniques include nanoindentation, Fourier transform infrared spectroscopy (FTIR), secondary ion spectroscopy and Auger spectroscopy.

7.2 CHARACTERISATION TECHNIQUES

Characterisation techniques are general methods used to check the outcome after the fabrication of a nanocomposite. After fabrication, a composite needs to be assessed in terms of the quality of the fabrication. In this regard, various techniques are employed to visualise and verify the properties of the composite. Broadly, characterisation techniques may be classified as destructive and non-destructive. In destructive characterisation methods like tensile testing, 3-point bend test, etc., the coating is made to fail under certain duress, applied as external load. This includes failure studies where the mechanism of failure or rupture under constant of varying modes of stress is studied. However, most of the characterisation tests use non-destructive type of tests where the evaluation does not harm the coating. In this chapter, most of the focus is specifically drawn towards non-destructive testing of coatings.

7.2.1 Appearance, Structure and Phase

The appearance of a coating is the first indication of its integrity. It should have a uniform colour and appearance with no discolouration or patches. The centre of the substrate and edges should have similar coverage indicating a uniform coating. A high-resolution camera is generally used to characterise the appearance. The structure on the other hand is used to characterise the features internally. This includes the phase, structure and elemental composition. All of this information together will help distinguish the microstructure of the coating.

X-ray diffraction (XRD) is perhaps the most elementary characterisation method used. It provides detailed information about the crystallographic structure, composition and phases present. It is an analysis of the interaction between radiation and matter where the radiation is in the form of x-rays. X-rays typically have a wavelength between 0.01 and 10 nm. They are shorter in wavelength than UV rays and longer than gamma rays. XRD uses Bragg's principle to analyse the crystallinity of materials. Figure 7.1 shows a schematic of the interaction of x-rays with an ideal crystalline lattice (denoted by solid points arranged in an array). The interplanar distance is calculated according to Bragg's law, given in Equation 7.1:

$$n\lambda = 2d \sin\theta \tag{7.1}$$

where d is the interplanar distance, λ is the wavelength of radiation, and θ is the angle of incidence, as shown in Figure 7.1.

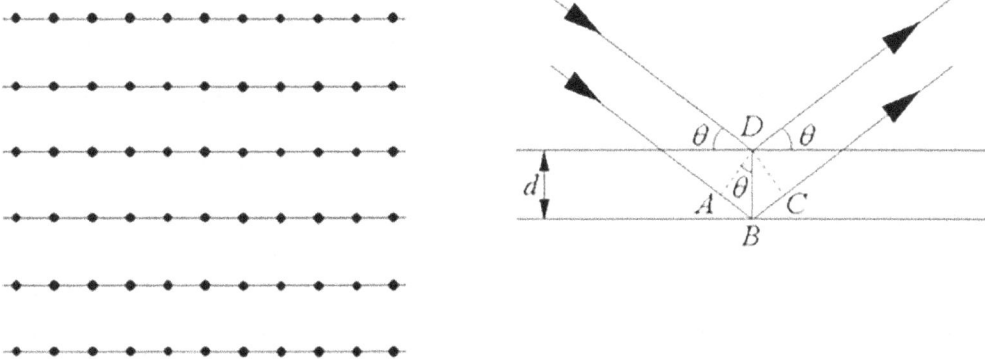

FIGURE 7.1 Illustration of Bragg's law in crystal lattice.

Although there are many variations of x-ray diffractometers, the most common type is called a powder diffractometer. It consists of an x-ray generator, a stage for samples and an x-ray detector. The x-rays generated are collimated and incident on the sample at an angle, θ, that is detected by the detector at an angle 2θ. The attachment of the sample holder and detector together is called the goniometer. The incident angle is constantly changed throughout the measurement. The output is a plot of 2θ vs counts or intensity of radiation detected. A perfectly crystalline sample will show sharp peaks with good intensity at predefined 2θ positions, whereas an amorphous sample shows low-angle broad peaks with no defined clarity of peaks. An XRD plot of a nanocomposite polymer will show somewhat broad peaks, albeit at the predefined 2θ positions. Most polymers are amorphous or semicrystalline in nature. However, only the XRD does not give justice to understanding the nanocomposite. For a complete picture, the microstructure gives further evidence.

The microstructure is high-magnification imaging of the structure. It gives first-hand information on the shape, size and integrity of the nanocomposite. There are basically three different techniques of microscopy imaging, viz. optical microscopy (OM), scanning electron microscopy (SEM) and transmission electron microscopy (TEM). While OM uses white-light radiation to image the structure, SEM and TEM use electron beams for imaging. OM has a limitation of imaging particles in the micrometre to the submicrometre range, while SEM and TEM can image in the nanometre and angstrom metre range. High-resolution TEM (HRTEM) can also resolve atomic features of angstrom and sub-picometre range. Usually, SEM and TEM are used to resolve the features of nanocomposites. However, while SEM is non-destructive, TEM is destructive that requires careful and laborious sample preparation.

The microstructure of the polymer matrix may be amorphous and yet have some microstructural features to it, like porosity, voids, etc., that are imaged usually by OM and SEM. The nanosized reinforcement is usually imaged by a high-resolution field emission SEM or an HRTEM. The information about the shape, size, agglomeration, defect chemistry, bonding, etc. is harnessed using this technique. For nanocomposite polymer coatings, the interaction between particle and matrix is important towards the integrity of the coating and its performance.

For the sake of illustration, let us now consider the case of CeO_2 nanoparticles dispersed in a conjugated PNC fabricated by chemical co-precipitation. Figure 7.2 shows the XRD plots of a polymer as well as the nanocomposite fabricated with 20 and 40 wt.% CeO_2 nanoparticles. In Figure 7.2i, the polymer matrix, in this case, is amorphous which increases in crystallinity due to the addition of the CeO_2 nanoparticles. Figure 7.2ii shows the (a) TEM and (b) HRTEM micrographs fabricated with CeO_2 nanoparticles and (c) TEM and (d) HRTEM images of 40 wt.% CeO_2 PNCs. The high-resolution micrographs show the crystallinity, shape and size of the CeO_2 nanoparticles. It even shows the atomic arrangement and d spacing in the CeO_2 particles.

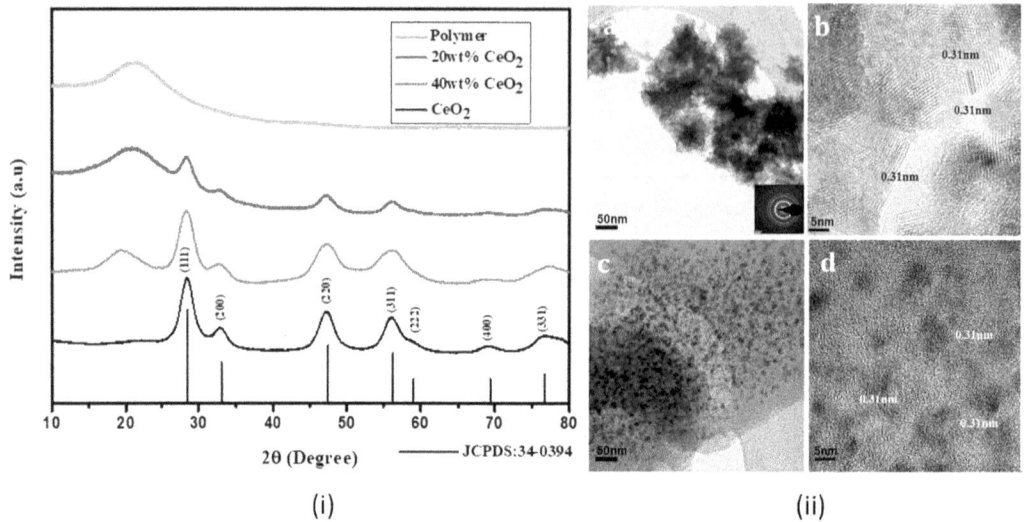

(i)   (ii)

FIGURE 7.2 (i) XRD plot, and (ii) microstructure (TEM and HRTEM) for PNC (CeO$_2$ nanoparticle dispersed in a conjugated polymer matrix) [3]. (Open Access.)

7.2.2 THERMAL CHARACTERISATION

Thermal characterisation involves a few basic techniques like thermogravimetric analysis (TGA), differential scanning calorimetry (DSC), differential thermal analysis (DTA), dynamic mechanical thermal analysis, thermal mechanical analysis, etc. [4]. Here, a small amount of the materials (a few micrograms) are heated gradually in a controlled atmosphere. The changes in weight, heat output, etc. are noted against the temperature. This provides an excellent tool, especially for polymer characterisation, like melting point glass transition, polymerisation, thermal degradation and oxidation reactions.

Thermal characteristics like curing temperature are essential parameters to determine the temperature required to heat the polymer without degradation of its properties. In this regard, DSC is helpful in determining isothermal curing temperature to set polymeric materials [5]. Similarly, Boparai et al. have used DSC and TGA together to determine the melting characteristic of a polymer composite [6]. It has been shown that with addition of reinforcement to a polymeric material, the thermal characteristics, like decomposition temperature, dehydration temperature, etc., change drastically with the increasing fibre loading [7]. With high cellulose fibre-loading content, the degradation property of the bio-polymer increases drastically with temperature. Further information about the change in heat capacity of the nanocomposite can also be measured by the combination of DTA and DSC [8]. Another important aspect is the film or coating interaction with the substrate material. With increasing temperature, the coating may at places react with the substrate to create heat spots [9].

There are other less frequently used techniques also like thermosonimetry, thermoluminescence, thermomanometry, thermomagnetometry, dynamic reflectance spectroscopy and emanation thermal analysis [10]. Newer methods of thermal characterisation include coupling TGA with mass spectroscopy to elucidate the thermal decomposition behaviour related to polymerisation mechanisms [11]. Another such technique is the high-resolution TGA that couples heating rate to weight, such that the temperature of the sample is not heated at a pre-determined heating rate, rather it is not raised till the weight loss/gain at an isothermal temperature is not completed [8]. Another such technique is temperature-modulated DSC, where the temperature scan is maintained by a sinusoidal heat flow. However, the analysis of such conjugated techniques is rather tricky and involves careful expertise [12].

7.2.3 RHEOLOGICAL CHARACTERISATION

The rheological characteristics of the material involve the characterisation of flow or deformation of the material in response to applied stress. This is most significant and prominent in polymeric materials that have a visco-elastic response towards applied stress. It is most commonly measured by a rheometer. A rheometer measures the resistance (to flow) of a material (usually fluid) when shear stress is applied. This is different from a viscometer which simply measures the viscosity of a fluid at certain parameters. Rheometers provide more data on the material structure and its elasticity.

Here, the most common parameters that are studied are viscosity, concentration and their relationship in different types of environments. Sometimes, their effect on changes in temperatures may be studied. For layered structures like nanoclay reinforcement in polymers, the information on the degree of intercalation, exfoliation and dispersion of a PNC is studied as a function of applied stress. Zhao *et al.* have shown that rheological measurements are complimentary information to traditional analysis techniques for PNC coatings [13]. Duggirala *et al.* have studied the rheological characterisation of cellulose and alginate-based polymers [14]. They have studied the effect of polymer concentration, temperature and autoclaving on the viscosity of the polymers. Abdel-Goad *et al.* have studied the melt rheology properties of expanded graphite in low-density polyethylene [15]. Here, the melt-state linear visco-elastic properties were evaluated by varying the nanofiller from 1 to 20 wt.%. For polymers and filler particles that have charges at their melt state, the dielectric properties including zeta potential are useful to be evaluated [16].

7.2.4 BIOLOGICAL CHARACTERISATION

Biological characterisation of PNCs is performed for biocompatible materials that are used in association with the human body. The characterisation may be done external (in-vivo) or internal (in-vitro) with respect to the human body. By rule of thumb, the materials are first characterised externally, that is, in-vivo before being used in in-vitro. The focus of this work will be to review the most common in-vivo techniques.

One of the most common methods for testing biological nanocomposite samples is potentiodynamic polarisation scanning. This is an electrochemical method of evaluating the response of the sample to body fluid in a controlled environment. Here, a small voltage is applied across the sample and its current measured over a period of time as the electrochemical reactions stabilise. Other corrosion testing methods include immersion testing, impedance spectroscopy, cyclic polarisation, etc., where the sample is tested in the body fluid and changes recorded over time [17].

7.2.5 OTHER SPECIFIC TECHNIQUES

Other specific characterisation techniques include methods like that show microstructure through atomic force microscopy, confocal microscopy, scanning electron transmission spectroscopy, etc. These are advanced characterisation techniques that also trace the surface details and morphology using different types of analytical tools and methods. The chemical composition may also be evaluated through FTIR spectroscopy, Raman spectroscopy, X-ray photoelectron spectroscopy, etc., where the chemical bonding between the atoms is detected to identify certain compounds. Other specific techniques may include methods like nanoindentation where specific material properties by indenting at the nanometric level may be identified.

To get an overall performance of a nanocomposite, all the characterisation techniques have to be synergistically coordinated. Quadri et al. have studied the corrosion behaviour of zinc oxide nanofiller embedded in polymeric substrates on mild steel [18]. They have synergistically investigated its structural properties through SEM and TEM, its optical properties through UV–Vis spectroscopy, thermal properties through TGA and corrosion properties through potentiodynamic polarisation, linear polarisation and impedance spectroscopy. They have concluded that addition of nanoparticles of ZnO improves the corrosion resistance of the coating. The coating behaves like a mixed-type inhibitor. The SEM micrographs show the effect of corrosion on the surface.

Similarly, most of the characterisation techniques act like different senses in our body, while some can 'see', other can 'touch', 'smell' and 'taste' to get an accurate picture of the structure-property-performance of the PNC. Completely relying on one type of characterisation technique will always lead to erroneous results.

7.3 APPLICATIONS OF POLYMER NANOCOMPOSITE COATINGS

7.3.1 BIOLOGICAL APPLICATIONS

Nanocomposite coating plays a very important role in biomedical applications due to the uniqueness of multifunctional nanofillers in the polymer matrix. The best property cannot be achieved in the polymer or nanofiller individually but combining these two shows the synergism properties in the PNC coating. Polymers and inorganic nanofillers are widely used for the coating on the substrate. Nanopolymer composite has several purposes for the development of human health in biomedical engineering. Nanocellulose and nanometal have been mainly used as a filler in nanocomposites which is widely used in biomedical applications like drug delivery, biosensor and diagnostics, medical implants, antimicrobial membrane, skin tissue repair, vascular graft and scaffolds as shown in Figure 7.3.

Nanocomposite helps humankind in several ways which is one of the large and fast-growing researches of nanotechnology in biomedical applications. Negroiu et al. [19] have prepared PNC

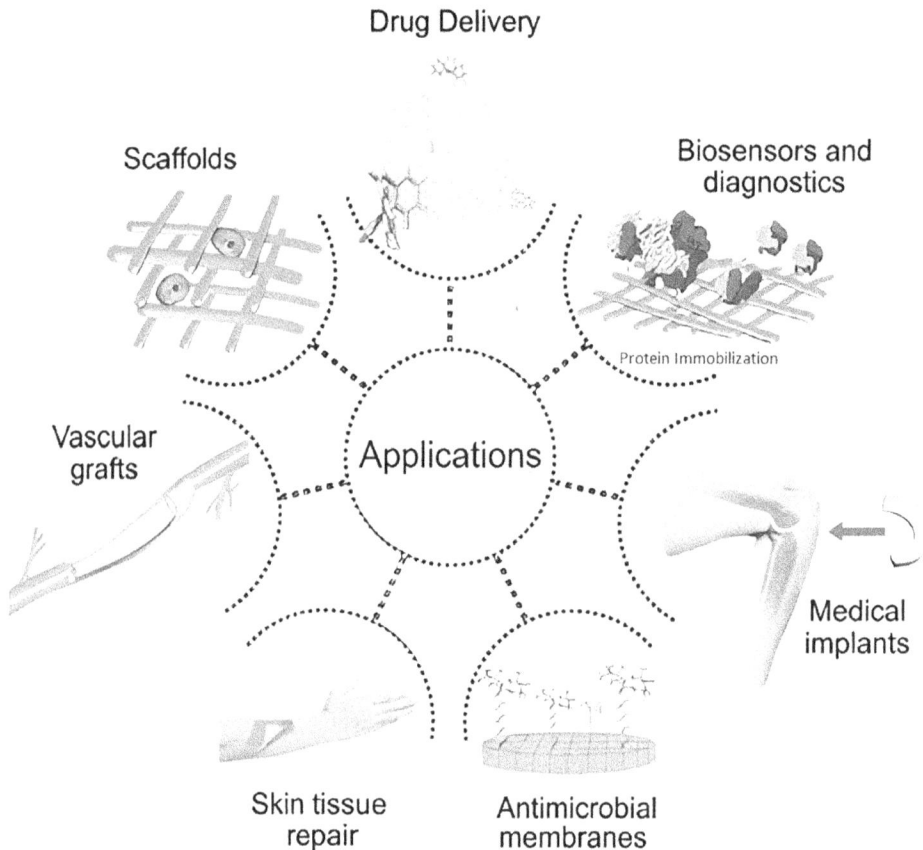

FIGURE 7.3 Application of cellulose nanoparticles in biomedical [21]. (Open Access.)

for scaffolds in dental or orthopaedic implantology which is made of hydroxyapatite and a sodium maleate copolymer on titanium substrate. Similarly, polytetrafluoroethylene (PTFE) with nanosilver (Ag) and gold (Au) fillers shows the good antibacterial efficiency at different bacteria growth [20]. The nanoparticles in polymer coatings have synergism effect in the dental and bone implantation. Nanocellulose in polymer is also used in biomedical application. Most common nanocelluloses are cellulose nanofibre and cellulose nanocrystals (CNC) which are extracted from wood and bacterial cellulose (BC) from bacteria. Cellulose nanoparticles are produced by mechanical methods such as high-pressure homogenisation, microfluidisation, high-intensity ultrasonication, grinding, steam explosion, cryocrushing with liquid nitrogen, etc. Chemical or enzymatic process treatments can also be done before mechanical processes to improve the reactivity of fibres [21]. CNCs are generally extracted from cellulose fibres by acid hydrolysis [22]. The copolymer/clay nanocomposites are also widely used in biomedical applications like drug delivery systems and regenerative medicine as reported by Murugesan and Scheibel [23]. For application in sensors and biosensors, the nanoparticles in polymer should show good electrical and catalytic properties which helps the timely detection of diseases like cancer [24]. Conducting polymers like polypyrrole, polyaniline, polythiophene and poly(3,4-ethylenedioxythiophene) are also used to design the sensors and biosensors. Addition of nanoparticles in conducting polymers utilised as biosensors [25]. Similarly, nanocomposite of graphene oxide and polyaniline nanofibres are used to control the glucose concentration for clinical diagnosis and the food industry [26]. For longer use of hip and knee implants, dental, etc., surface coating of biomedical devices is required which improves the biocompatibility and avoid the unfavourable effect like toxicity, infection, irritation or inflammation. Nowadays, self-healable hydrogels, nanocellulose are used as biomaterials in medical applications which has ability to repair the physical defects and traumas [27,28].

7.3.2 Superhydrophobic Application

Superhydrophobic surfaces are now trending in research and development in the last two decades for their excellent water-repellent behaviour which is the best concept to keep away dirt, fouling, fogging and icing from surfaces. Superhydrophobic surfaces and coatings nanostructures and their nanocomposites with metals, ceramics and polymers are highly used in anticorrosion, self-cleaning and oil-water separation etc. PNC superhydrophobic materials are produced by dispersing nanoparticles (particle size between 1 and 100nm) in a polymer matrix. It (contact angle larger than 150°) can be formed by forming hierarchical micro/ nanoscale binary structures and by chemical modification of materials with low surface energy [29]. Ibrahim et al. [30] have studied the superhydrophobic property of smart self-cleaning coating films made of polystyrene encapsulated modified nano-SiO_2. Saji [31] has reviewed superhydrophobic surfaces and coatings of carbon nanostructures and their metals, ceramics, and polymers nanocomposites and also discuss different applications of nanocomposites in oil separation, sensors, anticorrosion, anti-icing, antibiofouling, etc. A number of reports are available on superhydrophobic nanocomposite with different nanofillers like SiO_2, Al_2O_3, SiC and Co. Table 7.1 summarises the different PNC coating in superhydrophobic applications.

7.3.3 Films and Coatings for Whole and Cut Fruits and Vegetables

Biopolymer-based films and coatings are used for improving the shelf-life of whole and cut fruits and vegetable. Biopolymers with different nanofillers are used for coating fruits and vegetables for their prolonged storage. Nanocomposite coatings on fruits and vegetable surfaces are applied by different methods like dipping, spraying and spreading by brush as shown in the Figure 7.4. Jafarzadeh et al. [42] have reviewed the biopolymer-based nanocomposite films and coatings for the increment of shelf-life of whole and cut fruits/vegetables and also study the progress in the development, design and application of bio-nanocomposite. Figure 7.5 shows the effect of coating on vegetables and its importance. Vegetables without coating have more bacterial growth, so it loses their

TABLE 7.1

PNC Coating in Superhydrophobic Applications

Matrix	Nanofillers	Preparative Method	Contact Angle (in degree)	Reference
Polystyrene	SiO_2	Miniemulsion	130°	[29]
Fluoropolymer	CNT, graphene	Drop casting suspensions	170°	[32]
Polystyrene and PVC	SiO_2	Spin coating	162°	[33]
Polystyrene	TiO_2	Blending ambient-cured	168.7°	[34]
Polysiloxane	Fe_3O_4	Coating	158.3^0	[35]
Polydimethylsiloxane	SiO_2	Ultrasonic stirring, brushing	150°	[36]
Polydimethylsiloxane	Al_2O_3/Clay	Ultrasonic stirring, spraying	164°	[37]
Polymethylhydrosiloxane (PMHS) and polystyrene (PS)	Hydrothermally synthesised Al_2O	Dip coating	171°	[38]
Tetraethylorthosilicate and glycidyloxypropyltrimethoxysilane	Al_2O_3	Dip coating	164°	[39]
Silicone polymer	Tungsten (VI) oxide (WO_3)	Dispersion	161°	[40]
Acrylic polyurethane	ZnO	Spraying and co-curing process	171°	[41]

Dipping

uncoated   Dipping   Evaporation   coated

Spraying

uncoated   Spraying   coated

Brushing

uncoated   Brushing   coated

FIGURE 7.4 Different method of edible coatings on vegetables and fruits [42]. (Copyright 2021. Reproduced with permission from Elsevier.)

freshness whereas vegetable with coating last for a long time. This indicates the coating increases the life of the fruits and vegetables for a long time. Different nanoparticles are embedded in the matrix to make an edible coating on the fruits. Table 7.2 shows the application of nanocomposites films and coatings in fruits and vegetable in detail.

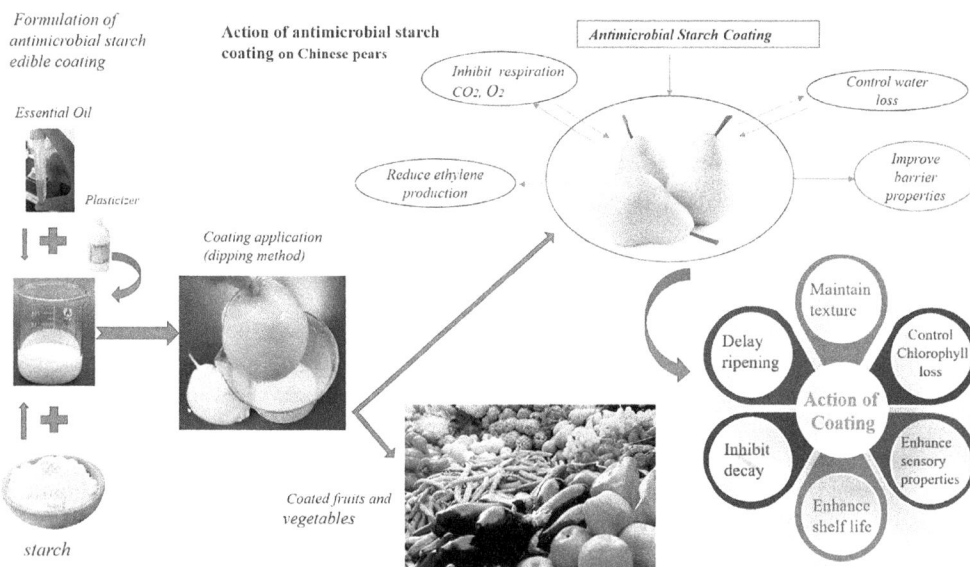

FIGURE 7.5 Effect of coating on vegetables [43]. (Copyright 2022. Reproduced with permission from Elsevier.)

TABLE 7.2

Application of Nanocomposite Films and Coatings in Fruits and Vegetables

Fruits/Vegetables	Nanoparticles	Matrix	Finding	Reference
Persimmon and tomato	Nano-ZnO	Carboxymethyl cellulose	Preserve the qualities and delaying black spot disease	[44]
Tomato	Nano-titanium	Chitosan	delayed ripening rate of the tomatoes and fewer quality changes	[45]
Orange	Nanoclay	Carnauba wax	Improve the nutritional quality and orange sensory acceptability. Reduce the orange weight loss and respiration rate	[46]
Cut papaya	Nano-ZnO	Chitosan	During the storage period, less microbial growth is coated as compared to uncoated papaya	[47]
Melon	Chitosan silver Nanoparticle	Chitosan	Improve the nutritional quality, decrease respiration rate and avoidance of softening.	[48]
Banana	Nano-ZnONP	Soy protein isolate	Delayed weight loss and ripening rate of bananas. Prevention of fungal growth during the storage period	[49]
Guava	Nano-ZnO	Alginate and chitosan	Shelf-life extended from 7 to 20 days	[50]
Table grapes	Fungal chitosan nanoparticles	Chitosan	Delaying the Ripening rate of table grapes, maintaining titratable acidity and moisture	[51]
Asparagus spears	Nano-Ag	Polyvinylpyrrolidone	Maintain the colour and texture, reduce weight loss, and increase the shelf-life to 10 days at 2°C	[52]
Mango	Bergamot essential oils, Nano-Ag and Nano-TiO	Polylactic acid	Maintain the quality of mango and increased the postharvest life to 15 days	[53]
Black grapes	Nano-Ag	Sodium alginate/gum acacia	Good at antimicrobial activity against various foodborne bacteria and increased the shelf-life	[54]

7.3.4 ANTICORROSIVE AND SELF-HEALING COATINGS

Corrosion deals with material failure and equipment shutdown which results into economic loss. Several efforts have been done to improve the corrosion resistance in the material by anticorrosion approaches. Surface coating is one of the most potential methods for anticorrosion of metal to slow down the corrosion rate. PNC coatings are widely used for controlling the corrosion, fouling and scratching of metallic materials. This is due to the fact that nanocrystalline structural coatings are able to modify the surface microstructure of interior or exterior surfaces of a given material at varying temperature ranges and improve their barrier protective performances [55]. There are various studies of PNC coating and its application. Mohammadzadeh et al. [56] have reviewed the different PNC coatings on inorganic and organic hybrid substrates. Table 7.3 shows different PNCs are used in anticorrosion application for different matrices and nanofillers. Anticorrosion property of nanocomposite coating can also be enhanced by hybridisation method where more than one filler is added in the matrix [36,55–57,63]. Hybrid nanocomposite is able to improve the self-healing ability of the coating to protect the material from the corrosion degradation. The self-healing polymer composite coating is widely used as coating for automobile and aerospace, biomedical and biomedical device, oil and gas industry, antistaining and scratch resistance coating as shown in Figure 7.6.

TABLE 7.3
PNCs in Anticorrosion Application

Matrix	Nanofillers	Preparative Method	Reference
Ethylene tetrafluoroethylene	PANi/CNT	Ultrasonic stirring, air spray	[36]
Polyaniline	Reduced graphene oxide and modified clay sodium montmorillonite (Na-MMT)	Dispersion and re-oxidised	[56,57]
PTFE	Ni-P-Al$_2$O$_3$	Electroless deposition	[58]
Epoxy resin	Oxides, clay	Ultrasonic stirring, solution dispersion	[59]
PANi	Fe$_2$O$_3$·NiO	precipitation–oxidation	[60]
PANi	Clay	Electrodeposition	[61]
Polyester	Clay	Electrostatic method	[62]

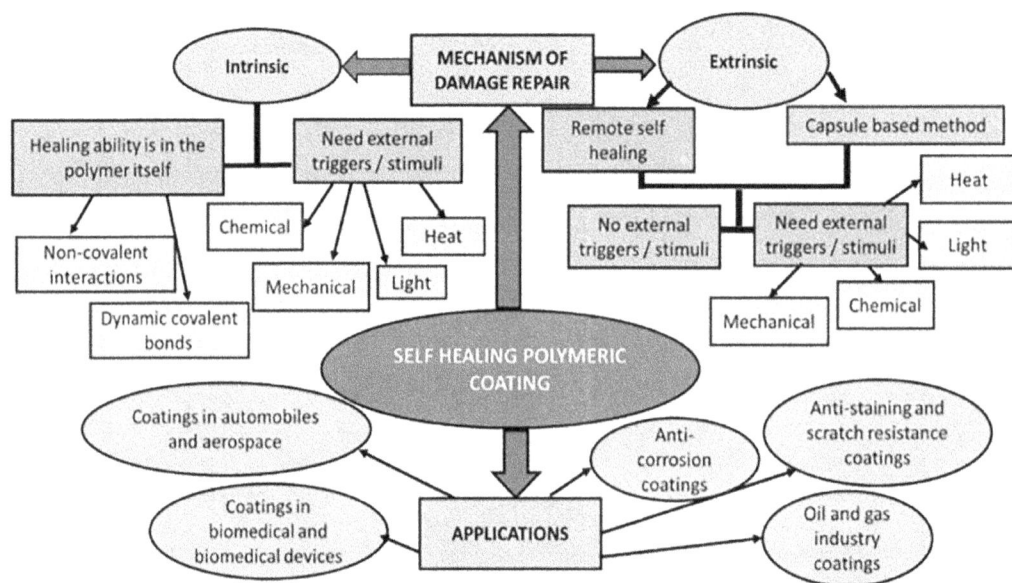

FIGURE 7.6 Mechanisms and applications of self-healing polymeric coatings [64]. (Open access.)

Kongparakul et al. [63] have prepared the self-healing hybrid nanocomposite made of self-healing microcapsules and organ silane-modified nanosilica in epoxy matrix. The self-healing may be the extrinsic and intrinsic mechanism. In extrinsic mechanism, the external healing agents such as microcapsules and vascular networks introduced the polymer matrix, job of which is to recover the damage. In the intrinsic mechanism, external stimuli are required for the chemical bonding of the matrix itself to recover the damage [62]. The details about the mechanisms are shown in Figure 7.6.

7.4 CONCLUSION AND FUTURE SCOPE

PNC is a revolutionary material in material science which has wide scope in the biomedical, food industry and automobile industries. This chapter deals with the different characterisation techniques of PNC coating and its potential applications. The salient points of the chapter may be summarised as follows:

- Characterisation techniques of PNC are highly required to analyse the mechanical, thermal and rheological property of the coating to get the knowledge of the effective coating and its integrity.
- Nanosized fillers in polymer show the best coating property. Low amount of nanofiller reinforcing in polymer increases the strength-to-weight ratio, surface area, surface energy and functionalities of the polymer composite.
- Hence, PNCs with unique architecture have several advantages over other conventional materials. In future, more emphasis should be given on various new metal nanoparticles in the polymer matrix for the future development of efficient self-healing materials, biomaterials, superhydrophobic application, anticorrosion and food packaging.

REFERENCES

1. N. Kyung Kwon, H. Kim, I. Kyung Han, T. Joo Shin, H. W. Lee, J. Park, S. Youn Kim, Enhanced mechanical properties of polymer nanocomposites using dopamine-modified polymers at nanoparticle surfaces in very low molecular weight polymers, *ACS Macro Letters*, 2018, 7(8), 962–967.
2. P. Nguyen-Tri, T. Anh Nguyen, P. Carriere, C. Ngo Xuan, Nanocomposite coatings: Preparation, characterization, properties, and applications, *International Journal of Corrosion*, 2018, doi: 10.1155/2018/4749501.
3. T. B. Atisme, C. Y. Yu, E. N. Tseng, Y. C. Chen, P. K. Shu, S. Y. Chen, Interface interactions in conjugated polymer composite with metal oxide nanoparticles, *Nanomaterials*, 2019, 9(11), 1534, doi: 10.3390/nano9111534.
4. C. E. Corcione, M. Frigione, Characterization of nanocomposites by thermal analysis, *Materials (Basel)*, 2012, 5(12), 2960–2980, doi: 10.3390/ma5122960.
5. M. R. Kamal, S. Sourour, Kinetics and thermal characterization of thermoset cure, *Polymer Engineering and Science*, 1973, 13(1), 59–64, doi: 10.1002/pen.760130110.
6. K. S. Boparai, R. Singh, F. Fabbrocino, F. Fraternali, Thermal characterization of recycled polymer for additive manufacturing applications, *Composites Part B: Engineering*, 2016, 106(1), 42–47, doi: 10.1016/j.compositesb.2016.09.009.
7. A. S. Singha, V. K. Thakur, Mechanical, morphological, and thermal characterization of compression-moulded polymer biocomposites, 2010, 87–97, doi: 10.1080/10236660903474506.
8. V. Mittal, Thermal characterization of filler and polymer nanocomposites. In *Characterization Techniques for Polymer Nanocomposites*, edited by V. Mittal, John Wiley & Sons, Weinheim, Germany, 2012-Technology & Engineering-378 doi:10.1002/9783527654505.ch2.
9. O. Okhamafe, P. York, Thermal characterization of drug/polymer and excipient/polymer interactions in some film coating formulation, *Journal of Pharmacy and Pharmacology*, 1989, 41(1), 1–6, doi: 10.1111/j.2042-7158.1989.tb06318.x.
10. W. W. Wendlandt, P. K. Gallagher, Chapter 1: Instrumentation. In *Thermal Characterization of Polymeric Materials*, edited by E. Turi, Academic Press, Inc., Elsevier Inc. , 1981, 1–90, doi:10.1016/B978-0-12-703780-6.50006-7.

11. J. L. de la Fuente, M. Ruiz-Bermejo, D. Nna-Mvondo, R. D. Minard, Further progress into the thermal characterization of HCN polymers, *Polymer Degradation and Stability*, 2014, 110, 241–251, doi: 10.1016/ j.polymdegradstab.2014.09.005.

12. G. B. McKenna, S. L. Simon, Handbook of thermal analysis and calorimetry, 2002, 3, 49–109, doi: 10.1016/S1573-4374(02)80005-X.

13. J. Zhao, A. B. Morgan, J. D. Harris, Rheological characterization of polystyrene–Clay nanocomposites to compare the degree of exfoliation and dispersion, *Polymer*, 2005, 46(20), 8641–8660, doi: 10.1016/j. polymer.2005.04.038.

14. S. Duggirala, P. P. Deluca, Rheological characterization of cellulosic and alginate polymers, *PDA Journal of Pharmaceutical Science and Technology*, 1996, 50(5), 290–296.

15. M. Abdel Goad, P. Pötschke, D. Zhou, J. E. Mark, G. Heinrich, Preparation and rheological characterization of polymer nanocomposites based on expanded graphite, *Journal of Macromolecular Science, Part A, Pure and Applied Chemistry*, 2007, 44(6), 591–598, doi: 10.1080/10601320701284840.

16. J. M. Thomassin, M. Trifkovic, W. Alkarmo, C. Detrembleur, C. Jérôme, C. Macosko, Poly (methyl methacrylate)/graphene oxide nanocomposites by a precipitation polymerization process and their dielectric and rheological characterization, *Macromolecules*, 2014, 47(6), 2149–2155, doi: 10.1021/ ma500164s.

17. A. Kausar, Corrosion prevention prospects of polymeric nanocomposites: A review, *Journal of Plastic Film & Sheeting*, 2019, 35(2), 181–202, doi: 10.1177/8756087918806027.

18. T. W. Quadri, L. O. Olasunkanmi, O. E. Fayemi, M. M. Solomon, E. E. Ebenso, Zinc oxide nanocomposites of selected polymers: Synthesis, characterization, and corrosion inhibition studies on mild steel in HCl solution, *ACS Omega*, 2017, 2(11), 8421–8437, doi:10.1021/acsomega.7b01385.

19. G. Negroiu, R. M. Piticescu, G. C. Chitanu, et al., Biocompatibility evaluation of a novel hydroxyapatite-polymer coating for medical implants, *Journal of Material Science: Materials in Medicine*, 2008, 19, 1537–1544, doi: 10.1007/s10856-007-3300-6.

20. V. Zaporojtchenko, R. Podschun, U. Schürmann, A. Kulkarni, F. Faupel, Physico-chemical and antimicrobial properties of co-sputtered Ag–Au/PTFE nanocomposite coatings, *Nanotechnology*, 2006, doi: 10.1088/0957-4484/17/19/020.

21. J. Rojas, M. Bedoya, Y. Ciro, Chapter 8: Current Trends in the Production of Cellulose Nanoparticles and Nanocomposites for Biomedical Applications. Intech, 2016, doi: 10.5772/61334.

22. Y. Habibi, L. A. Lucia, O. J. Rojas. Cellulose nanocrystals: Chemistry, self-assembly, and applications, *Chemical Reviews*, 2010, 110(6), 3479–500.

23. S. Murugesan, T. Scheibel, Copolymer/clay nanocomposites for biomedical applications, *Advanced Functional Materials*, 2020, 30, 1–28.

24. A. Ramanavicius, Y. Oztekin, A. Ramanaviciene, Electrochemical formation of polypyrrole-based layer for immunosensor design, *Sensors and Actuators B: Chemical*, 2014, 197, 237–243.

25. A. Ramanaviciene, A. Kausaite, S. Tautkus, A. Ramanavicius, Biocompatibility of polypyrrole particles: An *in vivo* study in mice, *Journal of Pharmacy and Pharmacology*, 2007, 59, 311–315.

26. A. Popov, R. Aukstakojyte, J. Gaidukevic, V. Lisyte, A. Kausaite-Minkstimiene, J. Barkauskas, A. Ramanaviciene, Reduced graphene oxide and polyaniline nanofibers nanocomposite for the development of an amperometric glucose biosensor, *Sensors (Basel)*, 2021, 21(3), 948.

27. L. Zhou, F. Chen, Z. Hou, Y. Chen, X. Luo, Injectable self-healing CuS nanoparticle complex hydrogels with antibacterial, anti-cancer, and wound healing properties, *Chemical Engineering Journal*, 2021, 409, 128224.

28. S. Akgöl, F. Ulucan-Karnak, C. I. Kuru, K. Kuşat. The usage of composite nanomaterials in biomedical engineering applications, *Biotechnology Bioengineering*, 2021, 118(8), 2906–2922.

29. R. Rioboo, M. Voue, A. Vaillant, D. Seveno, J. Conti, A. I. Bondar, D. A. Ivanovand, J. De Coninck, Superhydrophobic surfaces from various polypropylenes, *Langmuir*, 2008, 24(17), 9508–9514, doi: 10.1021/la801283j.

30. S. Ibrahim, A. Labeeb, A. F. Mabied, O. Soliman, A. Ward, S. L. Abd-El-Messieh, A. A. Abdelhakim, Synthesis of super-hydrophobic polymer nanocomposites as a smart self-cleaning coating films, *Polymer Composite*, 2016, doi: 10.1002/pc.24023.

31. V. S. Saji, Carbon nanostructure-based superhydrophobic surfaces and coatings, *Nanotechnology Reviews*, 2021, 10, 518–571.

32. A. Asthana, T. Maitra, R. Büchel, M. K. Tiwari, D. Poulikakos, Multifunctional superhydrophobic polymer/carbon nanocomposites: Graphene, carbon nanotubes, or carbon black? *ACS Applied Materials & Interfaces*, 2014, 6(11), 8859–8867.

33. W. Wang, Preparation of new polymer nanocomposites and analysis of their superhydrophobic properties, *IOP Conf. Series: Earth and Environmental Science*, 2021, doi: 10.1088/1755-1315/714/3/032072.
34. X. Ding, S. Zhou, G. Gu, L. Wu, A facile and large area fabrication method of superhydrophobic self-cleaning fluorinated polysiloxane/TiO_2 nanocomposite coatings with long-term durability, *Journal of Materials Chemistry*, 2011, 21(17), 6161–6164.
35. X. Ding, S. Zhou, G. Gu, L. Wu, Facile fabrication of superhydrophobic polysiloxane/magnetite nanocomposite coatings with electromagnetic shielding property, *Journal of Coatings Technology and Research*, 2011, 8(6), 757–764.
36. D. S. Facio, M. J. Mosquera, Simple strategy for producing superhydrophobic nanocomposite coatings in situ on a building substrate, *ACS Applied Materials & Interfaces*, 2013, 5(15), 7517–7526.
37. R. Yuan, S. Wu, H. Wang, et al., Facile fabrication approach for a novel multifunctional superamphiphobic coating based on chemically grafted montmorillonite/Al_2O_3-polydimethylsiloxane binary nanocomposite, *Journal of Polymer Research*, 2017, 24(4), 59.
38. R. S. Sutar, S. Nagappan, A. K. Bhosale, K. K. Sadasivuni, K. H. Park, C. S. Ha, S. S. Latthe, Superhydrophobic Al_2O_3–polymer composite coating for self-cleaning applications, *Coatings*, 202, doi: 10.3390/ coatings11101162.
39. M. Khodaei, S. Shadmani, Superhydrophobicity on aluminium through reactive-etching and TEOS/ GPTMS/nano-Al_2O_3 silane-based nanocomposite coating, *Surface and Coatings Technology*, 2019, 374, 1078–1090, doi: 10.1016/j.surfcoat.2019.06.074.
40. S. Dixon, N. Noor, S. Sathasivam, Y. Lu, I. Parkin, Synthesis of superhydrophobic polymer/tungsten (VI) oxide nanocomposite thin films, *European Journal of Chemistry*, 2016, 7(2), 139–145.
41. C. Xie, C. Li, Y. Xie, Z. Cao, S. Li, J. Zhao, M. Wang, ZnO/acrylic polyurethane nanocomposite superhydrophobic coating on aluminum substrate obtained via spraying and co-curing for the control of marine biofouling, *Surfaces and Interfaces*, 2021, doi: 10.1016/j.surfin.2020.100833.
42. S. Jafarzadeh, A. Mohammadi Nafchi, A. Salehabadi, N. Oladzad-abbasabadi, S. M. Jafari, Application of bio-nanocomposite films and edible coatings for extending the shelf life of fresh fruits and vegetables, *Advances in Colloid and Interface Science*, 2021, doi: 10.1016/j.cis.2021.102405.
43. W. Oyom, Z. Zhang, Y. Bi, R. Tahergorabi, Application of starch-based coatings incorporated with antimicrobial agents for preservation of fruits and vegetables: A review, *Progress in Organic Coatings*, 2022, doi: 10.1016/j.porgcoat.2022.106800.
44. M. N. Mooktida Saekow, W. Tongdeesoontorn, Y. Hamauzu, Effect of carboxymethyl cellulose coating containing ZnO-nanoparticles for prolonging shelf life of persimmon and tomato fruit, *Journal of Food Science and Technology*, 2019, 5, 41–48.
45. P. Kaewklin, U. Siripatrawan, A. Suwanagul, Y. S. Lee, Active packaging from chitosan titanium dioxide nanocomposite film for prolonging storage life of tomato fruit, *International Journal of Biological Macromolecules*, 2018, 112, 523–529.
46. E. Motamedi, J. Nasiri, T. R. Malidarreh, S. Kalantari, M. R. Naghavi, M. Safari, Performance of carnauba wax-nanoclay emulsion coatings on postharvest quality of 'Valencia' orange fruit, *Scientia Horticulturae (Amsterdam)*, 2018, 240, 170–178, doi: 10.1016/j.scienta.2018.06.002.
47. M. Lavinia, S. N. Hibaturrahman, H. Harinata, A. A. Wardana, Antimicrobial activity and application of nanocomposite coating from chitosan and ZnO nanoparticle to inhibit microbial growth on fresh-cut papaya, *Food Research*, 2020, 4, 307–311.
48. G. Ortiz-Duarte, L. E. Pérez-Cabrera, F. Artés-Hernández, G. B. Martínez-Hernández, Ag-chitosan nanocomposites in edible coatings affect the quality of fresh-cut melon, *Postharvest Biology and Technology*, 2019, 147, 174–184.
49. J. Li, Q. Sun, Y. Sun, B. Chen, X. Wu, T. Le, Improvement of banana postharvest quality using a novel soybean protein isolate/cinnamaldehyde/zinc oxide bionanocomposite coating strategy, *Scientia Horticulturae (Amsterdam)*, 2019, 258:108786.
50. B. J. Arroyo, A. C. Bezerra, L. L. Oliveira, S. J. Arroyo, E. A. Melo, A. M. P. Santos, Antimicrobial active edible coating of alginate and chitosan added ZnO nanoparticles applied in guavas (Psidium guajava L.), *Food Chemical*, 2020, 309, 125566.
51. N. F. Castelo Branco Melo, B. L. de MendonçaSoares, K. Marques Diniz, C. Ferreira Leal, D. Canto, M. A. P. Flores, et al., Effects of fungal chitosan nanoparticles as eco-friendly edible coatings on the quality of postharvest table grapes, *Postharvest Biology and Technology*, 2018, 139, 56–66.
52. J. An, M. Zhang, S. Wang, J. Tang, Physical, chemical and microbiological changes in stored green asparagus spears as affected by coating of silver nanoparticles-PVP. *LWT - Food Science and Technology*, 2008, 41(6), 1100–1107.

53. H. Chi, S. Song, M. Luo, C. Zhang, W. Li, L Li, et al., Effect of PLA nanocomposite films containing bergamot essential oil, TiO_2 nanoparticles, and Ag nanoparticles on shelf life of mangoes, *Scientia Horticulturae (Amsterdam)*, 2019, 249, 192–198.

54. V. Kanikireddy, K. Kanny, Y. Padma, R. Velchuri, G. Ravi, B. Jagan Mohan Reddy, et al., Development of alginate-gum acacia-Ag0 nanocomposites via green process for inactivation of foodborne bacteria and impact on shelf life of black grapes (Vitis vinifera), *Journal of Applied Polymer Science*, 2019, 136(15), doi: 10.1002/app.4733.

55. W. F. Huang, Y. L. Xiao, Z. Huang, G. P. Tsui, K. W. Yeung, C. Y. Tang, Q. Liu, Super-hydrophobic polyaniline-TiO_2 hierarchical nanocomposite as anticorrosion coating, *Materials Letters*, 2020, 258, doi: 10.1016/j.matlet.2019.126822.

56. A. Mohammadzadeh, H. G. Taleghani, M. Soleimani Lashkenari, Preparation and comparative study of anticorrosion nanocomposites of polyaniline/graphene oxide/clay coating, *Journal of Materials Research and Technology*, 2021, 13, 2325–2335, doi: 10.1016/j.jmrt.2021.05.098.

57. S. Kesarwani, R. K. Verma, A critical review on synthesis, characterization and multifunctional applications of reduced graphene oxide (rGO)/composites, *Nano*, 2021, 16(9), 2130008.

58. A. Sharma, A. K. Singh, Electroless Ni-P-PTFE-Al_2O_3 dispersion nanocomposite coating for corrosion and wear resistance, *Journal of Materials Engineering and Performance*, 2014, 23(1), 142–151.

59. T. A. Nguyen, H. Nguyen, T. V. Nguyen, H. Tai, X. Shi, Effect of nanoparticles on the thermal and mechanical properties of epoxy coatings, *Journal of Nanoscience and Nanotechnology*, 2016, 16(9), 9874–9881.

60. D. N. Nguyen, T. T. Ngo, Study on synthesis and anticorrosion properties of polymer nanocomposites based on superparamagnetic $Fe_2O_3 \cdot NiO$ nanoparticle and polyaniline, *Synthetic Metals*, 2009, 159(9–10), 831–834, doi: 10.1016/j.synthmet.2009.01.020.

61. M. Shabani-Nooshabadi, S. M. Ghoreishi, Y. Jafari, N. Kashanizadeh, Electrodeposition of polyaniline-montmorillonite nanocomposite coatings on 316L stainless steel for corrosion prevention, *Journal of Polymer Research*, 2014, 21(4), 416, doi: 10.1007/s10965-014-0416-5.

62. A. Golgoon, M. Aliofhazraei, M. Toorani, M. Moradi, A. S. Rouhaghdam, Corrosion and wear properties of nanoclay polyester nanocomposite coatings fabricated by electrostatic method, *Procedia Materials Science*, 2015, 11, 536–541.

63. S. Kongparakul, S. Kornprasert, P. Suriya, D. Le, C. Samart, N. Chantarasiri, P. Prasassarakich, G. Guan, Self-healing hybrid nanocomposite anticorrosive coating from epoxy/modified nanosilica/perfluorooctyl triethoxysilane, *Progress in Organic Coatings*, 2017, 104, 173–179.

64. N. N. F Nik Md Noordin Kahar, A. F. Osman, E. Alosime, N. Arsat, N. A. Mohammad Azman, A. Syamsir, Z. Itam, Z. A. Abdul Hamid, The versatility of polymeric materials as self-healing agents for various types of applications: A review, *Polymers*, 2021, 13, 1194, doi: 10.3390/polym13081194.

8 Recent Trends in Nanometric Dispersed Polymer Composites

Rahul Samanta, Atul Bandyopadhyay, Abhijit Mondal,
Apurba Das, Arijit Sinha, and Gurudas Mandal

CONTENTS

8.1 INTRODUCTION

In the modern technology-oriented world, polymer nanocomposites' synthesis has become a blooming science of polymer nanotechnology. Insertion of nanometric inorganic compounds into polymers makes its properties more improved. After successful insertion, it carries a lot of potential for use in various sectors or industries such as the medical sector, food packaging sector, automobile sector, solar energy industries, and many more. Applications of polymer nanocomposites depend on the inorganic material that is present in polymers [1]. As processed food becomes a key aspect of human food habits and that also increases the chances of cancer, this chapter is mainly highlighted the impact of nanotechnology on the two most impactful categories of human life, i.e., (a) food packaging industry and (b) medical sector in order to fight against cancer.

DOI: 10.1201/9781003343912-8

In terms of food packaging, mechanical and chemical properties can help to enhance through the use of Nanometric Dispersed Polymer Composites (NDPC) like antimicrobial properties, gas barriers, and many more. In recent years, nanofillers like graphene nanoplates (GNP), nanoclays montmorillonite (NMT), carbon nanotubes (CNT), and kaolinite are commonly used to enhance its properties [2]. Barrier properties' improvements are suitable especially for benefiting biopolymers which consist of limited intrinsic barrier properties in general. Moreover, (NDPC) with surface coatings are more useful in the food packaging industry as modulating surface affinity in case of food packaging leads toward different pastes and liquids, for example, in order to obtain waterproof and or repellent paper-based packaging [3] or for enabling easy-to-empty features [4]. In the near future, the possibilities of polymer composites' development via various matrixes can be proposed in order to evaluate its potential in application and several fabrication techniques that can improve the properties of the polymer composite [5].

In medical science, NDPC can have a great impact, especially in cancer detection, and its proper diagnosis is grounded mainly on the early-stage identification of unfamiliar features of abnormal tissues like typical mutations in genetic materials [6–8]. Also, morphological properties such as adhesive properties, distinct in structure, and elasticity, along with some biochemical processes, indicate the development of cancer tissues. In this regard, biosensor has been emerging as the fascinating and most exciting techniques for cancer diagnosis and for that inclusion of NDPCs in biosensor analytical devices. It commonly comprises a physicochemical detector and a biological component [9]. Biosensor enables the attachment of chemicals of interest such as chemicals, macromolecules, toxins, metabolites, toxins, etc., in biologic materials like tissue, organoids, cells, surfaces, organs, etc., or any such related sample in order to a probe and their quantification and fruitful identification. Nanotechnology-embedded biosensors make early diagnosis and detection of cancer demon possible and that can decrease the demise rate significantly. The study mainly focused on the advantages of NDPC over existing materials or compounds that help to boost the food packaging and medical sector comprehensively.

8.2 PROCESSING OF NDPC

Physical properties of nanocomposites largely depend on the degree of dispersion of nanoparticles in polymer matrices. The influence of dispersion degree impacts the rheological and mechanical properties of nanomaterials. Dispersion transfer of load by interfacial bonding to the reinforcement form matrix causes enhancement in mechanical properties of NDPC. This is due to the absence of chemical bonding between particles and polymers. According to Yan et al. [10], fine particles are attributed to combine together and form bond aggregates. That produces agglomerates and larger structure which tends to strengthen their properties [11]. Repulsion between constituents can effectively be controlled with the use of pH concentration or by a concentration of electrolyte in the solvent cast system [12]. For example, SiO_2 nanoparticle surfaces that are grafted by L-lactide in PLLA matrix significantly enhance bonding between matrix and filler attributes to higher tensile strength and toughness.

8.3 MECHANISM OF NDPC

In NDPC, a synthetic route is generally preferred as it enables suitable conditions for polymerization. The primary mechanism of NDPC depends on the compatibility of phases that separate blends of polymers. It leads to reduction in the interfacial tension between particles coalescence and phases during melt processing [13]. On the other hand, the interactions between nanoparticles and the reacting groups rely on the chemical structure of the polymer. In Figure 8.1, interactions of polymers have been showed that helps to understand the mechanism of NDPC.

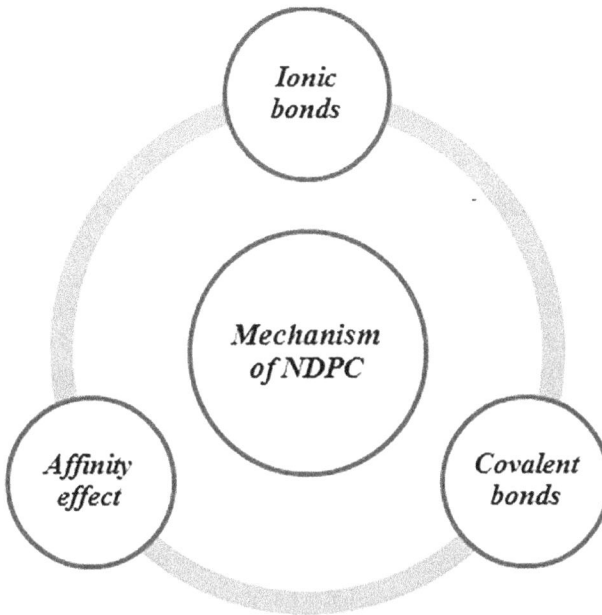

FIGURE 8.1 Mechanism of NDPC.

These interactions in polymers have been discussed below.

Affinity effect: In NDPC, surface affinity in polymer chains effectively forms very strong interactions. It ensures the homogeneous distribution of nanoparticles into the nanometric dispersed polymer matrix. Polymers, surfactant vesicles, reverse micelles, zeolites, and glass can be used for preparing NDPC as the substrate. It can prevent aggregation and hold accurate shape, size, and structure of the surface [14].

Covalent bonds: Amine groups or amide, sulfonation groups, and carboxyl groups can make a covalent bond with a combination of groups on the nanoparticles and hydroxyl.

Ionic bonds: In case of an ionic bond, a Columbic attraction between negative and positive charges helps in combining. In general, nanoparticles and a selected polymer chain have opposite charges, and that tend to form the ionic interaction between them that stabilizes the system of nanocomposites.

8.4 RECENT TRENDS AND APPLICATIONS

8.4.1 Recent Trends of NDPC

NDPC has become a recent trend setter in terms of its applications in several fields such as the medical sector, food packaging industry, automobile industry, solar energy industry, and many more industries. The food packaging industry and medical industry especially to boost biosensors and their application to detect and diagnose cancer at very early stages acquired NDPC the most. On the other hand, automotive applications have got benefits of NDPC as mechanical and electronic properties of it can be enhanced, and thus, wear resistance, thermal isolation, and flame retardancy have become a key interest among researchers. Low-weight stronger materials increase the fuel efficiency of automobiles, and thus, it has become a major boost in this industry.

In the solar energy industry, contributions toward efficient and multifunctionality photovoltaic cells and panel NDPC and nanocoatings have shown effective relevance in order to make a protective layer on the surface of solar panels. In this process, energy harvesting has become easier and research and development on organic photovoltaic have already been started as well [15]. Further, in order to provide self-cleaning effects, nanocoatings and nanotextured surfaces have become a key aspect of solar panels [16,17]. In this sector, NDPC plays a significant role to enable these properties that enlighten the path of modern nanotechnology.

8.4.2 APPLICATION OF NDPC IN THE MEDICAL SECTOR AND FOOD PACKAGING INDUSTRY

8.4.2.1 Application of NDPC in Medical Science in Order to Fight the Cancer Demon

In recent years, it has been reported that the second cause of demise is cancer and it has been rising day by day. As a result, cancer detection at early stages has become difficult and its treatment process becomes late, which causes the survival percentage of patients uncertain and unclear. Thus, early detection of cancer has become an essential issue for its therapy. Detection of cancer-based biomarkers can effectively increase the early detection of it and subsequent treatment. According to the researchers, nanomaterial-based nano-biosensors to detect cancer at early stages can be an excellent tool especially to diagnose disease and detect a molecular abnormality.

8.4.2.1.1 Biosensors: An Elixir in Medical Science

A biosensor can be described as a device that is used for detecting a biological analyte, be it natural or within the human body (basically biological in origin). Biosensors provide information like whether or not the analyte is still present and at what level it is transduced into an electrical signal which can be displayed, amplified, and analyzed. For example, analytes included antibodies, antigens, enzymes, nucleic acid, and metabolic components like glucose or other biological components. In medical science, biosensors can be used for monitoring levels of glucose in the blood in diabetics, detecting pathogens, and diagnosing and monitoring cancer.

Figure 8.2 indicates number of biomarkers available to detect different types of cancers in the market. On the other hand, Table 8.1 shows commonly used biomarkers to detect cancers at various human body parts like breasts, lungs, prostate, and many more types of cancers. In order to detect at early stages beside other professionals, the military personnel are very keen on the development of biosensors as it helps to counter bioterrorism devices. Bioterrorism devices can significantly detect elements of harmful chemicals and biological warfare in order to avoid potential infection or exposure or both. In the foreseeable future, biosensors can be included in chip-scale devices placed on the biological parts in origin (animal or human body) to monitor abnormalities, and vital signs, or even signal a call in order to accumulate help in an emergency [18]. Theoretical perspectives show an enormous application of biosensors. In real, biosensors act as tumor biomarkers and detect cancer cells or stages by measuring levels of certain proteins, viz. antibodies, antigen, and enzymes conveyed and/or secreted by tumor cells. Previously, it was predicted that biosensors have the potential

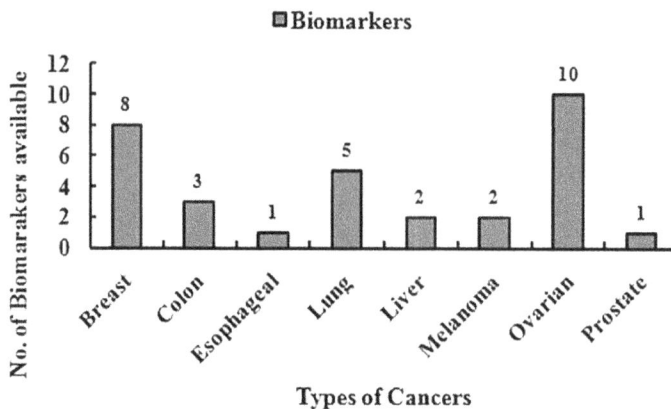

FIGURE 8.2 Number of biomarkers available to detect different types of cancers in the market.

TABLE 8.1

Commonly Used Biomarkers for Cancer Cell Detection

Breast	Colon	Esophageal	Lung	Liver	Melanoma	Ovarian	Prostate
BRCA1	EGF	SCC	CA 19-9	CEA	NY-ESO-1	CA 125	PSA
BRCA2	CEA		SCC	AFP	Tyrosinase	TAG 72	
NY-BR1	p53		CEA			CEA	
CA 15-3			NSE			HCG	
CA 125			NY-ESO-1			p53	
ING-1						CA 549	
CA 27.29						CASA	
CEA						CA 19-9	
						MCA	
						MOV-1	

to detect whether a tumor is present or not, whether it is cancerous or benign, and whether therapy or treatment has been effective in eliminating or reducing cancerous cells. As cancerous cells are involved in multiple biomarkers, a biosensor with the detecting ability of multiple analytes can be proved very useful, especially in the diagnosis and monitoring of cancer. It also helps with financial resources along with monitoring and diagnosis of cancerous cells [19].

Three main components are used in biosensors, viz. (a) a signal transducer, (b) a recognition element, and (c) a signal processor. The recognition element component detects a "signal" from the human body for cancer cells as an analyte form. After that, the signal transducer converts it to an electrical output from biological signal. After that, the result display process has been initiated by the signal processor component. NDPC not only enabled new technology for biosensors but also ensured a high detection rate that increases the early detection of this fatal disease at its early stages.

8.4.2.1.2   Biosensor and Nanotechnology: A Revolution in Medical History

In the modern day, a rapidly growing field: "Nanotechnology," especially nanometric dispersed polymer composites, has an enormous impact on biosensors. It also helps in diagnosis, prognosis, and more importantly monitoring of cancerous cells. According to the report in almost 60% of cases, cancer has been diagnosed after the tumor has metastasized which makes it much more difficult to treat and it become deadly [8]. In order to detect the cancerous cells at early stages and to increase the survival rates of patients, the application of nanotechnology for developing biosensors can improve the chances. Figure 8.3 indicates that if cancer tissues have been detected and diagnosed at early stages, the survival percentage of patients is very high.

For example, magnetic resonance imaging (MRI) is used to detect cancerous cells nowadays. Nowadays, it is used in the diagnosis and monitoring of cancerous cells very commonly. One of the key drawbacks of MRI is that entities of smaller sizes (less than a few centimeters) cannot be detected. NDPCs-enabled biosensors can provide precise and more sensitive measurements of cancerous cells. Nanomaterials such as buckyballs, liposomes, dendrimers, and CNT have been used for improving cancer imaging [20]. Additionally, NDPC enables smaller sensors that effectively translate into better access to doctors. Moreover, the detection of cancerous tissues becomes more powerful, and specific signals are significantly enhanced as well. In this process, not only cancer makers can be detected at early stages also it reduces cost [21].

□ I(A)   □ II(A)   □ III(A)   □ IV(A)

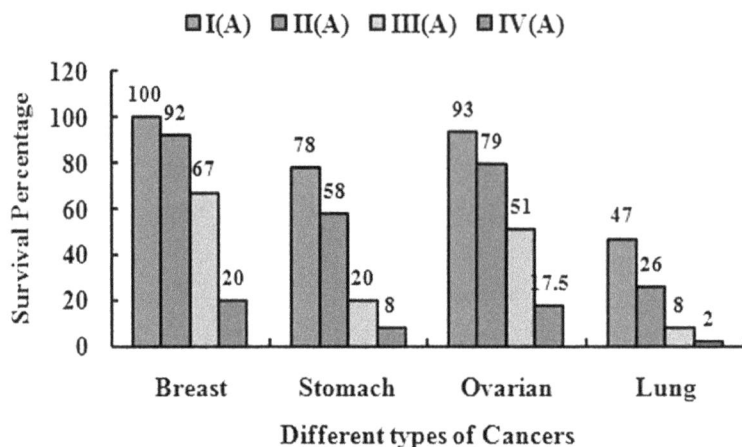

FIGURE 8.3   Survival rate of cancer patients, detected at various stages.

8.4.2.2   Application of NDPC in the Food Packaging Industry

A strong drive has been raised in the food packaging industry to make a solution of sustainable development by recycling food packages. In those instances, NDPC changed the course of the food packaging by enabling high-efficiency recyclable products compared to existing traditional plastic with desired mechanical properties and barriers. Moreover, the addition of biologically active or organic elements improves the mechanical and chemical properties of packaging products, and in that way, the economy gets a major boost as dealing with unwanted plastic becomes easier. Due to nanometer size, dispersion of nanometric dispersed composite polymers leads to an enhancement in strength and modulus [17]. Also, increased water repellent and gas permeability make it more suitable in the food packaging industry. Properties like antioxidant and antimicrobial food industry get triggered as well. The introduction of NDPC in the food industry not only depends on its mechanical and chemical properties but also on customer acceptability also plays a significant role.

8.5   OPPORTUNITIES AND PERSPECTIVES

8.5.1   Opportunities Nanotechnology Brings to Food Packaging and Medical Science

8.5.1.1   Effect of Nanometric Dispersed Polymer Composites on Biosensors

A nanoparticle present in NDPC is of the size of 1–100 nm in diameter. Nanoparticle size and surface-to-volume ratio are inversely proportional to each other, and hence, untouched body parts are also can be accessed with nanoparticles. Nanochannels, nanocantilevers, and nanowires have been exploited to detect cancer-specific molecular events and certainly improve signal transduction [22]. Wang developed a nanowire-based biosensor that can detect micro-RNAs (miRNAs) [23]. According to them, miRNAs are important regulators in terms of gene expression, and those are associated with cancer tissue development. Conventional methods of miRNA detection like northern blot analysis have very low sensitivity and are costly and time-consuming. Nanotechnology is attributed to the introduction of an inexpensive, sensitive, and easy-to-use biosensor in order to detect cancer-related miRNAs.

A comparison between normal human astrocyte cells and cancer patient's astrocyte cells detected by biochips can be a key component that are used as an attachment in surface-enhanced laser detectors of cancer biomarkers. In this concern, single-walled carbon nanotubes (SWCNTs) have an upper hand as it greatly enhances the capabilities of detection of electrochemical biosensors, and also, they can provide much higher sensitivity to enzymatic reactions [24]. Principle of detection of SWCNT, SWCNTs have enhanced activity toward hydrogen peroxide and NADH61 that have been utilized in immunosensors and nucleic acid-based sensors for cancerous tissue biomarkers in order to enhance signal detection and followed by transduction [25,26].

8.5.1.2 Recent Advancements in Nanometric Dispersed Polymer Composites to Boost Biosensors

Kim et al. reported the latest advancement in cancerous tissue detection with the help of nanoparticles [27]. Some very common nanomaterials such as CNT, graphene, Ag, Au, Pt, and Fe_3O_4 and their applications have been reported along with some newly emerged nanomaterials like up-conversion nanoparticles, quantum dots, metal–organic frameworks, and inorganics (ZnO, MoS_2). These nanoparticles can effectively detect and diagnose cancer biomarkers related to prostate, breast, colon, and lung cancer. On the other hand, Andrei et al. have demonstrated a magnetoresistive biosensor with the use of an InSb-based semiconductor channel [28]. A target antigen and Fe_3O_4 nanoparticle were linked with each other through a capture antibody in the reagent followed by immobilization of detection antibody on an InSb channel. On the other hand, a selectively bound target antigen was attached to the antigen capture antibody–nanoparticle complex. The detected magnetic nanoparticles showed resistance due to Lorentz's force as it generated a stray magnetic field which attached with a change. The number of included nanoparticles on the sensor surface was proportional to the antigen concentration and it can be determined by measurement of the sensing channel's magnetoresistance level. This antigen–antibody reaction sensor can effectively detect extremely low amounts of liver cancer antigen (concentration level: 1 pg/mL). The systematic test report also confirmed the good reusability and selectivity of this biosensor.

In recent times, researchers made efforts for applying aluminum composites in the field of medical science, especially for cancer tissue detection and its diagnosis [29–31]. Moreover, surface optimization of aluminum alloy and its composites by using environment-friendly, low-cost, and easy methods is one of the major prospects in the field of research [32–34]. On the other hand, existing surface engineering processes such as hot dipping, gas plasma, and vacuum-based methods are very costly that are very much unsuitable for cancer tissue detection cases or they affect the environment badly, especially processes like chromate coating and anodizing [35–38]. Moreover, in oxidation methods, an alkaline solution is used in an eco-friendly method, unlike anodizing methods. Also, the plasma electrolytic oxidation (PEO) method helps in producing more continuous and thicker on the aluminum surface [39–42].

Interrelationship of several technologies can contribute to clinical nanodiagnostics. As PEO is a low-cost method as well due to the non-requirement of any special conditions like a vacuum or high temperature. it has an upper hand to attract industrial applications in terms of biomarker makers [43,44]. PEO makes coating on aluminum, its alloys, and its composites possible. Amiri et al. reported that PEO belongs to an important coating method on aluminum surfaces [45]. They have investigated the potential of aluminum oxide coatings, coated using the PEO method in the medical industry. In this instance, the authors used electrolytes of sodium silicate, potassium hydroxide, sodium tetra phosphate, and sodium aluminates to coat an Al 2XXX series alloy. The results indicated that 500 V is the appropriate voltage for achieving consistent coating with thickness. The addition of sodium silicate into electrolyte solution can invite porosity and properties like non-adhesion to the material. However, tetra sodium pyrophosphate penetrated the metal and coating interface and that increased the adhesion of the coating. The authors found the optimum solution composed of 10, 3, and 3 g/L of tetra sodium pyrophosphate, KOH, and sodium aluminates, respectively, for PEO coating. In order to control the coating process and to get a coating uniformity, DC pulsed coating has been recommended for biosensor applications. Also, at 1000 Hz frequency coating on the substrate becomes ideal under a duty cycle of 30%.

In the modern day, despite recent progress in medical science especially in the therapy of cancer cells, most cases remain diagnosed late. After tumors have metastasized, it has been detected that leaving the patient helpless and grim prognosis. However, early detection of this disease can make the treatment possible at early stages that can reduce the burden of cancer from the heart of patients and their families significantly. In this regard, NDPC in biosensors is in a unique position in order to transform cancer diagnostics. In order to introduce new types of biosensors and medical imaging methods with higher accuracy and sensitivity of recognition, NDPC can play a major role. The authors examined the in vivo diagnostic and in vitro applications of nanoparticles, and some other nano-devices that in the foreseeable future may have an impact on the medical field. Future developments in the new generation of biosensors after the inclusion of NDPC in it may lead to multifunctional detection and followed by treatment of cancer tissues. Recent developments show that NDPC have also implemented advanced technology in optical biosensors in the form of surface-enhanced Raman scattering (SERS). As per the opinion of Jain with SERS, medical science can achieve a degree of multiplicity far better than the existing methods [46]. They have also claimed that without any interference SERS at once can measure up to 20 biomarkers.

8.5.2 Perspectives on NDPC Especially for Medical Sector and Food Packaging Industry

8.5.2.1 Perspectives on NDPC in Terms of the Medical Sector

Nowadays, advancement in a microfluidic laboratory-on-a-chip (LOC) device helps us to understand how NDPCs can effectively improve diagnosis and patient care. LOC technology removes the complexity of laboratory-based testing and enables an easy-to-use, low-cost, portable system that can very easily be used by concerned physicians or patients. The LOC method is incorporated with DNA hybridization arrays and immunoassays that have been tested for their ability in order to identify the individuals who might have a high risk of cancer. Quantum dots also can be considered as another key application of nanotechnology. It is luminescent nanocrystals that have similar properties to optical biosensors. The use of quantum dots in cancer detection can become a revolution as it emits light of different wavelengths, spectral width, and intensity that allows early detection and diagnosis of multiple unique cancer tissues. Howarth et al. reported that quantum dots are able to track entire cells once they move in the environment [47]. Monitoring cancer cell development, drug therapy, and cancer metastasis can be possible at the early stages by tracking those with the help of quantum dots. In the case of cancer cell detection, NDPC with advanced technology can detect very small diameters which are very essential and as the allure of quantum dots consists of multimodality, high stability, and small size (~50–100 units) in diameter detection technology, it is best suited for biological applications. Quantum dots effectively deliver therapeutic agents to some specific target sites in order to improve pharmaceutical effectiveness with very less side effects. In a similar way, dendrimers can be used as drug delivery systems that can target strategies effectively that have also emerged from the nanotechnology field.

Soda et al. observed that treatment success of fatal diseases like cancer depends upon early-stage diagnosis and critical analysis [48]. For use in immunodiagnostic and bioseparation applications, authors have bioengineered multifunctional core-shell structures embedded with poly-3-hydroxybutyrate (PHB) core which was densely coated with protein functions. The authors reported that bioengineering of Escherichia coli for self-assembled PHB can positively impact the protein A-derived antibody-binding Z domain and codisplay a ferritin-derived iron-binding peptide. According to the researchers using biological self-assembly systems, super-paramagnetic core-shell structures can be derived especially in the case of sensitive and specific electrochemical detection of cancer biomarkers. On the other hand, it was reported that magnesium-based materials can be considered promising biodegradable metals, especially for applications in orthopedic bone implants [49]. Magnesium materials exhibit similar elastic modulus and density to that of bone, excellent osteogenic properties, and biodegradability. In biosensors, use of Mg-based NDPC can effectively eliminate the limitations of existing implant materials like the need for a second surgery and stress shielding. This state-of-the-art review especially focuses on the Mg-additive manufacturing of biodegradable materials and their mechanical and chemical properties.

In cancer tissue detection, composites based on polyoxometalates (POMs) and Mucin 1 protein (MUC1) have grabbed the attention of the modern world. Due to excellent redox activities, the outstanding capability of proton and electron transport, and multitudinous architectures, POM has extensively become a research topic in recent few years [50]. MUCI is also known as a tumor marker of higher sensitivity that can detect cancer tissue at early stages. Besides existing copper-based metal–organic frameworks which are decorated with platinum and gold nanoparticles (AuPt, NPS), and hybrid nanocomposite that consists of MOFs, enzyme-free CHA and alloy NPs also can provide a higher platform for early cancer detection and diagnosis [51]. Manna et al. observed that existing widespread "label-free" biosensors to detect cancer tissues employ optoelectronic phenomena such as piezoelectric techniques or surface plasmon resonance techniques [52]. In the foreseeable future, efforts to eliminate tough challenges after the early detection of cancer cells can be overcome using nanotechnology, especially NDPC in the biosensor category [53].

8.5.2.2 Perspectives on NDPC in Terms of Food Packaging Industry

Nanotechnology showed appreciable promise in the form of NDPC for applications in the food packaging industry. In this technology-dependent era, the main focus has become sustainability and recyclability of materials. In this instance, NDPC shows an effective result that raises hope among the food packaging industry. Several materials from renewable resources tend to be increased to make the packages recyclable and it contributes to decreasing the cost of packaging as well [54]. Reinforcement of biomaterials provides higher toughness and desired barrier that leads to higher efficiency. Nanofillers with homogeneous distribution can help in achieving desired properties. Recent research indicates that nanoclays are most effective in terms of enhancing mechanical and chemical properties. Moreover, layer-by-layer surface grafting also impacts its properties positively. In the foreseeable future, NDPC can become the most desired component in the food packaging industry.

8.6 CHALLENGES AND LIMITATIONS

Nanotechnology has made huge achievements in recent yeast due to its flammability, water repellent properties, and many more. However, usage of environmental security has significantly been ignored by humans. Multiple pieces of evidence have been demonstrated that nanomaterials can perform various influences on different organisms. A key challenge is environmental safety as it deals with chemicals, and unnecessary use of it affects the environment badly [55]. For example, nano-TiO_2 increases the risk of environmental sustainability as it is a very harmful chemical, and after proper processing, it becomes usable. For that, it is indeed a necessity for researchers to take care of those factors in order to protect the environment. On the other hand, the normalization of nanomaterials and their proper control need to be established, and after that, challenges related to environmental safety can be overcome. Research on NDPC has been initiated and those are in the primary stages mostly. Advancement in the nanotechnology field needs to be triggered to mitigate the challenges soon.

8.7 IMPORTANCE OF THE STUDY

The importance of the study lies on the application of NDPC in the modern world at various sectors like automobile, medical, food packaging, and many more with an eco-friendly manner. Promotion of green polymer composites that are not responsible of any kind of environmental hazard makes the scope of the study paramount. In this process, bioreinforcement is one of the key aspects as it helps to maintain the motto of biodegradable and eco-friendly product [56]. It consists of unique properties that grows it demand significantly. For example, nowadays, using high-magnification extrusion foaming process foams are prepared where 85% is air bubble and in remaining 15% polylactic acid and starch is used [57]. This kind of innovations and technical changes attributes the world toward greenery, and in foreseeable future, there will be negligible environmental hazard. For that, the book focuses on the scientific mechanisms and frameworks that lead manufacturing and processing of materials eco-friendly.

8.8 CONCLUSIONS

In the nanotechnology world, nanometric dispersed polymer composites (NDPCs) gain trust among the researchers, and thus, they are very keen to develop mechanical and chemical properties of it. As it covers a wide range of areas like polymeric bio-nanomaterials, nano-electronics, and nano-composites, it can be applied for several applications such as the food packaging industry, drug delivery systems, other medical segments, and many more. In comparison to existing microfilled polymer composites, NDPC display far better mechanical and chemical properties. One of the key reasons for it is the very large interface between macromolecules and nanosized heterogeneities of neat polymers which provides large reinforcement, high conductivity, and low flammability at very low nanofillers' concentration. In the food packaging industry and medical sector especially in biosensors, application of NDPC has become a key component for its dispersion at nanometer size. As it enhances strength and modulus and decreases the permeability of gas and most importantly increases the resistance of the water for the food packaging industry, it has become the most wanted component. Further, according to their requirements, the food packaging industry can add biologically active elements to it as well. In the medical sector especially in biosensors to diagnose and detect cancer at early stages also NDPC make a huge impact as it consists of biofunctional properties that enable the huge potential for application in the medical industry. Recent trends in NDPC effectively boost human lifestyle in terms of enabling technology and upgrading medical science.

REFERENCES

1. Arora A, Padua GW. Nanocomposites in food packaging. *Journal of Food Science*. 2010;75(1):R43–9.
2. Mihindukulasuriya SD, Lim LT. Nanotechnology development in food packaging: a review. *Trends in Food Science & Technology*. 2014;40(2):149–67.
3. Anand S, Paxson AT, Dhiman R, Smith JD, Varanasi KK. Enhanced condensation on lubricant-impregnated nanotextured surfaces. *ACS Nano*. 2012;6(11):10122–9.
4. Sims L, Egelhaaf HJ, Hauch JA, Kogler FR, Steim R. Plastic solar cells. *Comprehensive Renewable Energy*. 2012;1:439–480.
5. Kesarwani S, Verma RK. A critical review on synthesis, characterization and multifunctional applications of reduced graphene oxide (rGO)/composites. *Nano*. 2021;16(09):2130008.
6. Iqbal J, Ginsburg O, Rochon PA, Sun P, Narod SA. Differences in breast cancer stage at diagnosis and cancer-specific survival by race and ethnicity in the United States. *JAMA*. 2015;313(2):165–73.
7. Gnerlich JL, Deshpande AD, Jeffe DB, Sweet A, White N, Margenthaler JA. Elevated breast cancer mortality in women younger than age 40 years compared with older women is attributed to poorer survival in early-stage disease. *Journal of the American College of Surgeons*. 2009;208(3):341–7.
8. Bianchi F, Nicassio F, Marzi M, Belloni E, Dall'Olio V, Bernard L, Pelosi G, Maisonneuve P, Veronesi G, Di Fiore PP. A serum circulating miRNA diagnostic test to identify asymptomatic high-risk individuals with early stage lung cancer. *EMBO Molecular Medicine*. 2011;3(8):495–503.
9. Onozato ML, Kovach AE, Yeap BY, Morales-Oyarvide V, Klepeis VE, Tammireddy S, Heist RS, Mark EJ, Dias-Santagata D, Iafrate AJ, Yagi Y. Tumor islands in resected early-stage lung adenocarcinomas are associated with unique clinicopathologic and molecular characteristics and worse prognosis. *The American Journal of Surgical Pathology*. 2013;37(2):287–94.
10. Yan S, Yin J, Yang Y, Dai Z, Ma J, Chen X. Surface-grafted silica linked with L-lactic acid oligomer: a novel nanofiller to improve the performance of biodegradable poly (L-lactide). *Polymer*. 2007;48(6):1688–94.
11. Schaefer DW, Kohls D, Feinblum E. Morphology of highly dispersing precipitated silica: impact of drying and sonication. *Journal of Inorganic and Organometallic Polymers and Materials*. 2012;22(3):617–23.
12. Oberdisse J, El Harrak A, Carrot G, Jestin J, Boué F. Structure and rheological properties of soft–hard nanocomposites: influence of aggregation and interfacial modification. *Polymer*. 2005;46(17):6695–705.

13. Paul DR, Robeson LM. Polymer nanotechnology: nanocomposites. *Polymer.* 2008;49(15):3187–204.

14. Yang B, Zhang J. Nanoparticles: synthesis in polymer substrates. In Contescu, C.I. (Ed.). *Dekker Encyclopedia of Nanoscience and Nanotechnology*, p. 5912, Second ed., Taylor & Francis, Boca Raton, 2009.

15. American Chemical Society. Solar cells that can face almost any direction and keep themselves clean. 2015. http://m.phys.org/news/2015-12-solar-cells.html.

16. Youssef AM. Polymer nanocomposites as a new trend for packaging applications. *Polymer-Plastics Technology and Engineering.* 2013;52(7):635–60. doi: 10.1080/03602559.2012.762673.

17. Lagashetty A, Venkataraman A. Polymer nanocomposites. *Resonance.* 2005;10(7):49–57.

18. Bohunicky B, Mousa SA. Biosensors: the new wave in cancer diagnosis. *Nanotechnology, science and applications.* 2011;4:1.

19. Tothill IE. Biosensors for cancer markers diagnosis. *Seminars in Cell & Developmental Biology.* 2009;20(1):55–62.

20. Grodzinski P, Silver M, Molnar LK. Nanotechnology for cancer diagnostics: promises and challenges. *Expert Review of Molecular Diagnostics.* 2006;6(3):307–18.

21. Banerjee HN, Verma M. Use of nanotechnology for the development of novel cancer biomarkers. *Expert Review of Molecular Diagnostics.* 2006;6(5):679–83.

22. Zhang GJ, Chua JH, Chee RE, Agarwal A, Wong SM. Label-free direct detection of MiRNAs with silicon nanowire biosensors. *Biosensors and Bioelectronics.* 2009;24(8):2504–8.

23. Wang J. Carbon-nanotube based electrochemical biosensors: a review. *Electroanalysis.* 2005;17:7–14.

24. Cai H, Cao X, Jiang Y, He P, Fang Y. Carbon nanotube-enhanced electrochemical DNA biosensor for DNA hybridization detection. *Analytical and Bioanalytical Chemistry.* 2003;375(2):287–93.

25. Yu X, Munge B, Patel V, et al. Carbon nanotube amplification strategies for highly sensitive immunodetection of cancer biomarkers. *Journal of the American Chemical Society.* 2006;128(34):11199–205.

26. Sharifianjazi F, Rad AJ, Esmaeilkhanian A, Niazvand F, Bakhtiari A, Bazli L, Abniki M, Irani M, Moghanian A. Biosensors and nanotechnology for cancer diagnosis (lung and bronchus, breast, prostate, and colon): a systematic review. *Biomedical Materials.* 2021; 17: 012002.

27. Kim SJ, Lee SW, Song JD, Kwon YW, Lee KJ, Koo HC. An InSb-based magnetoresistive biosensor using Fe_3O_4 nanoparticles. *Sensors and Actuators B: Chemical.* 2018;255:2894–9.

28. Andrei VA, Radulescu C, Malinovschi V, Marin A, Coaca E, Mihalache M, Mihailescu CN, Dulama ID, Teodorescu S, Bucurica IA. Aluminum oxide ceramic coatings on 316L austenitic steel obtained by plasma electrolysis oxidation using a pulsed unipolar power supply. *Coatings.* 2020;10(4):318.

29. Tajzad I, Ghasali E. Production methods of CNT-reinforced Al matrix composites: a review. *Journal of Composites and Compounds.* 2020;2(2):1–9.

30. Zhang K, Jang HW, Van Le Q. Production methods of ceramic-reinforced Al-Li matrix composites: a review. *Journal of Composites and Compounds.* 2020;2(3):77–84.

31. Jazi EH, Esalmi-Farsani R, Borhani G, Jazi FS. Synthesis and Characterization of In Situ Al-$Al_{13}Fe_4$-Al_2O_3-TiB_2 nanocomposite powder by mechanical alloying and subsequent heat treatment. *Synthesis and Reactivity in Inorganic, Metal-Organic, and Nano-Metal Chemistry.* 2014;44(2):177–84.

32. Rino JJ, Chandramohan D, Sucitharan KS, Jebin VD. An overview on development of aluminium metal matrix composites with hybrid reinforcement. *International Journal of Science Research.* 2012; 1(3).

33. Fattahi M, Vaferi K, Vajdi M, Moghanlou FS, Namini AS, Asl MS. Aluminum nitride as an alternative ceramic for fabrication of microchannel heat exchangers: a numerical study. *Ceramics International.* 2020;46(8):11647–57.

34. Nayebi B, Bahmani A, Asl MS, Rasooli A, Kakroudi MG, Shokouhimehr M. Characteristics of dynamically formed oxide films in aluminum–calcium foamable alloys. *Journal of Alloys and Compounds.* 2016;655:433–41.

35. Chlanda A, Oberbek P, Heljak M, Kijeńska-Gawrońska E, Bolek T, Gloc M, John Ł, Janeta M, Woźniak MJ. Fabrication, multi-scale characterization and in-vitro evaluation of porous hybrid bioactive glass polymer-coated scaffolds for bone tissue engineering. *Materials Science and Engineering: C.* 2019;94:516–23.

36. Asl MS, Nayebi B, Shokouhimehr M. TEM characterization of spark plasma sintered ZrB_2–SiC–graphene nanocomposite. *Ceramics International.* 2018;44(13):15269–73.

37. Delbari SA, Nayebi B, Ghasali E, Shokouhimehr M, Asl MS. Spark plasma sintering of TiN ceramics codoped with SiC and CNT. *Ceramics International.* 2019;45(3):3207–16.

38. Targhi VT, Omidvar H, Hadavi SM, Sharifianjazi F. Microstructure and hot corrosion behavior of hot dip siliconized coating on Ni-base superalloy IN738LC. *Materials Research Express.* 2020;7(5):056527.

39. Egorkin VS, Gnedenkov SV, Sinebryukhov SL, Vyaliy IE, Gnedenkov AS, Chizhikov RG. Increasing thickness and protective properties of PEO-coatings on aluminum alloy. *Surface and coatings Technology*. 2018;334:29–42.

40. Kasalica B, Petković-Benazzouz M, Sarvan M, Belča I, Maksimović B, Misailović B, Popović Z. Mechanisms of plasma electrolytic oxidation of aluminum at the multi-hour timescales. *Surface and Coatings Technology*. 2020;390:125681.

41. Kikuchi T, Taniguchi T, Suzuki RO, Natsui S. Fabrication of a plasma electrolytic oxidation/anodic aluminum oxide multi-layer film via one-step anodizing aluminum in ammonium carbonate. *Thin Solid Films*. 2020;697:137799.

42. Wang S, Liu X, Yin X, Du N. Influence of electrolyte components on the microstructure and growth mechanism of plasma electrolytic oxidation coatings on 1060 aluminum alloy. *Surface and Coatings Technology*. 2020;381:125214.

43. Angulakshmi N, Dhanalakshmi RB, Kathiresan M, Zhou Y, Stephan AM. The suppression of lithium dendrites by a triazine-based porous organic polymer-laden PEO-based electrolyte and its application for all-solid-state lithium batteries. *Materials Chemistry Frontiers*. 2020;4(3):933–40.

44. Hussein RO, Northwood DO, Nie X. The effect of processing parameters and substrate composition on the corrosion resistance of plasma electrolytic oxidation (PEO) coated magnesium alloys. *Surface and Coatings Technology*. 2013;237:357–68.

45. Amiri M, Padervand S, Targhi VT, Khoei SM. Investigation of aluminum oxide coatings created by electrolytic plasma method in different potential regimes. *Journal of Composites and Compounds*. 2020;2(4):115–22.

46. Jain KK. Applications of nano-biotechnology in clinical diagnostics. *Clinical Chemistry*. 2007;53(11):2002–9.

47. Howarth M, Takao K, Hayashi Y, Ting AY. Targeting quantum dots to surface proteins in living cells with biotin ligase. *Proceedings of the National Academy of Sciences of the United States of America*. 2005;102(21):7583–8.

48. Soda N, Gonzaga ZJ, Chen S, Koo KM, Nguyen NT, Shiddiky MJ, Rehm BH. Bioengineered polymer nanobeads for isolation and electrochemical detection of cancer biomarkers. *ACS Applied Materials & Interfaces*. 2021;13(27):31418–30.

49. Rasooly A, Jacobson J. Development of biosensors for cancer clinical testing. *Biosensors and Bioelectronics*. 2006;21(10):1851–8.

50. Zamani Y, Ghazanfari H, Erabi G, Moghanian A, Fakić B, Hosseini SM, Mahammod BP. A review of additive manufacturing of Mg-based alloys and composite implants. *Journal of Composites and Compounds*. 2021;3(6):71–83.

51. Tajgardoon R, Zarnegaryan A, Elhamifar D. A Lindqvist type hexamolybdate [Mo6O19]-modified graphene oxide hybrid catalyst: highly efficient for the synthesis of benzimidazoles. *Journal of Photochemistry and Photobiology A: Chemistry*. 2022;430:113960.

52. Manna K, Mukherjee N, Chatterjee N, Saha KD. Cancer diagnosis by biosensor-based devices: types and challenges. In Raju Khan, Arpana Parihar, and Sunil K. Sanghi (Eds.), *Biosensor Based Advanced Cancer Diagnostics* (pp. 353–73). Academic Press, Elsevier B.V., Amsterdam, 2022.

53. Li Y, Hou L, Liu Z, Lu W, Zhao M, Xiao H, Hu T, Zheng Z, Jia J, Wu H. A sensitive electrochemical MUC1 sensing platform based on electroactive Cu-MOFs decorated by AuPt nanoparticles. *Journal of the Electrochemical Society*. 2020;167(8):087502.

54. Sarfraz J, Gulin-Sarfraz T, Nilsen-Nygaard J, Pettersen MK. Nanocomposites for food packaging applications: an overview. *Nanomaterials*. 2020;11(1). doi: 10.doi.org/10.3390/nano11010010.

55. Yao D, Chen Z, Zhao K, Yang Q, Zhang W. Limitation and challenge faced to the researches on environmental risk of nanotechnology. *Procedia Environmental Sciences*. 2013;18:149–56. doi: 10.1016/j.proenv.2013.04.020.

56. Kudumula K. Scope of polymer nano-composite in bio-medical applications. *Journal of Mechanical and Civil Engineering*. 2016;13:18–21.

57. britannica.com, "An Overview of Foam Rubber", Accessed on 8th October, 2022, from: www.britannica.com/science/foam-rubber.

9 Dispersive Techniques and Methods to Synchronize Host–Guest Interactions of Nanocomposite-Reinforced Polymers

D. Lakshmi, M. Infanta Diana, P. Adlin Helen, and P. Christopher Selvin

CONTENTS

9.1 INTRODUCTION

Nanocomposite (NC): Polymer blend is a successful twosome finding application in wide range of technological fields starting from coatings, adhesives, ICs, drug delivery systems, aerospace and medical instruments. But the homogenous dispersion or blend of particle-polymer is challenging over its attractiveness. The amount of nanoparticle and quality such as absence of defects/strain/composition, morphological factors, surface energy and structure determine the overall output and suitability of the polymer: NC for target applications. Also, it is to be noted that synergistic effects of polymer-particle are different when compared to the individual form of it. To facilitate the homogenous dispersion and impactful improvement in mechanical properties of the polymers, rather than hand mixing, advanced and efficient techniques are in practice. Optimizing the dispersion of nanofiller inside the matrix varies from one nanofiller to another, and therefore, it is quite

DOI: 10.1201/9781003343912-9

complex to draw any conclusion on finding a method to determine the right amount of nanoparticles to be dispersed without paving the way for agglomeration inside the matrix and to achieve enhancement in the mechanical properties [1,2]. However, the quantity of dispersion can be assessed and ensured with the help of EDS analysis. This method entails connecting the crystalline and tensile characteristics of the nanocomposite polymer with the data acquired through image processing techniques [3]. Amid techniques, finding a suitable solvent to disperse the nanofillers inside the polymer matrix is essential in achieving the spatial distribution of nanoparticles inside the matrix inhibiting the formation of agglomeration [4–6]. This document briefly discusses about the different techniques for uniform distribution of NC inside polymer matrix and their impact on different properties of the polymer along with few case studies.

9.2   IMPACT OF DISPERSION IN NC: POLYMER MATRIX

Starting from simple hand blend to high technological processing for dispersion of nanoparticles in polymer chain, depending on the degree of dispersion, a heavy impact on physicochemical aspects such as crystallization, glass transition temperature, absorption properties, morphology, bands and vibration, roughness, stability, thermal/ electrical conductivity, refractive index and many other parameters is altered. Once the compatibility is achieved, the nanofillers begin to operate as surface active centers and play an important role in determining mechanical properties of the composite polymer since the nanoparticles are quite hard to break loose from the matrix [7–11]. Schematic illustration of different procedures involved on polymer: nanocomposite preparation is given in Figure 9.1. To ensure uniformity, a strong dispersion of NC in polymer matrix is demanded whereas few applications such as electrical conductivity are affected by high degree of dispersion which will be discussed later.

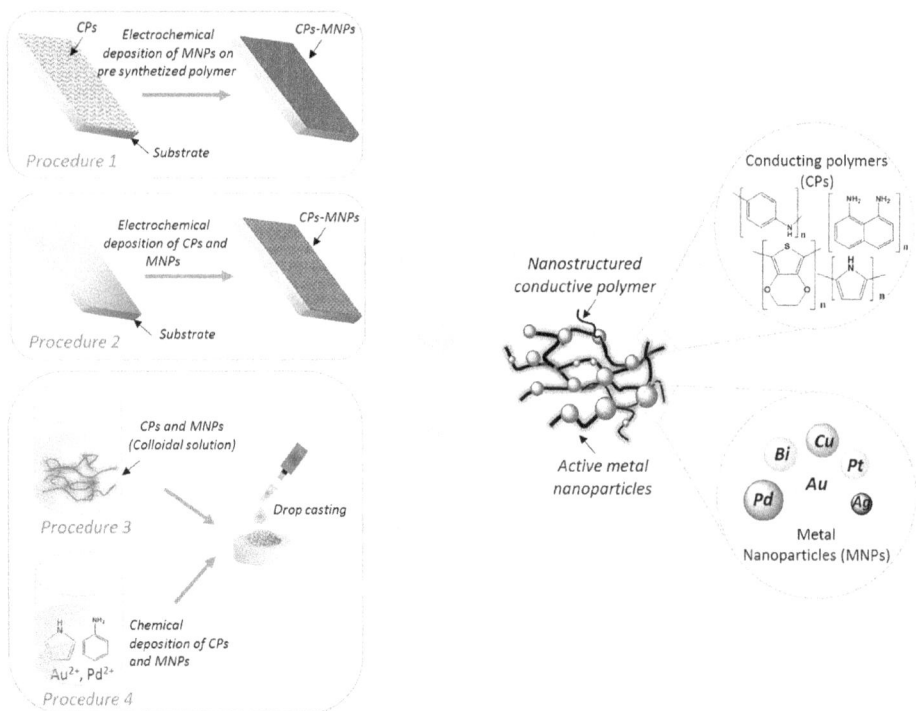

FIGURE 9.1   Schematic illustration of different procedures involved on polymer: nanocomposite preparation [10].

9.3 DISPERSION TECHNIQUES

To overcome the issue of agglomeration by the nanoparticles inside the polymer matrix, different dispersion techniques are adopted to prepare homogenous nanocomposite polymers to be reinforced with improved mechanical and other desired properties. Some of the best-known dispersion techniques involve ultrasonication, mechanical stirring, in-situ polymerization and solution mixing. These methods are chosen according to the aspect ratio of nanofillers and the viscosity of the polymers for different applications.

9.3.1 Ultrasonication Method

One of the versatile and widely used methods to prepare the nanocomposite polymer is ultrasonication method, also known as acoustic cavitation. The energy generated by the ultrasonic waves prohibits the formation of agglomeration of nanoparticles inside the polymer matrix which in turn greatly influences the crystallinity and glass transition temperature of the nanocomposite polymers. It is to be noted that the frequency and duration of sonication play a major role in synthesizing homogenous nanocomposite polymer where polymer density and quantity of nanoparticles play a key role [12]. The schematic shown in Figure 9.2 indicates the steps involved in ultrasonication method of polymer nanocomposite dispersion.

9.3.2 Mechanical Stirring

Nanocomposite polymers prepared via mechanical stirring are considered to be one of the simplest and easy-to-fabricate methods. This method involves dispersing the nanoparticles into the polymer matrix with shear force and does not require the need to find the suitable solvents for dispersing the nanoparticles. This method is often combined with ultrasonication or TRM process (three-roll mixing) for better results. In this method, when the shear force is increased at high loadings

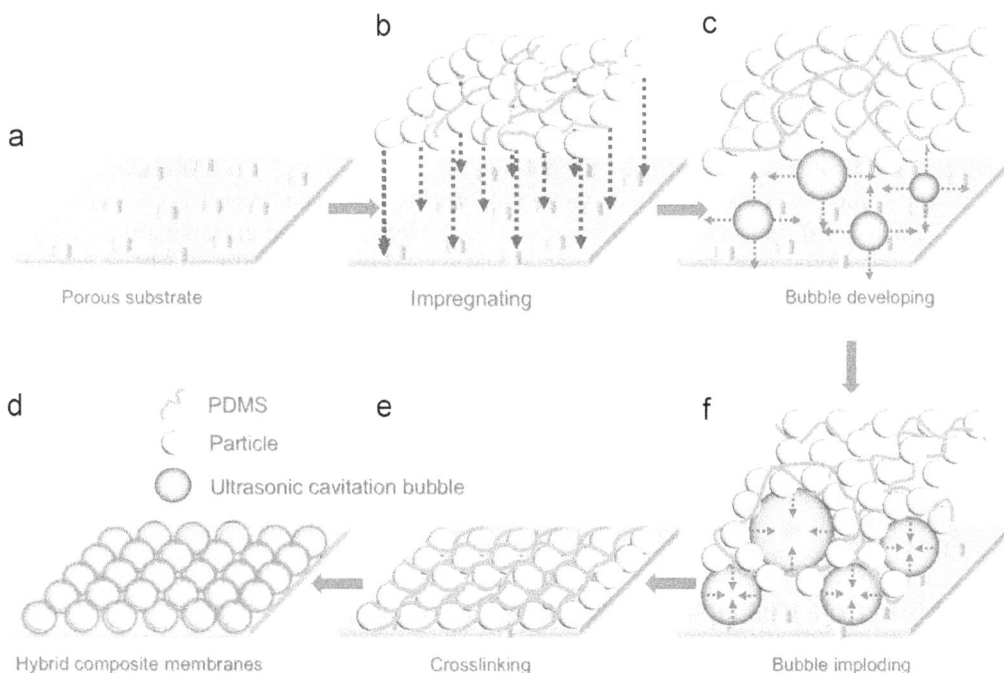

FIGURE 9.2 In-situ sonication assembly of polymer: nanocomposite [9]. (Reproduced with permission.)

of nanoparticles, the structure of the polymer chains is damaged, thereby greatly affecting the mechanical strength of the nanocomposite polymer [13].

9.3.3 SOLVENT MIXING

The tendencies to agglomerate by nanoparticles are pretty much avoided in the nanocomposite polymers prepared by solvent mixing method. The surface energy of the nanoparticles gets reduced when dispersed in a suitable solvent before mixing with the polymer solution. Due to this, the interfacial bond between the polymer and the filler is greatly enhanced [14].

9.3.4 IN-SITU POLYMERIZATION

This method is considered to be one of the viable methods to improve the compatibility of the nanoparticles with the polymer matrix as it is considered to be the most influential factor greatly impacting the mechanical strength of the nanocomposite polymers. Here, the samples are prepared by dispersing the nanoparticles into the liquid monomer followed by polymerization induced by either heating or exposure to radiation. However, the scalability of this method is quite difficult as there are limitations like changes in viscosity of the polymer solutions, and this could eventually lead to aggregate formation at higher particle loadings even after achieving the compatibility between the nanoparticles and the polymer matrix [15].

9.4 PERCENTAGE OF DISPERSIONS WITH DIFFERENT TECHNIQUES AND METHODOLOGIES

When the filler content exceeds the ideal proportion, the polymer matrix crystallites, obstructs the polymer chain movement, indicating the nanofiller's nucleation capacity. This has an impact on the polymer's overall characteristics since an increase in crystallinity causes a loss in tensile strength due to the formation of agglomeration. This could even lead to debonding of nanoparticles from the polymer matrix. For example, the amount required to enhance the mechanical properties of polystyrene polymer with MoS_2 is 0.002 wt.%. At this ultralow level of MoS_2 filler, the enhancement in the mechanical properties such as the tensile strength, elongation break and toughness was achieved. Interestingly, the amount of MoS_2 filler required for enhancement in mechanical properties is different for different polymer host designed for variety of applications. It is to be noted that the surface modification of pure MoS_2 filler and dispersion through melt processing technique could be the detrimental reasons behind this reinforcement at low-level loadings [16].

Also, when the composite polymer is subjected to mechanical stress, deformation zones are created inside the polymer matrix, and the nanoparticles present inside the matrix absorb a lot of impact energy in order to achieve the effect of toughening and strengthening at the same time. As the size of the nanoparticle decreases, the hydrogen bond and polarization force increase, substantially contributing the nanoparticle's interface factor to the bonding with the polymer matrix. This interfacial bonding of filler with matrix ensures the improvement in the composite material's mechanical quality [17].

9.5 IMPACT OF DISPERSION ON DIFFERENT PROPERTIES

As mentioned earlier, the degree and depth of dispersion impose a huge impact on physicochemical, optical, electrical, electrochemical and almost all the natural properties of both polymer and filler nanocomposites. A brief overview on these aspects is discussed here.

9.5.1 Mechanical Properties

It is a basic criterion to incorporate nanoparticles to enhance the mechanical properties such as tensile strength, crystallization, heat resistance temperature and notch impact strength of the polymers. To ensure the facilitating role of nanoparticles in polymer matrix as homogenous mixture, different dispersion techniques have been practiced as discussed above. One of the positive aspects of preferring nanoparticles over other elastomers or microfillers [18] is their high surface area which can cover complete working area of the polymer, strengthen the bonding, prevent crack formation and improved interparticle interaction. But the downside is unavoidable agglomeration in polymer-particle blend or aggregation in particle-particle contacts. Agglomerations such as soft and hard are unavoidable in nanoparticles since several internal (electrostatic attraction or van der Waals force) and external factors (adsorption, thermodynamic reasons) contribute to this phenomena. There are different technical snags that could be mentioned such as choice of dispersing techniques, fillers and surfactants while preferring for certain applications.

To maintain the virtue of the polymer along with strengthening of its mechanical aspects for particular application cannot go with a universal method. One example is opting of surfactants for well-distributed nanoparticles may not be compatible with biomedical applications due to possible chemical toxicity issues. Hence, this becomes a challenging task and hence solvents play a smoothening role here to maintain the particle distribution. However, after evaporation of the solvent, the agglomeration can again take place which is inevitable in most of the cases. But when the homogeneity of the polymer matrix NC improves, the mechanical properties are greatly enhanced [18]. Regarding the choice of fillers, the common one is carbon-based materials with different dimensions such as spherical, rod, plates, etc. Materials like silica, titania, alumina and $CaCO_3$ also share the podium. But the target application of the reinforced polymer material can be a crucial role in preferring a particular filler. For example, conductive property of carbon material is a limitation for dielectric applications whereas loss of transparency of polymers due to the addition of colored fillers is useless, etc. However, it is important to stick to the defect free NPs and smaller quantity of NP fillers to avoid the detrimental effects of the supplements. Table 9.1 provides a comparative view of mechanical/electrochemical and other parameters associated with one another.

9.5.2 Electrochemical and Electrical Properties

Analysis of electrochemical/electrical properties of any material contains the preliminary electrode preparation which includes polymer binder and a solvent like NMP. Report [19] suggests that rather than coating and drying technique of electrode, rate of mixing of binder/solvent determines the dispersion of nanoparticles in an electrode system. Addition of required quantity of solvent in batches rather than in single step ensures better dispersion. Figure 9.3 compares the one-step with two-step dispersion process. But in the case of electrodes, highly distributed nanoparticles lose the close connectivity between two particles which end up in poor redox property and electrical conductivity [19]. Still, it is a widely accepted fact that inclusion of salt and ceramic filler in a polymer electrolyte material improves the mechanical aspects and ionic conductivity of the polymer matrix. Addition of surfactants/surface modification by introducing inorganic fillers into organic compounds is another effective approach in gel polymer electrolyte category which promotes the Lewis acid-base interactions thereby improving free charge particles. Further, addition of metal oxide nanoparticles improves the homogenous distribution of pores and surface roughness in gel polymer electrolyte. The formation of ion-ceramic complex through salt dissociation is a profit while expecting better storage properties by means of better host–guest interactions. Also, by improving the viscous nature of the polymer, these well-distributed nanocomposite fillers improve the electrolyte–electrode interface with less significant passive reactions at the junction. But in few cases, interfacial resistance between ceramic filler and polymer is not helpful for better ionic conductivity [20].

TABLE 9.1

Comparison of Various Polymers and Fillers Combination on Different Properties of Polymer: NC

S. No.	Polymer	Filler (np)	Method for Dispersion	% of NP	Mechanical		Others
					Increment in Tensile		
					Strength		
1.	Hybrid glass	Al$_2$O$_3$ and graphene nanoplatelets	Sonication	1.5%	17% [22]		bending/flexural: low Hardness: 14 VHN (Vickers hardness number)
2.	Epoxy resin with methyl ethyl ketone solvent	TiO$_2$ NP	Ultradual sonication	10%	50% [23]		Young's modulus: 28% higher for 20% of TiO$_2$
3	Polylactic acid (PLA)	ZnO NP	Electro-spinning and ultrasonication	0.5%	55.33% [24]		Elongation at break: 28.43% lesser than PLA
4.	PLGA	Titania	Sonication	30/70→ceramic/polymer ratio	6 times [18]		-
5.	PVC	CaCO$_3$ NP	-	10%	123% [25]		Notch impact strength: 133% than pure PVC

(Continued)

TABLE 9.1 (Continued)

Comparison of Various Polymers and Fillers Combination on Different Properties of Polymer: NC

Electrical/ Electrochemical

S. No.	Polymer	Filler (np)	Method for Dispersion	% of NP	Electrochemical/Electrical Aspects	Application
6.	PVDF-HFP/PMMA	TiO_2	Mechanical stirring	5 wt.%	C_{sp}: 163.56 mAh/g Bulk resistance: low [26]	Energy storage
7.	PEO	LLZO Salt: LiTFSI	Stirring+ sonication	10 wt.%	conductivity: 1.36×10^{-5} S/cm [20]	Energy storage
8.	PMMA	TiO_2 and ZnO	Solvation-unfolding-swelling	TiO_2: 2 wt.% ZnO: 3 wt.%	conductivity: 17.54×10^{-4} mho/cm (TiO_2) 18.41×10^{-4} mho/cm (ZnO) [12]	Electrochromic devices

Optical

S. No.	Polymer	Filler (np)	Method for Dispersion	% of NP	Band Gap	Application
9.	LDPE	1. TiO_2 2. Al_2O_3 3. TiO_2-Al_2O_3	Freeze granulation and low-temperature extrusion	20 wt.%	1. 3.25 eV 2. 3.27 eV 3. 3.29 eV [21]	Multiple
10.	PVA	Fe NP	In-situ formation	3 wt.%	2.8 eV [27]	Multiple

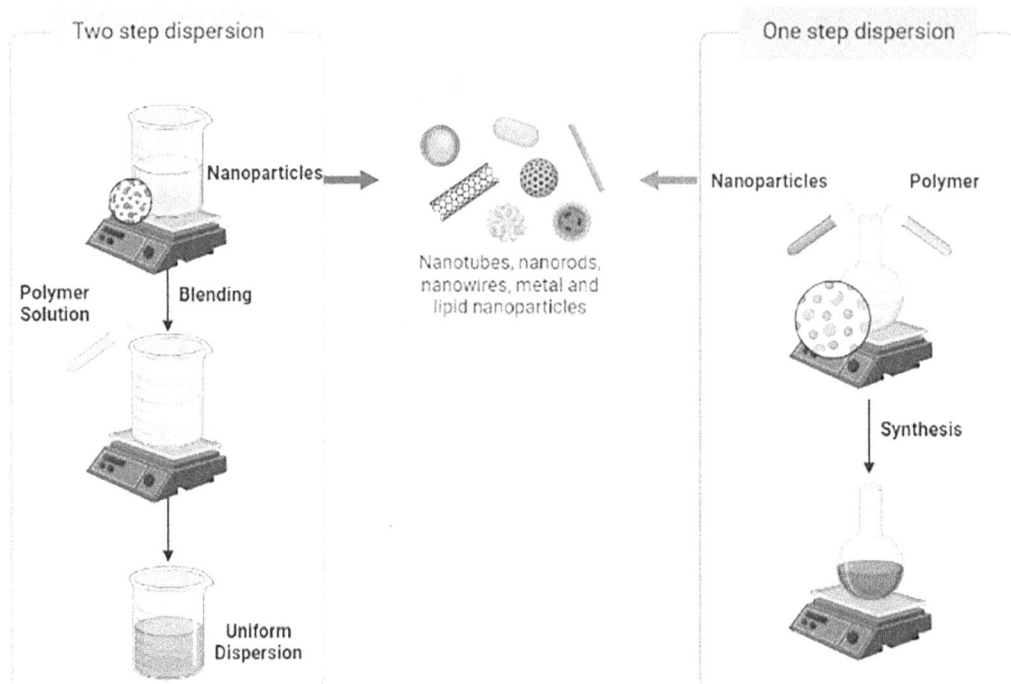

FIGURE 9.3 One-step vs. two-step dispersion process.

9.5.3 OPTICAL AND ENERGY GAP

Incorporation of suitable NC in a polymer matrix can tune the wide range optical properties of the polymer such as absorption, transparency, photoluminescence, coloration properties, energy gap, refractive indices, etc. Especially, direct solvent blending, direct melt blending, in-situ formation of particles inside polymer and core-shell arrangements of particle-polymer methods ensure great control over particle size, homogenous dispersion, multispecies combination and reliable outputs. Other hybrid methods also facilitate bulk production of polymer: NC [21]. A comparative illustration is given for PVA (with and without PVP) polymer with respect to different fillers and concentration in Figure 9.4 for band gap values. A definite improvement in band gap value with effective dispersion of fillers into the PVA base indicates the impact of the process, whereas pure PVA holds a wide band gap of ~6.27 eV and PVP is ~3.82 eV.

9.6 CASE STUDIES INVOLVING THE EFFECTIVE DISPERSION OF VARIOUS NANOMATERIALS

Scientific community has explored varieties of polymer: nanocomposite by different dispersion techniques for different applications. Herein, we compile few notable research works on this topic and present briefly.

9.6.1 PANORAMIC VIEW ON NANOFILLERS INTO POLYMER MATRIX

The ultralarge interfacial area and nanoscopic size of the fillers differentiate polymer nanocomposites from regular composites. Figure 9.5 depicts the classification of nanofillers based on their dimensions. Through the dominance of interfacial regions generated by nanoscale phase dimensions, mechanical and physical properties that are not present in standard polymer composites have

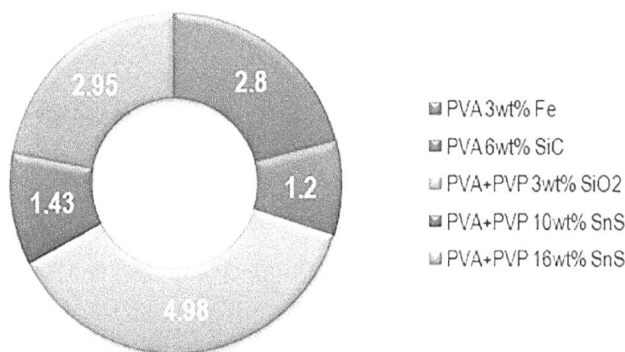

FIGURE 9.4 Comparative illustration of PVA on band gap value as simple or PVP blend with respect to different fillers concentration (data labels inside the plot indicate band gap values) [18,28–31].

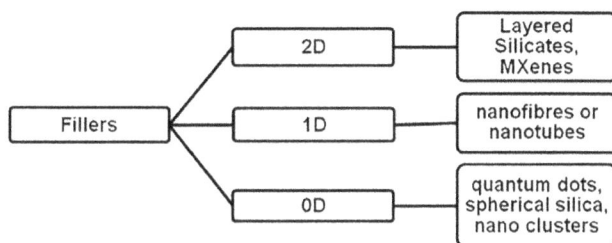

FIGURE 9.5 Range of nanofillers based on dimensions.

been developed. Because polymers and nanofillers have a large interfacial area, functionalization of nanofillers is used to modify interfacial states in polymer nanocomposites to improve mechanical and physical properties [32]. Grafting polymer molecular chains of graphene to functionalized graphene/polymer nanocomposites dramatically enhanced interfacial shear strength and interfacial mode fracture toughness. Various approaches for characterization of polymer nanocomposites as well as mechanical and physical properties of polymer nanocomposites are thoroughly examined.

Layered silicates, which belong to the 2:1 phyllosilicate structural family, are one of the examples of sheet-like nanofillers commonly employed in polymer nanocomposites. The most common layered silicates include hectorite, montmorillonite and saponite. Two-dimensional layers make up their crystal lattice. Two exterior silica tetrahedrons are fused to a central octahedral sheet of alumina or magnesia so that the oxygen ions of the octahedral sheet belong to the tetrahedral sheets. Direct physical mixing of polymer and layered silicates does not yield polymer nanocomposites. This is similar to the separation of layered silicates into separate phases that may occur in many polymer mixes. Poor physical interaction between inorganic fillers and polymers causes low mechanical and physical characteristics in immiscible systems, which are common in many traditional filled polymers. Strong interactions between layered silicates and polymers, on the other hand, may cause inorganic nanophases to disperse at the nanoscale level in polymer matrices. As a result, polymer nanocomposites will have mechanical and physical qualities that microcomposites do not possess [33].

In pure layered silicates, hydrated Na or K ions are abundant. In their basic form, layered silicates are only soluble with hydrophilic polymers [34]. Hydrophilic silicate surfaces should be changed to organophilic silicate surfaces to allow for the intercalation of other polymers when combining layered silicates with other polymers. A significant interlayer gap is typically formed when hydrated cations are exchanged using organic cationic surfactants such as alkylammonium or alkylphosphonium (onium) [35,36].

Organosilicates with alkylammonium or alkylphosphonium cations lower surface energy and improve the wetting capabilities of polymer matrices, allowing for greater interlayer spacing. Alkylammonium or alkylphosphonium cations, on the other hand, have functional groups that can react with polymer matrices or, in certain situations, trigger monomer polymerization to change the inorganic clay-polymer matrix interface adhesion [37].

Expansive graphite has a multilayer structure composed mostly of parallel boards that can buckle and deform, resulting in numerous holes ranging in diameter from roughly 10 nm to 10 m. Exfoliated graphite has graphite sheets that are several hundred nanometers thick. Graphite flakes of varying sizes form expanded graphite in an identical way, despite their differing expansion ratios [38].

Multilayer compounds, namely, $NiTe_2$, Bi_2Te_3, MoS_2, WS_2, $TaSe_2$, $MoSe_2$, $MoTe_2$ and $NbSe_2$ can be dispersed easily in common solvents. As individual flakes or films, these stacked compounds can be placed in the polymer matrices. These materials are exfoliated into separate layers, and polymer nanocomposites are created by combining them with polymer solutions [39].

The alignment of CNTs has also been used to prepare nanofibers. SWCNTs are distributed in 102% sulfuric acid, which may protonate sidewalls and enable fiber drawing [40]. Drawing and spinning CNT fibers straight from a CVD reaction chamber is one way. After that, the CNT fibers can be postimpregnated with epoxy resin to create a nanocomposite. These fibers have strength of around 1 GPa, which is substantially lower than a single nanotube's. As a result, greater CNT-polymer interfacial adhesion is required for CNTs to properly distribute the load.

Many research works aim to create spherical nanoparticles with adjustable sizes. The influence of particle size on their qualities is the major driving reason behind this interest. Nanofillers can provide polymer matrices features that microsized filler where other fillers fail. For example, adding nanoparticles to polymers at the right concentrations can improve strength and modulus while maintaining ductility, but adding microsized fillers is unlikely to achieve this balance.

The hydrolysis of tetraethylortho-silicate in ethanol followed by condensation of dispersed phase material may be used to produce monodispersed silica particles with a consistent size distribution [41]. To create high-performance polymer nanocomposites, good dispersion of nanoparticles in polymers is essential, and dispersants are sometimes employed to increase the dispersibility of nanoparticles. In the formation of a slurry containing nano-ZnO particles by ball milling, for example, two types of dispersants are utilized. One of the dispersants is a block polymer with an ultra-high molecular weight that has a great affinity for pigments; another dispersant is polyacrylate sodium, which has a molecular weight of only a few thousand [42].

Since the relevance of nanofiller–matrix interfaces in polymer nanocomposites has been recognized, specialized surface modification based on the particular purposes and properties of the polymer system is always required. The whole concept is predicated on the formation of a massive interfacial area between the polymer matrices and nanoscale fillers. However, the interfacial area is not the only factor influencing the properties of polymer nanocomposites; interfacial interaction is also significant in defining nanocomposites' features. Polymer nanocomposites with specific properties can be developed after appropriate surface modification for a number of technological applications. Figure 9.6 illustrates the schematic of different surface modification techniques for polymer: nanocomposites.

Silica nanotubes are used to strengthen polyimide (PI) sheets. The coupling agent (-aminopropyltriethoxy silane, KH 550) is used to promote the affinity of silica and polyimide and hence their interfacial adhesion. It should be emphasized that the proper amount of coupling agent promotes tensile strength and modulus [43].

9.6.2 Mechanical Properties of Highly Studied Polymer: Nanocomposites

The mechanical characteristics of carbon nanotubes or nanofibers are quite high. The mechanical characteristics of the resulting nanocomposites once carbon nanotubes or nanofibers are introduced

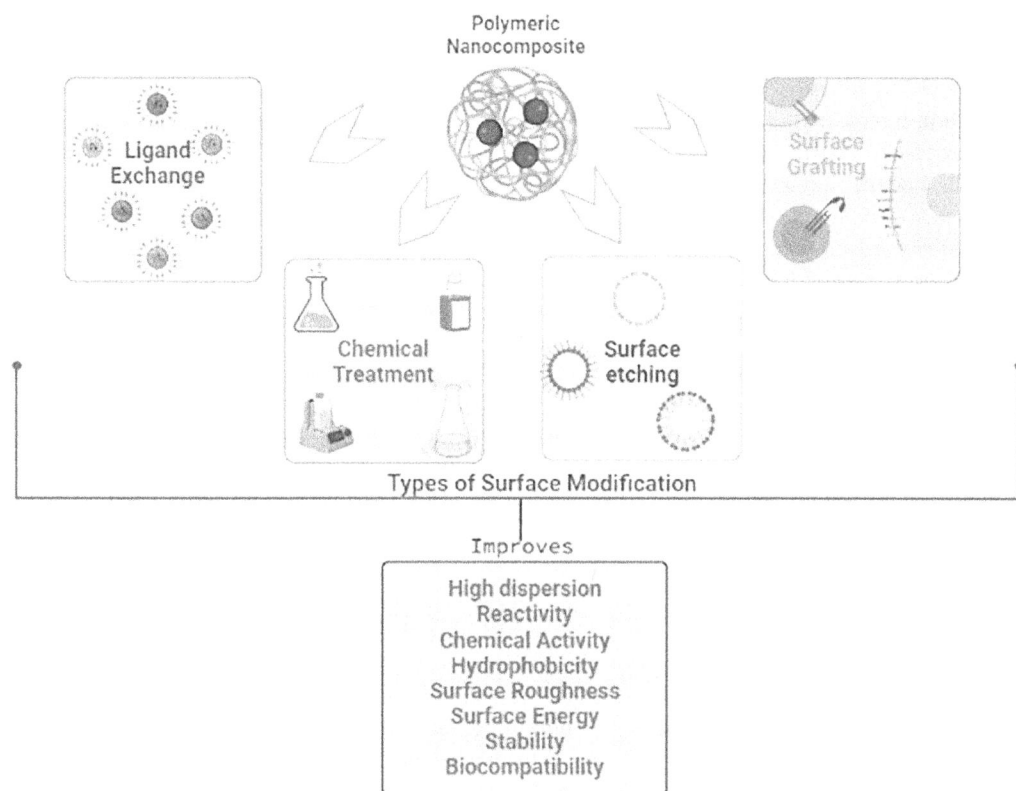

FIGURE 9.6 Schematic illustration of various surface modification techniques.

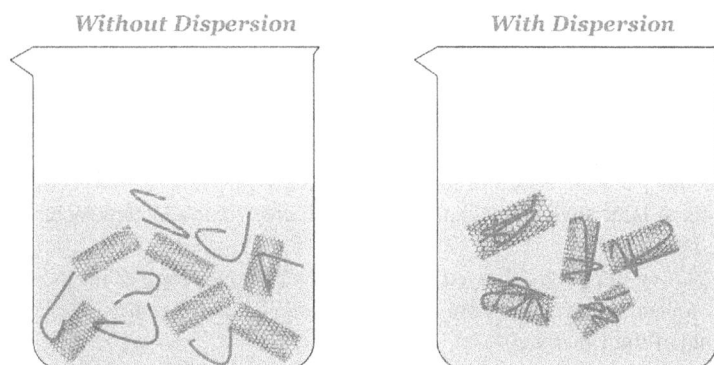

FIGURE 9.7 Effect of dispersion in CNT: polymer matrix.

into polymers are therefore of interest. Typical comparison on the effect of dispersion in blending the polymer with CNT is shown in Figure 9.7.

For CNT/polymer nanocomposites, optimal CNT length and interface strength would lead to optimal fracture toughness. The ideal diameter of CNTs may occur in the transition from interfacial debonding to CNT breaking, and smaller CNTs did not necessarily impart a better fracture toughness on their reinforced composites. Short fiber-reinforced polymer composites have similar outcomes to those of long fiber-reinforced polymer composites. In order to achieve high fracture toughness in short-fiber composites, good interface bonding and a short-fiber aspect ratio are required. The strength of a discontinuous polymer composite is often proportional to the strength of the interface adhesion.

Carbon-nanotube/polystyrene composites: There is a 25% rise in tensile strength and a 36%–42% increase in elastic stiffness when only 1 wt.% of nanotubes is introduced. Nanotube pullout, nanotube fracture and nanotube crack-bridging have all been seen in nanotube composites, just as they have in standard fiber composites. The short-fiber composite theory is used to show that 10 wt.% carbon fibers are required to produce the same increase in elastic modulus as 1 wt.% CNTs [44].

Carbon-nanotube/PVA: The DMA is used to calculate the tensile elastic modulus and damping characteristics of PVA composite films as a function of CNT concentration and temperature. The experimental data have been used to predict a nanotube elastic modulus of 150 MPa using the theory developed for short-fiber homogenous composites [45].

WCNT/poly(ethersulfone)(PES)/epoxy composites: The three-roll milling technique at cryogenic temperature is adopted to prepare MWCNT/poly(ethersulfone)(PES)/epoxy composites. The addition of MWCNTs at the right concentrations improved cryogenic tensile characteristics. However, as the CNT content rises to 1.0 wt.% and beyond, the tensile strength and failure strain decrease due to CNT aggregation at high concentrations. The fracture toughness rises as the CNT content increases [46].

The function of particle/matrix interfacial adhesion in composite strength is also of interest. When the surfaces of silica nanoparticles are modified with aminobutyric acid to improve filler/matrix adhesion, the tensile strength of silica-filled nylon 6 nanocomposites increases in a bell shape with increasing silica content. It is also proven that at 5 wt.% silica, the tensile strength reaches its maximum value [47]. The separation-to-aggregation process of inorganic particles with increasing particle content helps explain this. When the particle surface is not changed, the strength falls as the particle load increases. The improved composites have high interfacial adhesion between the silica particles and the polymer matrix, as well as good dispersion. As a result, when the external stress is applied to nanocomposites, the stress transfer at the interface should be better for the silica-modified nanocomposite, increasing composite strength.

9.7 CONCLUSION

The role and impact of different fillers in various polymers have been extensively analyzed with respect to dispersion techniques. The following key points are highlighted from this review:

- Simple and binary combinations of fillers into single or polymer blends regulate the different properties of later such as mechanical, optical, electrical and electrochemical aspects of polymer: NC.
- The homogenous dispersion achieved via different techniques such as ultrasonic, mechanical stirring, solvent blend, in-situ polymerization, etc. leads to greater potential of the systems than that of their originals.
- Few case studies provide an overview of several nanoscale fillers with two-dimensional, one-dimensional and zero-dimensional morphologies that are employed in polymer nanocomposites, as a fundamental overview of how polymer nanocomposites are processed by utilizing various techniques.
- Adding a little amount of nanoscale fillers to polymers improves their mechanical and physical characteristics. However, particular fillers do not play specific role when employed in different polymers or in different concentrations.
- The mode of dispersion also modifies the explicit properties of nanomaterials by altering the bandgap, crystallinity, thermal stability, glass transition temperature, etc.
- There is no universal procedure to be adapted for preparing composites with specific fillers or polymers for a particular application whereas all these combinations require trail and optimization process to achieve the success.

REFERENCES

1. Kashiwagi T, Fagan J, Douglas JF, Yamamoto K, Heckert AN, Leigh SD, Obrzut J, Du F, Lin-Gibson S, Mu M, Winey KI, Haggenmueller R (2007) Relationship between dispersion metric and properties of PMMA/SWNT nanocomposites. *Polymer.* 48(16): 4855–4866.
2. Sadasivuni KK, Rattan S, Waseem S, Brahme SK, Kondawar SB, Ghosh S, Das AP, Chakraborty PK, Adhikari J, Saha P, Mazumdar P, "Silver nanoparticles and its polymer nanocomposites—synthesis, optimization, biomedical usage, and its various applications. polymer nanocomposites in biomedical engineering". In *Lecture Notes in Bioengineering.* Springer, Cham, pp. 331–373, 2019.
3. George G, Dev AP, Asok NN, Anoop MS, Anandhan S (2021) Dispersion analysis of nanofifillers and its relationship to the properties of the nanocomposites. *Mater Today: Proc.* 47(15): 5104–5109.
4. Haveriku S, Meucci M, Badalassi M, Cardelli C, Ruggeri G, Pucci A (2021) Optimization of the mechanical properties of polyolefin composites loaded with mineral fillers for flame retardant cables. *Micro.* 1: 102–119.
5. Helen A, Ajith K, Diana MI, Lakshmi D, Selvin PC (2022) Chitosan based biopolymer electrolyte reinforced with V_2O_5 filler for magnesium batteries: an inclusive investigation. *J Mater Sci: Mater Electron.* 33: 3925–3937.
6. Kesarwani S, Verma RK (2021) A critical review on synthesis, characterization and multifunctional applications of reduced graphene oxide (rGO)/composites. *Nano.* 16(09): 213008.
7. Hong RY, Chen Q, "*Dispersion of Inorganic Nanoparticles in Polymer Matrices: Challenges and Solutions*" Part of the Organic-Inorganic Hybrid Nanomaterials, Springer, Cham. vol. 267, pp. 1–38, 2014.
8. Fenin KA, Akinlabi ET, Perry N (2019) Quantification of nanoparticle dispersion within polymer matrix using gap statistics. *Mater Res Express.* 6: 075310.
9. Kumar S, Nehra M, Dilbaghi N, Tankeshwar K, Kim KH (2018) Recent advances and remaining challenges for polymeric nanocomposites in healthcare applications. *Prog Polym Sci.* 80: 1–38.
10. Rhazi ME, Majid S, Elbasri M, Salih FE, Oularbi L, Lafdi K (2018) Recent progress in nanocomposites based on conducting polymer: application as electrochemical sensors. *Int Nano Lett.* 8: 79–99.
11. Chang CJ, Xu L, Huang Q, Shi J (2012) Quantitative characterization and modeling strategy of nanoparticle dispersion in polymer composites. *IIE Trans.* 44(7): 523–533.
12. Goyat MS, Gosh PK (2018) Impact of ultrasonic assisted triangular lattice like arranged dispersion of nanoparticles on physical and mechanical properties of epoxy-TiO_2 nanocomposites. *Ultrason Sonochem.* 42: 141–154.
13. Gao Y, Wang Q, Wang J (2014) Synthesis of highly efficient flame retardant high-density polyethylene nanocomposites with inorgano-layered double hydroxides as nanofiller using solvent mixing method. *ACS Appl Mater Interfaces.* 6(7): 5094–5104. https://doi.org/10..1021/am500265a.
14. Tajik S, Beitollahi H, Nejad FG et al. (2021) Recent developments in polymer nanocomposite-based electrochemical sensors for detecting environmental pollutants. *Ind Eng Chem Res.* 60(3): 1112–1136.
15. Rodriguez CLC, Nunes MABS, Garcia PS, Fechine GJM (2021) Molybdenum disulfide as a filler for a polymeric matrix at an ultralow content: polystyrene case. *Polym Test.* 93: 106882.
16. Pfeifer S, Bandaru PR (2014) A methodology for quantitatively characterizing the dispersion of nanostructures in polymers and composites. *Mater Res Lett.* 2(3): 166–175.
17. Soliman TS, Vshivkov SA (2019) Effect of Fe nanoparticles on the structure and optical properties of polyvinyl alcohol nanocomposite films. *J Non-Cryst Solids.* 519: 119452.
18. Ahmed H, Abduljalil HM, Hashim A (2019) Structural, optical and electronic properties of novel (PVA–MgO)/SiC nanocomposites films for humidity sensors. *Trans Electr Electron Mater.* 20: 218–232.
19. Okail MA, Alsaleh NA, Farouk WM et al. (2021) Effect of dispersion of alumina nanoparticles and graphene nanoplatelets on microstructural and mechanical characteristics of hybrid carbon/glass fibers reinforced polymer composite. *J Mater Res Tech.* 14: 2624–2637.
20. Zhang R, Lan W, Ji T et al. (2021) Development of polylactic acid/ZnO composite membranes prepared by ultrasonication and electrospinning for food packaging. *LWT.* 135: 110072.
21. Choudhary S (2018) Characterization of amorphous silica nanofiller effect on the structural, morphological, optical, thermal, dielectric and electrical properties of PVA–PVP blend based polymer nanocomposites for their flexible nanodielectric applications. *J Mater Sci: Mater Electron.* 29: 10517–10534.
22. Badawi A (2020) Engineering the optical properties of PVA/PVP polymeric blend in situ using tin sulfide for optoelectronics. *Appl Phys A.* 126: 335.
23. Rajesh K, Crasta V, Kumar NBR et al. (2019) Structural, optical, mechanical and dielectric properties of titanium dioxide doped PVA/PVP nanocomposite. *J Polym Res.* 26: 99.
24. Liu H, Webster TJ (2011) Enhanced biological and mechanical properties of well-dispersed nanophase ceramics in polymer composites: from 2D to 3D printed structures. *Mater Sci Eng C.* 31: 77–89.

25. Kuratani K, Ishibashi K, Komod Y, Hidema R, Suzuki H, Kobayashi H (2019) Controlling of dispersion state of particles in slurry and electrochemical properties of electrodes. *J Electrochem Soc.* 166(4): A501–A506.

26. Liu T, Zhong J, "Effect of dispersion of nano-inorganic particles on the properties of polymer nanocomposites", *IOP Conf. Ser: Mater Sci Eng.* vol. 563, p. 022026, 2019.

27. Huy VPH, So S, Hur J (2021) Inorganic fillers in composite gel polymer electrolytes for high-performance lithium and non-lithium polymer batteries. *Nanomaterials (Basel).* 11(3): 614.

28. Din Mir Mehraj Ud, Häusler M, Fischer S M et al. (2021) Role of filler content and morphology in LLZO/PEO membranes. *Front Energy Res.* 9: 711610.

29. Pullanjiot N, Swaminathan S (2019) Enhanced electrochemical properties of metal oxide interspersed polymer gel electrolyte for QSDSSC application. *Sol Energy.* 186: 37–45.

30. Angastiniotis NC, Christopoulos S, Petallidou KC et al. (2021) Controlling the optical properties of nanostructured oxide-based polymer films. *Sci Rep.* 11: 16009.

31. Biswas M, Ray SS, "Recent progress in synthesis and evaluation of polymer-montmorillonite nanocomposites". In: *New Polymerization Techniques and Synthetic Methodologies*, Part of Advances in Polymer Science book series, vol. 155, pp. 167–221, 2001.

32. Alexandre M, Dubois P (2000) Polymer-layered silicate nanocomposites: preparation, properties and uses of a new class of materials. *Mater Sci Eng: R Reports.* 28: 1–63.

33. Karimi A, Wan Daud WMA (2017) Materials, preparation, and characterization of PVA/MMT nanocomposite hydrogels: a review. *Poly Comp.* 38: 1086–1102.

34. Tsunoji N, Fukuda M, Yoshida K et al. (2013) Characterization of layered silicate HUS-5 and formation of novel nanoporous silica through transformation of HUS-5 ion-exchanged with alkylammonium cations. *J Mater Chem A.* 1: 9680–9688.

35. Madejová J, Barlog M, Jankovic L et al. (2021) Comparative study of alkylammonium-and alkylphosphonium-based analogues of organo-montmorillonites. *Appl Clay Sci.* 200: 105894.

36. Gomes CS, Rautureau M, "General data on clay science, crystallochemistry and systematics of clay minerals, clay typologies, and clay properties and applications". In *Minerals latu sensu and Human Health*. Springer, Cham. pp. 195–269, 2021.

37. Huang A, Wang H, Ellingham T et al. (2019) An improved technique for dispersion of natural graphite particles in thermoplastic polyurethane by sub-critical gas-assisted processing. *Compos Sci Technol.* 182: 107783.

38. Coleman JN, Lotya M, O'Neill A et al. (2011) Two-dimensional nanosheets produced by liquid exfoliation of layered materials. *Science.* 42: 568–571.

39. Zestos AG (2018) Carbon nanoelectrodes for the electrochemical detection of neurotransmitters. *Int J Electrochem.* 2018: 3679627.

40. Patil MB (2020) Hybrid nanocomposite membranes of poly (vinyl alcohol) and cerium oxide for pervaporation dehydration of ethanol at their azeotropic point. *Int J Adv Sci Eng.* 6: 1472–1475.

41. Xiong M, Gu G, You B et al. (2010) Preparation and characterization of poly(styrenebutylacrylate) latex/nano-ZnO nanocomposites. *J Appl Poly Sci.* 90: 1923–1931.

42. Li YQ, Pan QY, Li M et al. (2007) Preparation and mechanical properties of novel polyimide/T-silica hybrid films. *Compos Sci Technol.* 67: 54–60.

43. Li Z, Zhang Y, Liang S et al. (2018) Effect of surfactant on microstructure, surface hydrophilicity, mechanical and thermal properties of different multi-walled carbon nanotube/polystyrene composites. *Mater Res Exp.* 5: 055035.

44. Yeung R, Zhu X, Gee T et al. (2020) Single and binary protein electroultrafiltration using poly (vinyl-alcohol)-carbon nanotube (PVA-CNT) composite membranes. *PLoS One*, 15: e0228973.

45. Yang JP, Chen ZK, Feng QP et al. (2012) Cryogenicmechanical behaviors of carbon nanotube reinforced composites based onmodified epoxy by poly(ethersulfone). *Compos Part B.* 43: 22–26.

46. Ou Y, Yang F, Yu Z (2015) A new conception on the toughness of nylon 6/silicananocomposite prepared via in situ polymerization. *J Polym Sci B Polym Phys.* 36: 789–795.

10 Application of Nanoparticle-Reinforced Polymeric Composites in Bioscience and Medicinal Disciplines
A Review

Shikha Rana and Mahavir Singh

CONTENTS

10.1 INTRODUCTION

Nanoparticle-polymer composites (NPCs) are a group of matrix-based "smart" substances which has numerous applications with desired mechanical, physiochemical, optical, and electrical properties [1]. NPC exists in 1, 2, or 3 nanodimensional phases of organic/inorganic nanofillers of different types like carbon nanotube (CNT), metal oxides, cellulose, hydroxyapatite, silica, graphene oxide (GO), etc. [2]. They have shown improved performance than other types of polymer composites which have micro- and macrodimensional fillers. The distinctive characteristics of NPC depend on the interplay between individual features of polymers and nanofilling which is also termed as "nano-effect" [3].

Generally, nanocomposites are synthesized by combining different phases of materials out of which any one intermixed phase is of nanodimension. Nanoparticles have shape and critical size and phase interactions-dependent optical, magnetic, and mechanical characteristics, and hence, synthesis of nanoparticles with desired shape and size becomes a need for smart material engineering. Nanocomposites are divided into three main groups depending on the used matrix type, i.e., polymer, metal, or ceramic-nanocomposites [4]. Among these types of nanocomposites, polymer-nanocomposites have maximum application in biomedical fields due to similarities between synthetic and natural polymer composites. The human body is made up of biopolymers like proteins

DOI: 10.1201/9781003343912-10

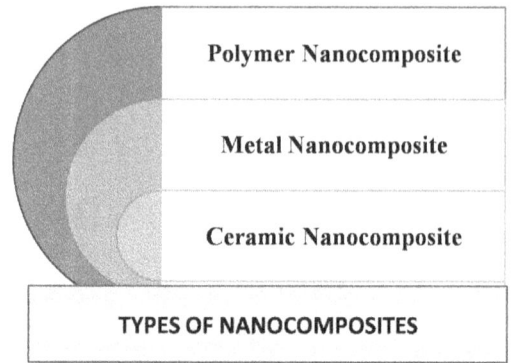

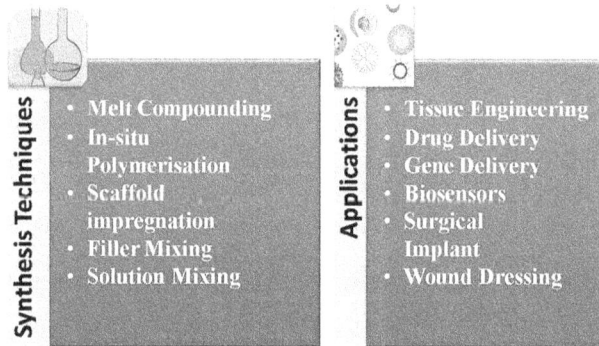

FIGURE 10.2 Synthesis techniques and applications of NPC.

and polysaccharides, and some of the synthetic/natural polymers like polyesters, etc. are found to possess the same characteristics [5]. Biomaterial-based NPC is accomplished with physiochemical and biological characteristics required for the preparation of smart bioengineered material. In the beginning, inorganic nanoparticles were embedded in the polymer matrix just to improve the basic characteristics of polymers [6]. Types of nanocomposites and inorganic fillers in polymer-based nanocomposites are listed in Figure 10.1. Encapsulation efficiency, loading capacity, degradation rate, and morphology of the natural or synthetic nanostructured polymeric substances play a vital role in NPC applications [7].

Here, this chapter presents basic aspects of the NPC synthesis, their desired properties, and their corresponding application in biomedical fields like externally guided drug delivery, bioimaging, tissue engineering, and hyperthermia as illustrated in Figure 10.2. It is mainly focused on the decisive synthesis techniques, and biofeatures of NPC are required for their influential applications in modern medical areas.

10.2 SYNTHESIS AND BIOLOGICAL ASPECTS

Polymers are long-chain matrices that are prepared by the polymerization of small blocks known as monomers. Polymers are compounded with other additives to get optimized and noninherent properties. NPC is generally synthesized by combining inorganic and organic substances as filler and matrix, respectively. Inorganic filler materials are generally functionalized to attain homogeneous mixing of filler in polymer matrices with the least agglomeration and better surface bonding among them [8]. Polymerization, various types of intercalation, and scaffold impregnation are some important types of synthesis techniques employed for NPC preparation as shown in Figure 10.2. In situ polymerization and melt, intercalation is the generally used synthesis technique for CNT and silicate-based NPC preparation [9]. Intercalation is one of the earliest used methods for NPC synthesis which involves the fabrication of an inorganic-layered polymer guest-host matrix either by chemical or mechanical routes. Solution mixing and in situ polymerization are two types of this method in which functionalized nanoparticle-based monomers are injected or mixed in the inorganic-layered structure and where these monomers show polymerization. In situ, polymerization can synthesize partially exfoliated NPC with the advantage of polymers grafting on the nanosurface of filler material. This method leads to the synthesis of simple intercalated to exfoliated, different types of NPC structures which vary with polymer penetration into silicate layer structure [10]. Melt intercalation is an eco-friendly high-temperature type of intercalation employed for the synthesis of thermoplastic NPC. This process needs to be optimized for good dispersion of filler in the polymer matrix, according to the type and compatibility of filler as high temperature sometimes destroys bonding between functionalized groups on the nanofiller surface. Template preparation includes easy in situ growth of inorganic-layered structure in a hydrophilic polymer solution. It is generally used for the synthesis of hydrophilic NPC at a large scale but is not favorable for clay- or silicate-based NPC [11]. The sol-gel process is a low-temperature two-step synthesis technique employed for NPC synthesis in which polymerization of colloidal solution of nanoparticles and monomers leads to gel formation of NPC. The sol-gel process leads to the formation of three-dimensional multiple matrices metal-based NPC. NPC can also be synthesized by direct mixing of polymers and nanofillers either by nonsolvent or solvent-mixing [12].

NPC is used in various biomedical applications as bioactive membranes, easily degradable products, and restorative implants due to their biocompatible characteristics. To enhance the compatibility between polymer matrices and native cells, NPC is functionalized with various bioactive molecules. Mohammad Shakir et al. fabricated a biodegradable and biocompatible NPC based on chitosan. They investigated mechanical, morphological, in vitro swelling, degradation, cytocompatibility, and hemocompatibility characteristics of the prepared samples. From the results, they confirmed that the samples have the potential to be used in tissue engineering as they have good mechanical strength and cytocompatibility, and also show in vitro enzymatic degradation [13]. F.S. Senatov et al. synthesized polyethylene-based NPC and investigated its mechanical and morphological characteristics, and in vitro and in vivo biocompatibility. They concluded that the prepared samples have excellent mechanical and biocompatible properties and are potentially applicable in cartilage replacement [14].

Several types of illnesses and contaminations in nature are caused by bacterial infections. NPC doped with ions of silver and gold-like metals is known for its antibacterial characteristics, which rely upon certain factors like surface oxidation and dispersion of particles. Silver-doped NPC not only possesses antibacterial characteristics but also helps in the identification and monitoring of bacteria-induced biochemical reactions. Many research articles are available for metal or metal oxide-doped NPC in which they have shown extraordinary antibacterial behavior against *S. aureus*, *E. coli*, *Pseudomonas aeruginosa*, and *Bacillus subtilis*-like bacteria [15]. Zeynep Kalaycıoğlu et al. fabricated cesium oxide-based chitosan NCP by solvent-casting technique. They found that the antibacterial behavior of the prepared samples depends directly on the nanoparticles' concentration [16]. Some of the latest synthesis techniques and applications of NPC are listed in Table 10.1.

Graphene and its derivatives have shown excellent antibacterial properties as nanofiller in NPC. Their antibacterial behavior is explained by chemical and physical mechanisms which involve direct bonding with bacterial membrane, generation of reactive oxygen species, indirect entrapment efficiency, and lipid extraction [26–28]. Ana M. Díez-Pascual et al. reviewed the antibacterial activity of graphene and its derivatives and provide a better understanding of possible mechanisms of antibacterial characteristics [29].

TABLE 10.1

A List of Latest NPC, Synthesis Details, and Applications

NPC Type	Synthesis Details	Application	Reference
PMMA, PGMA and PMSEMA @BaTiO$_3$	Surface-initiated reversible addition–fragmentation chain transfer polymerization	Energy storage	[17]
BN/PEI@TiO$_2$	Liquid-phase exfoliation by spin coating	Energy storage	[18]
Starch/PVA@copper/chitosan/ cotton NPC	Gamma-radiation grafted by padding	Antibacterial studies	[19]
PVA@ZnO$_2$/MWCNT	Ultrasonically dispersed	Food packaging	[20]
TiO$_2$-PVDF/PMMA	Phase inversion	Cross-flow filtration pilot photocatalytic waste water treatment	[21]
Conjugated microporous polymers CMP@TiO$_2$	In situ Sonogashira polymerization	Cytocompatibility and antibacterial studies	[22]
PE@TiO2/ZnO	Homogeneous mixing	Biomedical applications	[23]
CsPbBr$_3$@polystyrene-block- poly(acrylic acid) (PS-b-PAA)	Microemulsion Self-assembling	Bioimaging	[24]
Glycidyl methacrylate (GMA)@Fe3O$_4$@TEOS	Graft polymerization	Cytotoxicity	[25]

10.3 APPLICATIONS

10.3.1 HYPERTHERMIA

Cancer is one of the leading death causes throughout the world, and oncologists are employing functionalized nanoparticles to enhance the efficiency rates of various treatments [30]. Among the treatment methods for cancer, hyperthermia is one earliest designed and most commonly used promising method to control and treat carcinoma cells. It includes the treatment of cancer cells with high temperatures. It is also found to enhance the efficiency of other cancer treatments like chemotherapy and radiation therapy. The temperature between 41°C and 46°C has proven to be fatal for cancer cells and decreases their viability. The working temperature range for hyperthermia is acquired by radiofrequency, ultrasound, and microwave heating of the tumor cells [31]. Among all these heating techniques, radiofrequency heating is the most effective one in terms of specific absorption rate and thermal penetration depth. It is categorized in three ways depending on the type, size, and location of the tumor local, regional, and whole-body heating. Local, regional, and whole-body heating are employed for near-surface, deeply located, and metastatic tumors, respectively [32]. Hyperthermia raises the temperature near the tumor site which results in an increase in the blood flow and oxygen delivery to the tumor site and hence increases the possibility of drug delivery and radiation killing too. Magnetically induced hyperthermia is the best nanoparticle-based hyperthermia as magnetically active nanoparticles are controlled externally by an applied magnetic field and could generate local heat in the required tumor site due to hysteresis loss in magnetic dipoles. Magnetic heat induction by magnetic nanoparticles is measured in terms of specific absorption rate (SAR) or specific loss power (SLP). To avoid agglomeration in magnetic nanoparticles, they are embedded in the biodegradable and biocompatible polymer matrices.

Rafael S. Moraes et al. prepared magnetoactive NPC of poly(butylene succinate) (PBS) by solvent-mixing technique and investigated physiochemical, magnetic, and induction heating characteristics of the prepared samples. They concluded that the samples have shown high SAR and heat-generation properties after mixing nanoparticles in the polymer matrix as required in magneto-hyperthermia [33]. Ana Barra et al. synthesized magnetic NPC by one-pot solvent-casting method in chitosan and reduced graphene oxide (rGO) poly matrix. They investigated the behavior of the prepared samples in hyperthermia and calculate the value of specific loss power. They found with the rise in rGO concentration, heating efficiency increases too [34]. Zhi Wei Lim and his coresearchers studied the chemo-drug-loading efficiency of prepared biomimetic magnetic NPC for combined hyperthermia and chemotherapy for the treatment of HepG2 liver cancer cells [35]. Mehdi Hadjianfar et al. investigated and synthesized a magnetic polycaprolactone (PCL)-based NPC by employing electro-spinning for thermal treatment. They analyzed the physiochemical and heat performance of the prepared samples and found them good for heat generation at the low-alternating magnetic frequency [36]. G.R. Iglesias and his coworkers prepared a triggered drug-release magnetic NPC for possible effective multimode treatment of tumors. They found that the kinetics of gemcitabine drug release got better with an increase in temperature and change in the magnetic field as shown in Figure 10.3, which confirms the potential use of the prepared sample in temperature-triggered chemotherapy [37].

Ricardo J.R. Matos and his coworkers prepared cellulose-based magnetoactive NPC for hyperthermia application by electro-spinning. They analyzed physicochemical, mechanical, cytotoxicity, and heating ability characteristics of the prepared samples and concluded that the samples have no cytotoxicity and showed better mechanical characteristics with nanoparticles doping into the polymer matrix [38]. Seyed Ashkan Moghadam Ziabari et al. fabricated magnetite-based NPC by coprecipitation method followed by mixing with bentonite. They studied the effect of change in the magnetic field on the synthesis process and specific loss of power [39]. Three-dimensional hydrophilic polymer structures are named hydrogels which possess high biocompatibility and good enzymatic degradability. Vanessa Zamora-Mora et al. fabricated a chitosan-based hydrogel for hyperthermia application by mixing biopolymer with magnetic nanoparticles [40]. Sayan Ganguly and his coworker reviewed the fabrication processes and physicochemical and magnetic characteristics of magnetoactive hydrogels for various biomedical applications [41].

10.3.2 Targeted Drug Delivery

Biomedicine includes and investigates the basics of biochemistry to improve the effectiveness of drugs and health quality. In biomedicine, the therapeutic efficiency of a drug depends on its specific

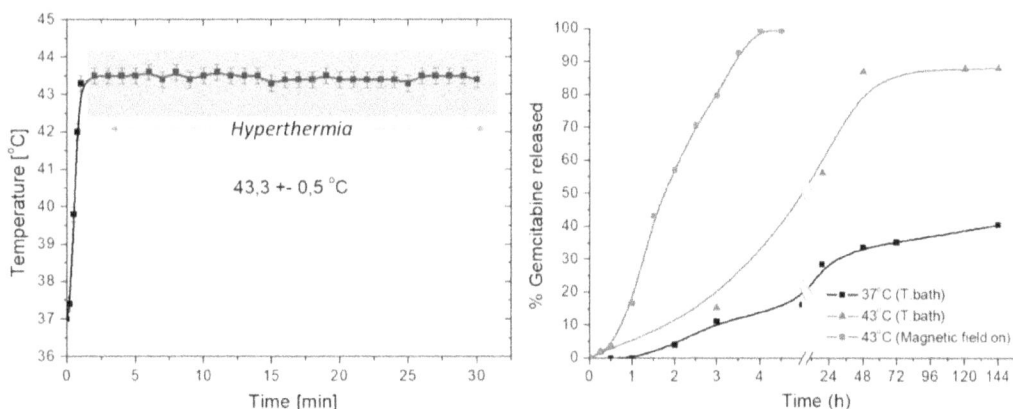

FIGURE 10.3 (a) Drug-release profile concerning change in temperature and time and (b) drug-release profile at different temperatures and in the presence of the magnetic field. (Reproduced from Reference [37] (Published under http://creativecommons.org/licenses/by/4.0/).)

delivery, release rate, time of exposure, and potential toxicity [42]. Conventionally injected drugs are less effective and have low targeted delivery rates. Generally, a drug molecule has to cross physiological and biochemical barriers to reach the chosen site. Targeted drug delivery (TDD) helps in the delivery of optimum drug dose at the specific site and decreases undesirable side effects. TDD is of two types: active targeted drug delivery (ATDD) and passive targeted drug delivery (PTDD). ATDD is a demand drug-release system in which triggers are used to govern the drug dose in a therapy. These stimuli involve temperature and pHs like internal factors and electric field, magnetic field, ultrasonic, and optical-like external factors as illustrated in Figure 10.4. This system can provide control on the drug-release phenomena in the human body but as the drug is injected intravenously; it leads to more drug accumulation and toxicity. PTDD is a sustained drug-release system in which nanofunctionalized groups are attached to drugs for site-specific drug delivery. This system includes composites of artificial or biopolymers with nanoparticles and provides less drug accumulation in the human body [43]. For efficient TDD, drug and carrier systems are prepared by doing an amalgamation of nanocarrier with drugs which provide easy control over drug movement inside the body and its specific delivery. The dynamical characteristics of a drug like solubility and activity are improved by conjugating drugs with nanocarriers.

NPC poses different features like inertness, biodegradability, and nontoxic nature which are required for purposeful TDD. NPC consists of either natural or synthetic polymers and provides more stability with increased hydrophilic kinetic behavior to the conjugated drugs. Protein and polysaccharide-based natural polymers are more favorable for in vivo TDD as they have biocompatible nature. The characteristics of NPC like solubility, drug loading, and encapsulation efficiency are easily tailored by the applicative area. Alginate, cellulose, starch, chitosan, gelatin, silk, and albumin-like natural biopolymers are employed for the preparation of NPC-based drug delivery vehicles [44]. Besides, its biocompatible and biodegradable nature NPC possesses good antibacterial, antimicrobial, and anticancer properties too. Poly(lactic-co-glycolic acid) (PLGA) is a nontoxic, hydrophilic, amorphous, and biodegradable polymer that possesses the property of sustained drug release attributed to these properties they are used in many therapeutic applications. Its degradation rate depends on polyglycolic acid ratio, composition, crystallinity, and adhesion [45]. In another study on PLGA, Paola Allavena and his coworkers synthesized PLGA-based NPC and investigated the prepared sample for in vivo drug delivery of doxorubicin at the tumor site in rats. They also confirmed effective in vitro drug delivery to monocytes by the prepared NPC [46].

Recently, Paula Ecaterina Florian et al. synthesized nano-clay-salecan-based NPC by polymerization method and investigated various physiochemical, cytotoxicity, cell proliferation, and internalization-like characteristics. They also studied the doxorubicin loading capacity of the prepared

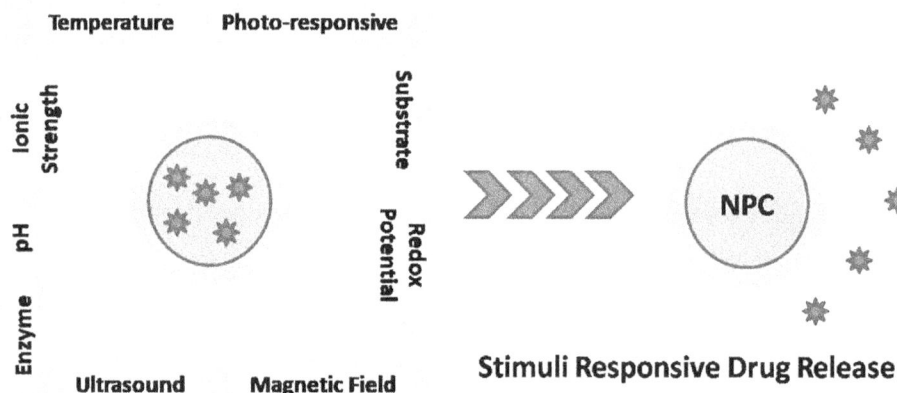

FIGURE 10.4   Illustrating a stimuli-responsive drug-release system based on NPC.

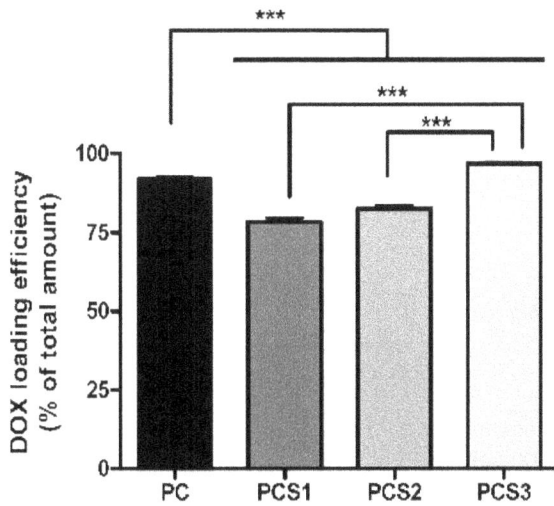

FIGURE 10.5 Drug-loading efficiency of all the prepared samples with different sale-can ratios. (Reproduced from Reference [47] (Published under http://creativecommons.org/licenses/by/4.0/).)

samples as shown in Figure 10.5 for all the prepared samples with different salecan ratios and found with salecan concentration drug-loading increases. From the obtained results, they also confirmed that the salecan ratio affects the microstructural properties, cell viability, and cytotoxicity of the samples and could act as a drug delivery vehicle for chemotherapeutic drugs [47].

10.3.3 BIOIMAGING

Bioimaging based on NPC provides new dimensions to molecular imaging and tremendously helped researchers in the cancer diagnosis. The nanosize of NPC helps in deep penetration and effective detection of the tumors. Recently, NPC of hydrophobic nature synthesized by various preparation techniques including nanoprecipitation and emulsification of a semiconducting polymer with nanoparticles is employed in the fluorescence imaging. Gaiying Lei and his coworkers have synthesized quantum dot-based NPC. They concluded that the prepared samples are capable to show better imaging due to easy cell endocytosis in the absence of transfection reagent [48]. Mojue Zhang et al. have synthesized a dual-mode operating NPC system by dehydration condensation. They found that the samples are of biocompatible and low toxic nature and could be used for in vivo photo thermal therapy-assisted bioimaging [49]. Tao Deng and his coworkers have synthesized carbon dot-based NPC and investigated in vivo and in vitro drug release by using the prepared samples [50]. Recently, L. Huang et al. has prepared a silica-based polymeric nanocomposite and explores their dispersibility and toxicity characteristics as fluorescent-based cell imaging material [51]. Magnetic material can be used in MRI-based imaging techniques and K. Mandal et al. has prepared such type of magnetically active NPC. They reported that the prepared samples have better colloidal and fluorescence stability with surface charges and amines surface groups which could be utilized for easy biomolecules labeling during imaging process [52].

10.3.4 TISSUE ENGINEERING

Tissue engineering (TEG) is an attractive branch of bioengineering since 1988 when this term was first coined in the USA. It presents an amalgamation of basic science concepts and engineering knowledge to treat damaged tissues and organs [53]. Also, NPCs are engineered substrates with desired mechanical strength and physiochemical properties which provide biodegradable 3-dimensional artificial cell support matrices for maintenance, restoration, and stimulated growth of tissues in the human body in TEG [54]. For in vitro treatment of defected and damaged tissues, biodegradable three-dimensional scaffolds are employed for cell culture growth as shown in Figure 10.6 and

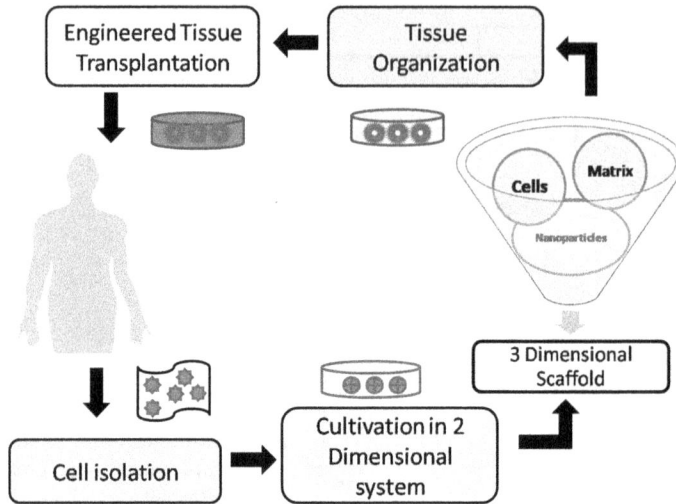

FIGURE 10.6   Schematic of NPC scaffold use in TEG.

later used as cell support matrices in the human body. TEG provides a transitional hybrid substitute of the inherent extracellular matrix with analogous mechanical characteristics. Biomaterials used as a substrate should be nontoxic and biodegradable, as after completion of the tissue repair process, these artificial substrates should be removed for remodeling of repaired tissues. Thus, the realization of hybrid material as a substrate in TEG is an impactful process that includes intense study of the basic characteristics and possible functionalization techniques. The required characteristics are fulfilled by nanofilled polymers which are also known as NPC. Many natural/synthetic biodegradable polymers like chitosan, CNT, gelatin, silk, PLGA, dendrimers, PVDF, hydroxyapatite, silicates, PCL, poly(lactic acid) (PLA), PBS, etc. are employed in the formation of NPC for TEG [55]. The synthesis of NPC-based scaffolds also depends on their degradation rate during the replacement stage, and this rate should be similar to the growth rate of new tissues for proper replacement of damaged tissues. Naturally biodegradable polymers have better biofunctionality and degradation rate but less mechanical strength; to overcome these disadvantages, synthetic polymers are incorporated in NPC scaffolds. Biodegradable scaffolds disintegrate into low-weight substructures by following various chemical/enzymatic reactions and are later on expelled out from the body. Their degradation rate is governed by polymer molecular weight, reaction medium conditions, bonding efficiency, and nanoparticle composition, shape, and size. NPC-based engineered scaffolds are prepared by various procedures like molding, casting, electro-spinning, etc. The different types of human organs are made up of diversified cellular matrices, so NPC-based scaffolds are specially designed to imitate native cells for proper cell proliferation [56].

Amir Sotoudeh et al. investigated viscoelastic characteristics of PLA-based nanocomposite for TEG application. The nonuniformity and anisotropy found in tissues lead to their arduous growth. To overcome these types of problems, biomimetic NPC is designed and implanted in TEG [57].

Graphene oxide is one of the types of oxidized two-dimensional graphene. It has a good solution stability, antibacterial activity, conductivity, and self-healing characteristics [58]. Muhammad Umar Aslam Khan and his coworkers studied an NPC consisting of hydroxyapatite and GO for bone TEG and found that this composite when coated with silver nanoparticles possessed better antibacterial and physiochemical properties. As shown in Figure 10.7, they found that with the increase in GO content for three different samples, mechanical properties got better which is attributed to better interfacial bonding between polymer matrix and nanoparticles. Also, they studied scaffold swelling and biodegradability for different media as given in Figure 10.8 and concluded that with GO concentration

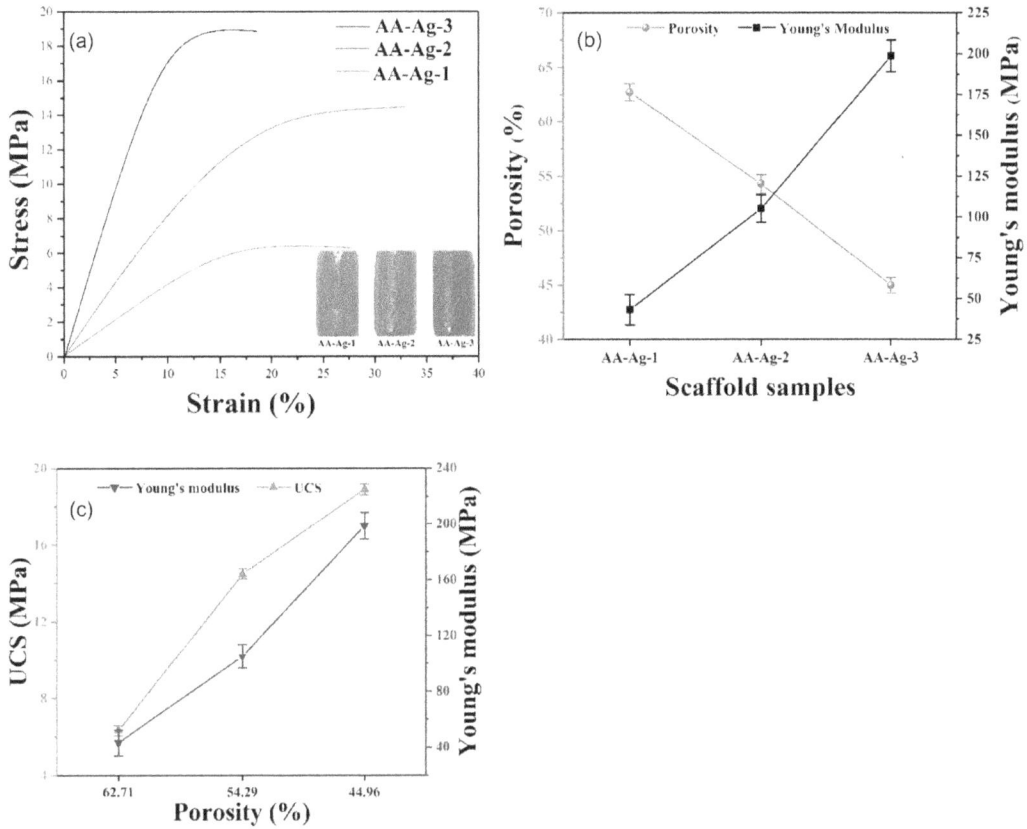

FIGURE 10.7 (a) Stress-strain curve, (b) porosity and Young's modulus curve, and (c) compression and Young's modulus curve for NPC scaffolds. (Reproduced from Reference [59]. (Published under http://creative-commons.org/licenses/by/4.0/).)

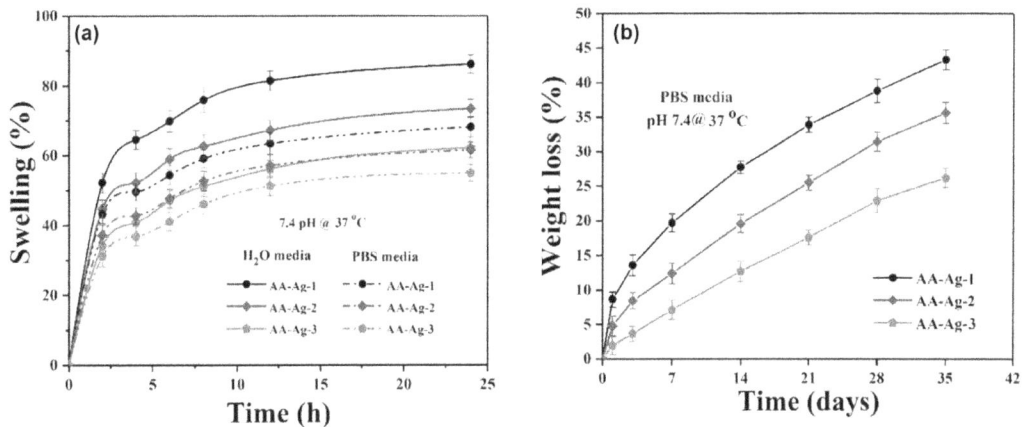

FIGURE 10.8 (a) Swelling and (b) degradation behavior of all the prepared NPC scaffolds in different media, respectively. (Reproduced from Reference [59]. (Published under http://creativecommons.org/licenses/by/4.0/).)

both the characteristics got better due to more functionality of the polymer matrix. The prepared NPC scaffold has shown better cell adhesion which shows its potential in bone TEG [59]. In another study, they employed the freeze-drying technique for hydroxyapatite and TiO_2-based NPC scaffold preparation. They investigated the physiochemical and biocompatible characteristics of the prepared samples and concluded their good cell attachment for MC3T3-E1 cells with better mechanical strength [60].

Magnetic nanomaterials are currently employed in all types of biomedical applications attributed to their easy separation and governance from the externally applied magnetic field [61,62]. Margarida M. Fernandes et al. have investigated a three-dimensional magnetically active NPC as a potential scaffold in bone tissue engineering. They used the solvent-casting technique to prepare a porous scaffold by mixing cobalt ferrite nanoparticles with PVDF in various ratios [63]. In the last few decades, metal oxides have shown exemplary applications in electronics, wastewater treatment, energy storage, and medical fields due to their low cost, surface exchange interactions, quantum confinement, and structural defects characteristics [64]. M. Rasoulianboroujeni and his coworkers have reported successful preparation of metal oxide and PLGA-based NPC scaffold by the 3D printing process. They investigated cell proliferation and hydrophilicity characteristics of the prepared samples and concluded their potential in bone TEG [65]. Balu Kolathupalayam Shanmugam et al. used the solvent-casting technique for the synthesis of TiO_2-based NPC scaffolds with different ratios of chitosan and sodium alginate. They investigated in vitro cell viability, biocompatibility, and microstructural, antibacterial, and thermal degradation characteristics of the prepared scaffolds and concluded that samples showed better cytotoxicity and antiproliferative characteristics which confirm its potential application as a scaffold in bone TEG [66].

Zahra Pedram Rad and his coworkers prepared PCL-based NPC with different ratios of polymer and nanofiller by employing the solvent-mixing method. They investigated the hydrophilicity, porosity, mechanical strength, cell viability, and antibacterial characteristics of the prepared samples and concluded that they have improved results with PCL addition in the nanocomposite of zein and gum arabic [67]. Rabia Kouser et al. have synthesized CNT-based NPC by solution mixing and investigated the microstructural, mechanical, biocompatible, and biocompatible characteristics of the samples. They found that with the addition of CNT, all these characteristics got better and confirm its potential application in TEG [68].

10.4 CONCLUSION

Nanoparticle-polymer composites exhibit a large surface-to-volume ratio, biocompatibility, biodegradability, and high conductivity, like characteristics due to embedded nanoparticles in the polymer matrix. This chapter focused on the basic synthesis techniques, biological aspects, and application of these composites in biomedical areas. As described in this chapter, many synthesis techniques are used for the fabrication of these composites with desired properties, but every method has to be optimized according to the desired application. The characteristics of NPC are tailored easily by amalgamating the electrical, catalytical, antibacterial, and magnetic behavior of nanoparticles with the biodegradability and biocompatibility of polymers. Recently, NPC is designed to perform with least toxicity in multimode treatments body as a stimuli-responsive system. Certainly, it can be concluded that NPCs have a strong influence on drug delivery and heat-generation systems and tissue growth, repair, and replacement processes. Therefore, extensive research is still needed in understanding the functionalization of nanoparticles and the surface interactions among nanoparticles and polymers.

REFERENCES

1. Y. Zare and I. Shabani, Polymer/metal nanocomposites for biomedical applications, *Mater. Sci. Eng. C* 60 (2016), pp. 195–203.
2. W. Qi, X. Zhang and H. Wang, Self-assembled polymer nanocomposites for biomedical application, *Curr. Opin. Colloid Interface Sci.* 35 (2018), pp. 36–41.

3. P.N. Kendre, M. Gite, S.P. Jain and A. Pote, Nanocomposite polymeric materials: State of the art in the development of biomedical drug delivery systems and devices, *Polym. Bull.* 79 (2021), pp. 9237–9265.

4. R. Xiong, A.M. Grant, R. Ma, S. Zhang and V. V. Tsukruk, Naturally-derived biopolymer nanocomposites: Interfacial design, properties and emerging applications, *Mater. Sci. Eng. R Reports* 125 (2018), pp. 1–41.

5. P. Matricardi, C. Di Meo, T. Coviello, W.E. Hennink and F. Alhaique, Interpenetrating polymer networks polysaccharide hydrogels for drug delivery and tissue engineering, *Adv. Drug Deliv. Rev.* 65 (2013), pp. 1172–1187.

6. U. Dubey, S. Kesarwani and R.K. Verma, Incorporation of graphene nanoplatelets/hydroxyapatite in PMMA bone cement for characterization and enhanced mechanical properties of biopolymer composites, *J. Thermoplast. Compos. Mater.* 0 (2022), pp. 1–31.

7. D. Bharatiya, S. Patra, B. Parhi and S.K. Swain, A materials science approach towards bioinspired polymeric nanocomposites: A comprehensive review, *Int. J. Polym. Mater. Polym. Biomater.* 0 (2021), pp. 1–16.

8. V. Mittal, *Synthesis Techniques for Polymer Nanocomposites*, WILEY-VCH, Weinheim; Baden-Wurttemberg, Germany, (2014), pp. 1–30.

9. C. Harito, D.V. Bavykin, B. Yuliarto, H.K. Dipojono and F.C. Walsh, Polymer nanocomposites having a high filler content: Synthesis, structures, properties, and applications, *Nanoscale* 11 (2019), pp. 4653–4682.

10. P.H.C. Camargo, K.G. Satyanarayana and F. Wypych, Nanocomposites: Synthesis, structure, properties and new application opportunities, *Mater. Res.* 12 (2009), pp. 1–39.

11. K.S. Kumar, V.B. Kumar and P. Paik, Recent advancement in functional core-shell nanoparticles of polymers: Synthesis, physical properties, and applications in medical biotechnology, *J. Nanoparticles* 2013 (2013), pp. 1–24.

12. S. Kalia, S. Kango, A. Kumar, Y. Haldorai, B. Kumari and R. Kumar, Magnetic polymer nanocomposites for environmental and biomedical applications, *Colloid Polym. Sci.* 292 (2014), pp. 2025–2052.

13. M. Shakir, I. Zia, A. Rehman and R. Ullah, Fabrication and characterization of nanoengineered biocompatible n-HA/chitosan-tamarind seed polysaccharide: Bio-inspired nanocomposites for bone tissue engineering, *Int. J. Biol. Macromol.* 111 (2018), pp. 903–916.

14. F.S. Senatov, M.V. Gorshenkov, S.D. Kaloshkin, V.V. Tcherdyntsev, N.Y. Anisimova, A.N. Kopylov et al., Biocompatible polymer composites based on ultrahigh molecular weight polyethylene perspective for cartilage defects replacement, *J. Alloys Compd.* 586 (2014), pp. S544–S547.

15. H. Moustafa, A.M. Youssef, N.A. Darwish and A.I. Abou-Kandil, Eco-friendly polymer composites for green packaging: Future vision and challenges, *Compos. Part B Eng.* 172 (2019), pp. 16–25.

16. Z. Kalaycıoğlu, N. Kahya, V. Adımcılar, H. Kaygusuz, E. Torlak, G. Akın-Evingür et al., Antibacterial nano cerium oxide/chitosan/cellulose acetate composite films as potential wound dressing, *Eur. Polym. J.* 133 (2020), pp. 1–7.

17. H. Li, L. Wang, Y. Zhu, P. Jiang and X. Huang, Tailoring the polarity of polymer shell on $BaTiO_3$ nanoparticle surface for improved energy storage performance of dielectric polymer nanocomposites, *Chinese Chem. Lett.* 32 (2021), pp. 2229–2232.

18. P. Wang, L. Yao, Z. Pan, S. Shi, J. Yu, Y. Zhou et al., Ultrahigh energy storage performance of layered polymer nanocomposites over a broad temperature range, *Adv. Mater.* 33 (2021), pp. 1–7.

19. S.N. Saleh, M.M. Khaffaga, N.M. Ali, M.S. Hassan, A.W.M. El-Naggar and A.G.M. Rabie, Antibacterial functionalization of cotton and cotton/polyester fabrics applying hybrid coating of copper/chitosan nanocomposites loaded polymer blends via gamma irradiation, *Int. J. Biol. Macromol.* 183 (2021), pp. 23–34.

20. Y.H. Wen, C.H. Tsou, M.R. de Guzman, D. Huang, Y.Q. Yu, C. Gao et al., Antibacterial nanocomposite films of poly(vinyl alcohol) modified with zinc oxide-doped multiwalled carbon nanotubes as food packaging, *Polym. Bull.* 79 (2022), pp. 3847–3866.

21. K.B. Errahmani, O. Benhabiles, S. Bellebia, Z. Bengharez, M. Goosen and H. Mahmoudi, Photocatalytic nanocomposite polymer-TiO_2 membranes for pollutant removal from wastewater, *Catalysts* 11 (2021), pp. 1–15.

22. Y. Wu, Y. Zang, L. Xu, J. Wang, H. Jia and F. Miao, Synthesis of high-performance conjugated microporous polymer/TiO_2 photocatalytic antibacterial nanocomposites, *Mater. Sci. Eng. C* 126 (2021), pp. 112121.

23. N.H. Harun, R.B.S.M.N. Mydin, S. Sreekantan, K.A. Saharuddin and A. Seeni, In vitro bio-interaction responses and hemocompatibility of nano-based linear low-density polyethylene polymer embedded with heterogeneous TiO_2/ZnO nanocomposites for biomedical applications, *J. Biomater. Sci. Polym. Ed.* 32 (2021), pp. 1301–1311.

24. S.K. Avugadda, A. Castelli, B. Dhanabalan, T. Fernandez, N. Silvestri, C. Collantes et al., Highly emitting perovskite nanocrystals with 2-year stability in water through an automated polymer encapsulation for bioimaging, *ACS Nano* 16 (2022), 13657–13666.

25. I.V. Korolkov, K. Ludzik, A.L. Kozlovskiy, M.S. Fadeev, A.E. Shumskaya, Y.G. Gorin et al., Immobilization of carboranes on Fe_3O_4-polymer nanocomposites for potential application in boron neutron cancer therapy, *Colloids Surfaces A Physicochem. Eng. Asp.* 601 (2020), pp. 125035.

26. R. Vivek, R. Thangam, S.R. Kumar, C. Rejeeth, S. Sivasubramanian, S. Vincent et al., HER2 targeted breast cancer therapy with switchable "off/on" multifunctional "smart" magnetic polymer core-shell nanocomposites, *ACS Appl. Mater. Interfaces* 8 (2016), pp. 2262–2279.

27. A. Gonsalves, P. Tambe, D. Le, D. Thakore, A.S. Wadajkar, J. Yang et al., Synthesis and characterization of a novel pH-responsive drug-releasing nanocomposite hydrogel for skin cancer therapy and wound healing, *J. Mater. Chem. B* 9 (2021), pp. 9533–9546.

28. S. Kesarwani, V.K. Patel, V.K. Singh and R.K. Verma, *Recent Trends in the Manufacturing of Reduced Graphene Oxide Modified Epoxy Nanocomposites as Advanced Functional Material*. AIP Publishing LLC, Melville, New York, (2022), pp. 9-1–9-34.

29. A.M. Díez-Pascual and J.A. Luceño-Sánchez, Antibacterial activity of polymer nanocomposites incorporating graphene and its derivatives: A state of art, *Polymers (Basel)*. 13 (2021), p. 2105.

30. N.E. Kazantseva, I.S. Smolkova, V. Babayan, J. Vilčáková, P. Smolka and P. Saha, Magnetic nanomaterials for arterial embolization and hyperthermia of parenchymal organs tumors: A review, *Nanomaterials* 11 (2021), pp. 1–28.

31. T.C. Lin, F.H. Lin and J.C. Lin, In vitro feasibility study of the use of a magnetic electrospun chitosan nanofiber composite for hyperthermia treatment of tumor cells, *Acta Biomater.* 8 (2012), pp. 2704–2711.

32. Z. Hedayatnasab, F. Abnisa and W.M.A.W. Daud, Review on magnetic nanoparticles for magnetic nanofluid hyperthermia application, *Mater. Des.* 123 (2017), pp. 174–196.

33. R.S. Moraes, V. Saez, J.A.R. Hernandez and F.G. de Souza Júnior, Hyperthermia system based on extrinsically magnetic poly (Butylene Succinate), *Macromol. Symp.* 381 (2018), pp. 1–7.

34. A. Barra, Z. Alves, N.M. Ferreira, M.A. Martins, H. Oliveira, L.P. Ferreira et al., Biocompatible chitosan-based composites with properties suitable for hyperthermia therapy, *J. Mater. Chem. B* 8 (2020), pp. 1256–1265.

35. Z.W. Lim, V.B. Varma, R. V. Ramanujan and A. Miserez, Magnetically responsive peptide coacervates for dual hyperthermia and chemotherapy treatments of liver cancer, *Acta Biomater.* 110 (2020), pp. 221–230.

36. M. Hadjianfar, D. Semnani and J. Varshosaz, An investigation on polycaprolactone/chitosan/Fe_3O_4 nanofibrous composite used for hyperthermia, *Polym. Adv. Technol.* 30 (2019), pp. 2729–2741.

37. G.R. Iglesias, F. Reyes-Ortega, B.L.C. Fernandez and Á.V. Delgado, Hyperthermia-triggered gemcitabine release from polymer-coated magnetite nanoparticles, *Polymers (Basel)*. 10 (2018), 269.

38. R.J.R. Matos, C.I.P. Chaparro, J.C. Silva, M.A. Valente, J.P. Borges and P.I.P. Soares, Electrospun composite cellulose acetate/iron oxide nanoparticles non-woven membranes for magnetic hyperthermia applications, *Carbohydr. Polym.* 198 (2018), pp. 9–16.

39. S.A. Moghadam Ziabari, M. Babamoradi, Z. Hajizadeh and A. Maleki, The effect of magnetic field on the magnetic and hyperthermia properties of bentonite/Fe_3O_4 nanocomposite, *Phys. B Condens. Matter* 588 (2020), pp. 412167.

40. V. Zamora-Mora, P.I.P. Soares, C. Echeverria, R. Hernández and C. Mijangos, Composite chitosan/agarose ferrogels for potential applications in magnetic hyperthermia, *Gels* 1 (2015), pp. 69–80.

41. S. Ganguly and S. Margel, Design of magnetic hydrogels for hyperthermia and drug delivery, *Polymers (Basel)*. 13 (2021), 4259.

42. F. Masood, Polymeric nanoparticles for targeted drug delivery system for cancer therapy, *Mater. Sci. Eng. C* 60 (2016), pp. 569–578.

43. S. Hossen, M.K. Hossain, M.K. Basher, M.N.H. Mia, M.T. Rahman and M.J. Uddin, Smart nanocarrier-based drug delivery systems for cancer therapy and toxicity studies : A review, *J. Adv. Res.* 15 (2019), pp. 1–18.

44. P. Jana, M. Shyam, S. Singh, V. Jayaprakash and A. Dev, Biodegradable polymers in drug delivery and oral vaccination, *Eur. Polym. J.* (2020), pp. 110155.

45. D.N. Kapoor, R. Kaur, R. Sharma and S. Dhawan, PLGA: A unique polymer for drug delivery, *Therapeutic Delivery*, 6 (2015), pp. 41–58.

46. P. Allavena, A. Palmioli, R. Avigni, M. Sironi, B. La Ferla and A. Maeda, PLGA based nanoparticles for the monocyte-mediated anti-tumor drug delivery system, *Journal of Biomedical Nanotechnology*, 16(2) (2020), pp.212–223.

47. P.E. Florian, M. Icriverzi, C.M. Ninciuleanu, E. Alexandrescu, B. Trica, S. Preda et al., Salecan-clay based polymer nanocomposites for chemotherapeutic drug delivery systems; Characterization and In Vitro Biocompatibility Studies. *Materials*, 13(23) (2020), p. 5389.

48. G. Lei, S. Yang, R. Cao, P. Zhou, H. Peng, R. Peng et al., In situ preparation of amphibious zno quantum dots with blue fluorescence based on hyperbranched polymers and their application in bio-imaging, *Polymers (Basel)*. 12 (2020), pp. 24–26.

49. M. Zhang, Y. Zou, Y. Zhong, G. Liao, C. Yu and Z. Xu, Polydopamine-based tumor-targeted multifunctional reagents for computer tomography/fluorescence dual-mode bioimaging-guided photothermal therapy, *ACS Appl. Bio Mater.* 2 (2019), pp. 630–637.

50. T. Deng, R. Zhang, J. Wang, X. Song, F. Bao, Y. Gu et al., Carbon dots-cluster-DOX nanocomposites fabricated by a co-self-assembly strategy for tumor-targeted bioimaging and therapy, *Part. Part. Syst. Charact.* 35 (2018), pp. 1–10.

51. L. Huang, S. Yu, W. Long, H. Huang, Y. Wen, F. Deng et al., The utilization of multifunctional organic dye with aggregation-induced emission feature to fabricate luminescent mesoporous silica nanoparticles based polymeric composites for controlled drug delivery, *Microporous Mesoporous Mater.* 308 (2020), p. 110520.

52. K. Mandal, D. Jana, B.K. Ghorai and N.R. Jana, AIEgen-conjugated magnetic nanoparticles as magnetic-fluorescent bioimaging probes, *ACS Appl. Nano Mater.* 2 (2019), pp. 3292–3299.

53. C.I. Idumah, Progress in polymer nanocomposites for bone regeneration and engineering, *Polym. Polym. Compos.* 29 (2021), pp. 509–527.

54. I. Armentano, M. Dottori, E. Fortunati, S. Mattioli and J.M. Kenny, Biodegradable polymer matrix nanocomposites for tissue engineering: A review, *Polym. Degrad. Stab.* 95 (2010), pp. 2126–2146.

55. T. Biswal, Biopolymers for tissue engineering applications: A review, *Mater. Today Proc.* 41 (2019), pp. 397–402.

56. K. Ghosal, C. Agatemor, Z. Špitálsky, S. Thomas and E. Kny, Electrospinning tissue engineering and wound dressing scaffolds from polymer-titanium dioxide nanocomposites, *Chem. Eng. J.* 358 (2019), pp. 1262–1278.

57. A. Sotoudeh, G. Darbemamieh, V. Goodarzi, S. Shojaei and A. Asefnejad, Tissue engineering needs new biomaterials: Poly(xylitol-dodecanedioic acid)–co-polylactic acid (PXDDA-co-PLA) and its nanocomposites, *Eur. Polym. J.* 152 (2021), p. 110469.

58. M. Silva, I.S. Pinho, J.A. Covas, N.M. Alves and M.C. Paiva, 3D printing of graphene-based polymeric nanocomposites for biomedical applications, *Funct. Compos. Mater.* 2 (2021), 1–21.

59. M.U.A. Khan, S.I. Abd Razak, H. Mehboob, M.R. Abdul Kadir, T.J.S. Anand, F. Inam et al., Synthesis and characterization of silver-coated polymeric scaffolds for bone tissue engineering: Antibacterial and in vitro evaluation of cytotoxicity and biocompatibility, *ACS Omega* 6 (2021), pp. 4335–4346.

60. M.U.A. Khan, S. Haider, S.A. Shah, S.I.A. Razak, S.A. Hassan, M.R.A. Kadir et al., Arabinoxylan-co-AA/HAp/TiO$_2$ nanocomposite scaffold a potential material for bone tissue engineering: An in vitro study, *Int. J. Biol. Macromol.* 151 (2020), pp. 584–594.

61. K.K. Kefeni, T.A.M. Msagati, T.T. Nkambule and B.B. Mamba, Spinel ferrite nanoparticles and nanocomposites for biomedical applications and their toxicity, *Mater. Sci. Eng. C* 107 (2020), pp. 110314.

62. S. Rana, A. Sharma, A. Kumar, S.S. Kanwar and M. Singh, Utility of silane-modified magnesium-based magnetic nanoparticles for efficient immobilization of *Bacillus thermoamylovorans* Lipase, *Appl. Biochem. Biotechnol.* 192 (2020), pp. 1029–1043.

63. M.M. Fernandes, D.M. Correia, C. Ribeiro, N. Castro, V. Correia and S. Lanceros-Mendez, Bioinspired three-dimensional magnetoactive scaffolds for bone tissue engineering, *ACS Appl. Mater. Interfaces* 11 (2019), pp. 45265–45275.

64. N. Ali, F. Ali, R. Khurshid, Ikramullah, Z. Ali, A. Afzal et al., TiO$_2$ nanoparticles and epoxy-TiO$_2$ nanocomposites: A review of synthesis, modification strategies, and photocatalytic potentialities, *J. Inorg. Organomet. Polym. Mater.* 30 (2020), pp. 4829–4846.

65. M. Rasoulianboroujeni, F. Fahimipour, P. Shah, K. Khoshroo, M. Tahriri, H. Eslami et al., Development of 3D-printed PLGA/TiO$_2$ nanocomposite scaffolds for bone tissue engineering applications, *Mater. Sci. Eng. C* 96 (2019), pp. 105–113.

66. B. Kolathupalayam Shanmugam, S. Rangaraj, K. Subramani, S. Srinivasan, W.K. Aicher and R. Venkatachalam, Biomimetic TiO$_2$-chitosan/sodium alginate blended nanocomposite scaffolds for tissue engineering applications, *Mater. Sci. Eng. C* 110 (2020), pp. 110710.

67. Z. Pedram Rad, J. Mokhtari and M. Abbasi, Fabrication and characterization of PCL/zein/gum arabic electrospun nanocomposite scaffold for skin tissue engineering, *Mater. Sci. Eng. C* 93 (2018), pp. 356–366.

68. R. Kouser, A. Vashist, M. Zafaryab, M.A. Rizvi and S. Ahmad, Biocompatible and mechanically robust nanocomposite hydrogels for potential applications in tissue engineering, *Mater. Sci. Eng. C* 84 (2018), pp. 168–179.

11 Bionanocomposites
Biological Aspects and Biomedical Applications

Jyoti Sharma, Vitthal L. Gole, and Mohammad Ali Siddiqui

CONTENTS

DOI: 10.1201/9781003343912-11

11.1   INTRODUCTION

Composites are materials that consist of two or more than two constituents, with distinct physical and chemical properties. In nanocomposites, the bonding between filler components and matrix takes place through ionic bonds, covalent bonds, hydrogen bonding and van der Walls forces and becomes single physical material after processing of two or more constituents finally having phase dimension of less than 100 nm [1,2]. In general, nanomaterials are substances whose size is in the "nano-meter" (nm) range, and at such a low size, they exhibit unique properties such as supermagnetism, superhydrophobicity and distinctly large surface area. The biomaterials are those materials that exhibit some biological activity by interacting with components of the living system.

In the current context, a biomaterial is a substance that has been created to take on a shape on its own or as a component of a complex system. It could be utilized to control or direct the progress of any therapy or diagnosis as a part of therapeutic treatment. This is done by controlling interactions with elements of biological systems. The aim of bionanomaterials is to provide a solution to the most important issues in medical field such as tissue replacement and surgeries. Since the inception of biomaterials, various developments have been taking place and are incorporated into various applicable areas. Consequently, there is a rise in applications in various domains, predominantly in the medical industry. Modern technological developments have improved access to knowledge of tissue generations, pathogenesis of diseases and prospects for the use of biomaterials. Different materials have been created with the intention of being compatible with the human body and acting as substitutes for both soft and hard tissues. The development of orthopedic materials from bioinert to third-generation materials has ensured that their biocompatibility meets the standards [3] and is being utilized for the replacement therapy.

11.1.1   History

Theng originally used the phrase "bio-nanocomposites" in 1970 [4]. It is sometimes referred to as "green composites," "bio composites," "nano composites," "nano bio composites," "bio hybrids" or "bio-based plastics" (bioplastics). Bionanocomposites are biohybrid materials made by mixing nano-scaled inorganic fillers with biopolymers (e.g., polysaccharides, proteins and nucleic acids) [1]. The main difficulty with bionanocomposites is controlling the unique features of each component to produce materials with superior performance. Better functional and structural materials are made possible due to the biocompatibility property of biopolymers as well as their inorganic counterpart's mechanical and thermal qualities. The production and properties of nanocomposites made by biodegradable polymers have been extensively studied in the past ten years. In this chapter, we will discuss the synthesis and characterization of bionanocomposites. A critical aspect of the properties of bionanocomposites will be discussed specially in terms of biological aspects.

11.1.2   Recent Research and Advances

The majority of research on bionanocomposites focuses on proteins, polysaccharides, polylactic acid (PLA) and polycaprolactone (PCL). Recent research, however, demonstrates that the formation of clay-polymer nanocomposites leads to efficient reinforcement of polymer as well as biopolymer matrices when using microfibrous clay minerals such as sepiolite and palygorskite. If we analyze the research study on nanocomposites, bionanomaterials, bionanopolymers, bionanocomposites and advanced nanocomposites like (surgical application-based bionanocomposites and living

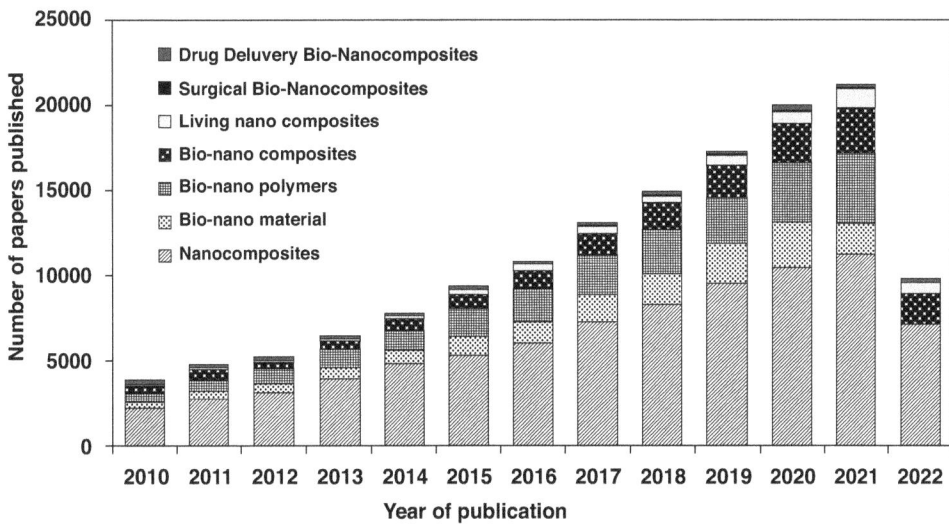

FIGURE 11.1 Number of research articles published in the context of various biological/medical aspects of bionanomaterial (2010–2022).

bionanocomposites, and bionanocomposites in drug delivery), we can see that an extensive research is being done in the field of biological and pharmaceutical for the fabrication of bionanomaterials and smart biomaterials. Data shown in Figure 11.1 are related with correspondence to the research papers published solely in the context of biomedical aspect of the bionanomaterial.

The research on nanomaterial is increasing year by year if we compare specific domains like drug-delivery and surgical nanocomposites; it is also gaining quantum from lower to higher value.

11.2 TYPES OF NANOCOMPOSITES

Composites can be defined as an ensemble of two phases: blended/mixed to obtain the desired properties and tailor-made for specific applications. Nanocomposites constitute a subgroup of this bigger domain of multifunctional materials in which one of the components is in the 1–100 nm size range. Figure 11.2 provides a brief classification of nanocomposites based on their framework and application. The arena of nanocomposites with unusual property combinations and design possibilities at a very low concentration of fillers has gained importance in recent years owing to a high matrix-to-filler interfacial area (the so-called nanoeffect) and greater aspect ratio. Based on the types of the framework that is used while synthesizing the nanocomposites, the following categories of nanocomposites originate [5–7].

11.2.1 FRAMEWORK-BASED

The framework of the nanocomposites is a combination of two things: matrix and reinforcement; matrix imparts structural stability and reinforcement imparts secondary and primary properties to the nanocomposites. By changing matrix or reinforcement type, we can have a whole different kind of nanocomposites. So, based on this, we have the following subcategories.

11.2.1.1 Matrix-Based

The reinforcement is typically implanted and evenly dispersed throughout the matrix, which is a monolithic material. Matrix materials include aluminum, magnesium, nickel, titanium and cobalt [1]. Based on the type of matrix composites, it can be further classified as follows.

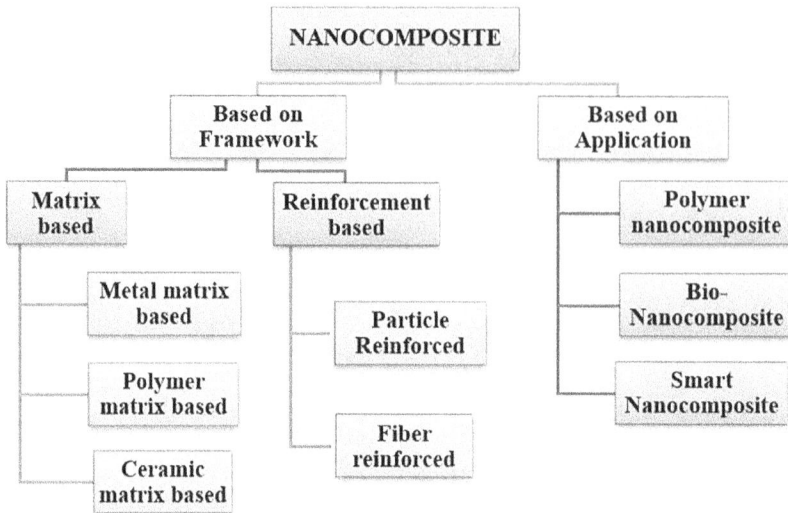

FIGURE 11.2 A brief classification of nanocomposites.

11.2.1.1.1 Metal Matrix Nanocomposites

In this type of composites, metal matrix is distributed with other metal, ceramic or organic substances. The metal matrix nanocomposites (MMNCs) that use copper, magnesium and aluminum as their matrix materials have received the greatest attention [8,9]. The qualities of the matrix metal are improved by the reinforcements/fillers. Comparing MMNCs to traditional engineering materials, MMNCs have superior physical and mechanical characteristics. This is because adding filler to the metallic matrix enhances the properties such as specific strength and stiffness along with wear, creep and fatigue resistance.

The features of MMNCs are greatly influenced by the distribution and characteristics of the filler particles. The most popular filler materials include silicon carbide, titanium oxide (TiO_2), aluminum oxide (Al_2O_3), B_4C, Y_2O_3, $Si_3 N_4$ and AlN [1,3,5–7]. Al_2O_3 reinforcement offers good wear resistance and compressive strength. One of the hardest known elements, boron carbide, imparts high elastic modulus and fracture toughness to the MMNCs. Silicon carbide (SiC) had been reported to increase the hardness of MMNCs, but it had little to no impact on their wear resistance [7].

11.2.1.1.1.1 Application of MMNCs
Fly ash-reinforced MMNCs are gaining interest as electromagnetic shielding effect improving material due to availability (as waste by-product from thermal power plants) and low cost. Automobile and aerospace sectors utilize aluminum- and titanium alloys-reinforced MMNCs in structural applications due to low density, good strength at high temperature and corrosion resistance along with better forming and joining properties. Carbon nanotube (CNT)-reinforced composites can replace carbon fiber composites owing to their high strength and stiffness also in high-temperature conditions. MMNCs are also applied in reduction of vibrations and noise in structural components due to their enhanced damping capacity. In literature, it is reported that powder-milled Mg–Al_2O_3 samples extruded at 350°C have shown a significant damping ability. The RHA-SiC-Al hybridization provided better mechanical properties in reduced heat treatment time [9,10]. However, as materials, they are expensive composites.

11.2.1.1.2 Polymer Matrix Nanocomposites

Polymer matrix nanocomposites (PMNCs) can be defined as hybrid materials that use polymer as matrix and nanomaterial as filler. In comparison to pure polymers, PMNCs had been reported to have good mechanical properties like tensile and compressive strength, yield stress and toughness. The gas barrier characteristics can also be improved 50–500 times, when nanofillers are dispersed into a polymer matrix. The addition and even distribution of the nanofillers inside the polymeric matrix improve the characteristics of PMNCs. By including magnetic elements, such as species based on iron, PMNCs can be further made magnetic.

11.2.1.1.2.1 Application of PMNCs   PMNCs have several uses in the fields of electronic, magnetic, optical and electrochromism due to their distinctive synergistic multifunctions that result from the combination of numerous components into one compatible entity. PMNCs have proven to be highly efficient, highly kinetic and recyclable nanoadsorbents for wastewater remediation applications [11]. PMNCs can be applied as membranes in the separation process (poly(amide-imide)/TiO_2 combination) and PET/clay as food-packaging materials. These nanocomposites are also working as bio and smart materials. Shape memory polymers/SiC are also the PMNCs used in medical devices for holding or releasing therapeutics drugs within blood vessels. Epoxide-carbon fibers-composite are applied in the bone fracture repair [12]. Urethanes or silicone rubber reinforced with nano-sized SiO_2 in catheters and thermosets reinforced with glass, carbon, or Kevlar fibers in prosthetic limbs create a new class of composites called bionanopolymer composites [12]. Photovoltaic (PV) cells and photodiodes, supercapacitors, printed conductors, light emitting diodes (LEDs) and field-effect transistors are further possible applications for PMNCs [13]. Research has been done on conjugated polymers that contain different nanoscale fillers for use in gas sensors, biosensors and chemical sensors, among other sensor applications. Some of the nanofillers employed are metal oxide nanowires, carbon nanotubes, and nanoscale gold, silver, nickel, copper, platinum and palladium particles [14]. Table 11.1 presents the type of matrix and reinforcements used in the synthesis of PMNCs. The discovery of carbon nanotubes coincided with the growing interest in graphene for use in electronics. As Moore's law limits are reached, graphene in sheet form may show potential as a replacement for silicon [15]. The promise for a new generation of hybrid materials, including photovoltaic, gas sensors, electrical devices, mechanics, optical displays, superconductors, and photoconductors, is offered by polymer semiconductor nanocomposites [16].

11.2.1.1.3 Ceramic Matrix Nanocomposites
Ceramic/metal fibers are enmeshed in a ceramic matrix to create ceramic matrix nanocomposites (CMNCs) (polycrystalline ceramic, glass or their mixture). In ceramic matrix composites,

TABLE 11.1
Matrix-Based Composites

Nanocomposites	Matrix	Reinforcements
Metal matrix nanocomposites	Metals like copper, magnesium and aluminum	Aluminum oxide (Al_2O_3), TiO_2, Silicon carbide (SiC), B_4C, Y_2O_3, Si_3N_4 and AlN
Polymer matrix nanocomposites	Polyethylene terephthalate (PET) Polyporpylene (PP) Nylons Ethylene vinyl alcohol	Ultrahigh molecular weight polyethylene (UHMWPE) fibers, glass, carbon, or graphite fibers, exotic fibers (boron), and ceramic fillers (calcium carbonate, fumed silica)
Ceramic matrix nanocomposites	Zirconia, silicon nitride, silicon carbide, aluminum nitride or alumina	Monocrystalline *whiskers* Carbon Liquid polymer

a ceramic substance, including carbon/carbon fibers, may make up the matrix. While the second component is generally a metal, the ceramic substance comes from the family of oxides like nitrides, borides and silicides. Both the components should be finely disseminated in one another for better optical, electrical and magnetic properties and for good corrosion resistance [17]. CMNCs should be designed with the help of binary phase diagram of the mixture in mind, and chemical reaction between the two components must be prevented. CMNCs are light in weight due to reduced density and reported to have high temperature and creep resistance, excellent resistant to thermal shock, corrosion and wear [18]. They provide high tensile and compressive strength, in comparison to conventional ceramics, so no sudden failure, and they are inert toward aggressive chemicals.

11.2.1.1.3.1 Applications of CMNCs Due to their high temperature-bearing capacity and strength retention property at high temperatures, they are used in heat-exchanging equipment and burner components. Turbine blades, combustion chambers, stator vanes and turbine engines are all made of coated silicon carbide fibers that are embedded in a ceramic matrix. For protection from intense heat, the aerospace industry uses coated ceramic tiles for space shuttle shielding, body flaps and shrouds [19]. CMNCs are also utilized in the nuclear power industry due to their high temperature-bearing capacity and inertness. Hypersonic vehicles and engine exhaust systems also utilize CMNCs for better component life and reduced weight and engine noise [20].

11.2.1.2 Reinforcement-Based

In this category of nanocomposites, the discontinuous phase known as reinforcement is included to give the matrix materials additional properties and functionality. The matrix is a ductile, or tough, material, and the reinforcement materials are strong with low densities. In general, reinforcements have an impact on the mechanical, physical or any other customized properties of the matrix material [5]. In the matrix phase, there may be fibers, sheets or particles (metal, ceramic or polymer).

The reinforcement boosts the total mechanical properties of the matrix while the matrix preserves the reinforcement to produce the desired form. Based on the type of reinforcement used, composites can be further classified as follows.

11.2.1.2.1 Particle-Reinforced Nanocomposites

A particulate composite comprises particles having any shape, size or configuration suspended in a matrix. The particles can be processed at nanoscale and can be used as a reinforced phase within the metal, glass, ceramic or polymer matrices. An anisotropic or orthotropic response can be seen in a particulate composite. Concrete and particle boards are examples of particulate composites. Flake and filled/skeletal particles are the two types of particulates employed in these kinds of composites [21].

These composites are employed in numerous situations where strength is of a great deal. The reinforcements, matrix materials, bond strength between the reinforcing materials and matrix, and weave type affect the quality of composites [22].

11.2.1.2.1.1 Applications of Particle-Reinforced Nanocomposites Aluminum metal matrix reinforced with nano-ZrO_2 has exceptional mechanical qualities, including high strength, hardness and fracture toughness [23]. Argon-degassed A356 alloy has a high tensile strength of 157.5 Mpa and is widely used in the making of chassis plate of racing cars. The polymer matrix can be reinforced by particles like carbon nanotubes, exfoliated clay platelets or carbon black nanoparticles to increase their properties like stiffness (modulus), strength, etc. A range of medical equipment, such as MRI and CT scanners, X-ray couches, mammography plates, etc. can be created using these particles [17,21].

11.2.1.2.2 Fiber-Reinforced Nanocomposites

Composites with fiber reinforcement that combine hard, brittle fibers with a softer, ductile matrix come under the category of fiber-reinforced nanocomposites (Table 11.2). The strength of composites is determined on the basis of adhesion of the fibers to the matrix. The physicochemical characteristics of the matrix material have an impact on adhesion. The length, diameter, concentration, strength, and direction of the fibers are significant in determining the mechanical properties of the fiber-reinforced composites. Two methods, namely [24],

a. employing nanofibers to reinforce nanocomposites, and
b. adding nanomaterial into the fiber-reinforced composites can be used to synthesize fiber-reinforced nanocomposites.

11.2.1.2.2.1 Applications of Fiber-Based Nanocomposites Construction and biomedical applications have been the principal uses of nanofiber-reinforced composites. When epoxy resins are strengthened with nanoparticles, the resulting materials are reported to have higher glass-transition temperatures, slightly increased glassy modulus, lower dielectric constant and noticeably higher mechanical properties [24]. Table 11.2 lists the various types of reinforcements applied and matrix materials in metal matrix-reinforced nanocomposites. In the realm of sports and games, as well as in the equipment like gas turbine rotators in nuclear power plants, they are employed as primary structures for aerospace vehicles, launch vehicles and spacecraft for space research. Using the freeze-granulation technique, a new nanocomposite of hydroxyapatite (HA) modified with PMMA and reinforced with multiwalled carbon nanotubes (MWCNTs) was created. This material is a new generation of biomedical implant coatings and bone cement [25].

11.2.2 Application-Based

Based on the application of nanocomposites, we have different kinds of nanocomposites, ranging from metal nanocomposites to magnetic core nanocomposites and polymeric nanocomposites, bionanocomposites and smart or artificial intelligence (AI)-based nanocomposites. Besides these, there are many kinds of nanocomposites today like hyperhygroscopic nanocomposites and thermo nanocomposites that come under the combined category of smart and advanced nanocomposites, we will focus on the following nanocomposites.

11.2.2.1 Bionanocomposites

The word bionanocomposites includes hybrid materials, i.e., bio-based matrix and nano-sized filler material. Bionanocomposites were developed from the fusion of biopolymers with inorganic solids to categorize a new emerging class of biohybrid materials with dimensions in the nanometer range (1–100 nm) [20]. They possess good mechanical, optical, barrier, properties etc., biocompatibility, antimicrobial activity, and biodegradability. They have wide area of application due to multidimensional properties. The most difficult task in bionanocomposites is to create materials with exceptional performance by deftly controlling the unique characteristics of its constituent parts.

TABLE 11.2
Reinforcement-Based Nanocomposites

Nanocomposites	Matrix	Reinforcements
Particle-reinforced nanocomposites	Metal matrixGlass matrix and ceramic matrix	SiO_2, ZrO_2 and carbon
Fiber-reinforced nanocomposites	Soft metals and clay, glass, boron and polymeric fibers	CNTs, SiO_n dust, naturals fibers

In recent years, natural polymers like starch, cellulose acetate, polylactic acid (PLA), etc., have seen widespread application. These materials are preferable over other materials from an environmental point of view. The food-packaging business uses optically transparent PLA-based bionanocomposite films that have been plasticized [26]. Natural cellulose acetate (CA) and nanohydroxyapatite (n-HA)-based artificial bone tissue scaffolds are used in in vitro bone regeneration studies [27]. However, the quest for and development of innovative, affordable materials for greener approach is a continuing activity. The general methods of production of bionanocomposites include the following:

a. Solution intercalation
b. In situ intercalative polymerization
c. Melt intercalation
d. Template synthesis

The properties that make bionanocomposites unique are as follows:

- Adaptability
- Reproducibility
- Self-repair
- Biodegradability
- Biocompatibility

We will discuss them in detail in the upcoming sections.

11.2.2.2 Smart Nanocomposites

Smart materials enable us to adapt to environmental changes by activating its functions. One or more physical and chemical properties of smart or intelligent materials can be controlled or altered by external inputs [28,29]. This response to stimuli can be divided into two categories: internal stimuli and exterior stimuli. pH, solvent, chemical recognition, and biological or enzymatic recognition are a few examples of the internal stimuli. The external stimuli, however, can be things like light, electric current, temperature and magnetic field [30]. External stimuli are possible to establish and control easily in comparison to internal stimuli. The active/smart materials provide large opportunity to use them on electrotechnological solutions. Multistimuli active nanocomposites have qualities that depend on the parameters of the matrix and the type of filler used, and they can react to one or more stimuli [31].

11.2.2.2.1 Applications of Smart Nanocomposites

Due to their responsive nature, smart materials are gaining interest and recognition in various fields of application. For oral colon-specific medication administration, smart nanocomposite hydrogels based on azo cross-linked graphene oxide are utilized. For high-temperature tribological applications, adaptive nanocomposite coatings with a titanium nitride diffusion barrier mask are used. In vitro studies of bacterial cellulose and magnetic nanoparticles reveal smart materials for efficient chronic wounds healings. A novel organoclay-reinforced UHMWPE nanocomposite coating is created and characterized for tribological applications. Cu-hemin metal-organic frameworks (MOFs) smart nanocomposites are developed for electrochemical glucose biosensing, and it is concluded that they will provide a practical solution of electrochemical sensing based on MOFs and biomolecules [32].

11.3 BIONANOCOMPOSITES

11.3.1 Classification of Bionanocomposites

Bionanocomposites have revolutionized medical science and have paved route to the smart hydrogels and regenerative and programmable tissues. Comparison of the research studies and results of synthesis and application of bionanocomposite concluded that the research and publications on these materials is ever increasing in last decades (Figure 11.1). Tons of advanced nanocomposites containing biocomponents as filler or as matrix are used extensively in the medical field. Based on the literature available, it can be classified as shown in Figure 11.3.

11.3.1.1 Bioactive Silicate-Based Nanocomposites

Synthetic silicate-based ultrathin (or 2D) nanocomposites exhibit significant anisotropy and functionality. It interacts with biological things in quite different manner than it does with their respective 3D macro-/micro- and nano-size ranges because it is reported to have high surface-to-volume ratio [33,34]. In biological and biomedical settings, these nanomaterials are used in imaging, therapy and disease-related diagnostics.

11.3.1.2 Mechanically Stiff Interpenetrating Networks (IPNs)

Cartilage is a tough tissue with great mechanical stiffness, and its simulation can be done by using IPNs, semi-IPNs and double networks [35]. Multiple types of polymers are composed of cartilages and muscles which form interpenetrating networks. By combining two or more polymers together, artificial mimicry can be obtained. Nowadays, industries are also conducting research to inject

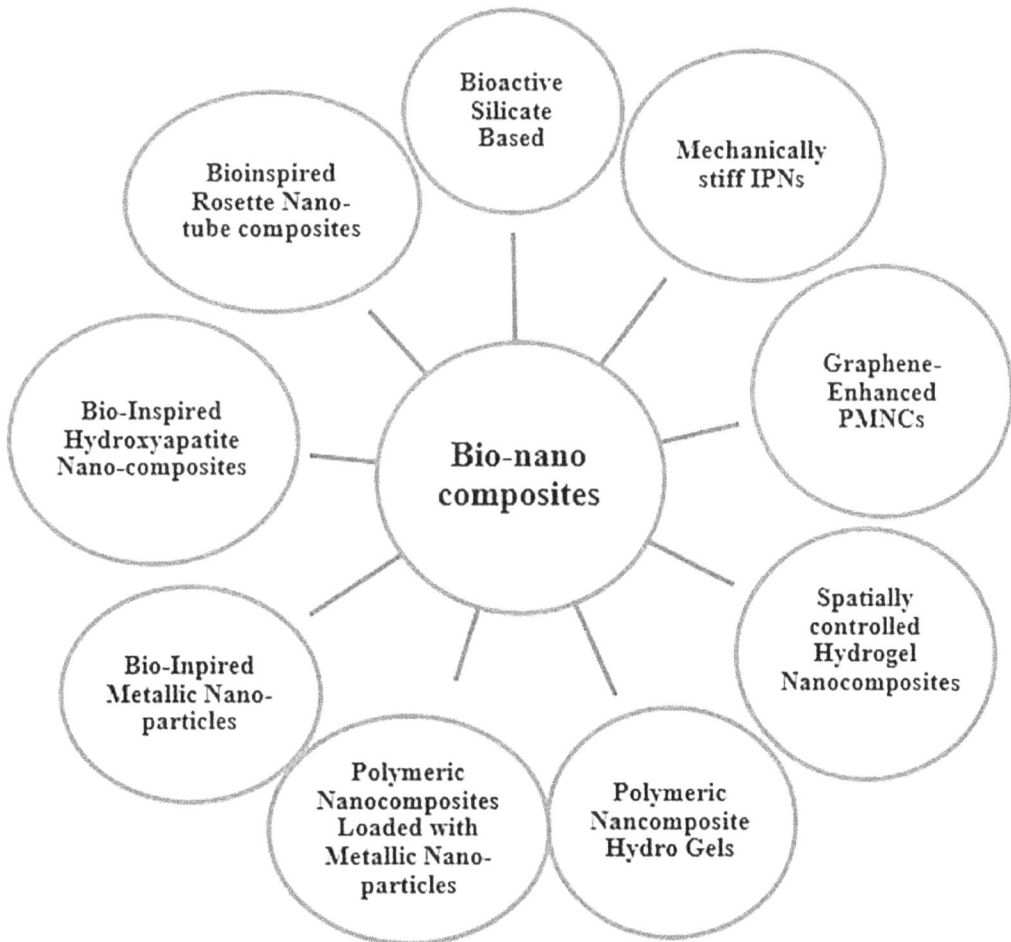

FIGURE 11.3 Classification of bionanocomposites.

bionanomachines in the human brain which can be adapted by the human brain and connected to internet, like, devices connect to each other using the concept of IoT.

11.3.1.3 Graphene-Enhanced Polymeric Nanocomposites

Graphene and its derivatives have excellent electrochemical, thermal and optical properties. Examples include reduced graphene oxide (rGO) and linear graphene oxide. According to the published reports, single-strand DNA and graphene oxide sheets interact covalently and noncovalently to self-assemble into three-dimensional structures [34]. The mechanical and electrical conductivity properties of graphene are still present in the final biomimetic hierarchical structure, and owing to these properties, it is utilized in tissue engineering. The ability of graphene to internalize cells and fill with biomolecules makes graphene oxide nanosheets to be utilized to transfer genes.

11.3.1.4 Spatially Controlled Hydrogel Nanocomposites

The goal of creating biomimetic hydrogel scaffolds is to mimic the intricacy of tissues by promoting encapsulated cells' proliferation and adherence. Microfabrication techniques like micromolding are utilized to control cellular microenvironments, and this will improve tissue complexity imitating. According to the published reports, bioengineered polymeric nanocomposites are employed to modulate the cellular activity [36].

11.3.1.5 Polymeric Nanocomposite Hydrogels

Hydrated three-dimensional polymeric networks filled with nanoparticles can be characterized as nanocomposite hydrogels. The hydrogels inflate after absorbing water, which hydrates cells in vivo and increases cell viability. Together with a biomimetic environment, this process enables the flow of nutrients and waste into an encapsulated cell population. Formulation of a mechanically stable and bioactive hydrogel is still a challenge [34,37]. Nanocomposite hydrogels are also incorporated with bioactivity using inorganic nanoparticles like hydroxyapatite and synthetic silicates. Electrical conductivity is provided to hydrogels by incorporating with carbon nanotubes and graphene oxide. Metals and their oxides (nanoparticles) provide antimicrobial and magnetic properties to hydrogels. Application of nanocomposite hydrogels is under research for tissue engineering, stem cell therapy, cancer research, immunomodulation, and cellular and molecular treatment [37].

11.3.1.6 Polymeric Nanocomposites Loaded with Metallic Nanoparticles

There are many methods for functional tissue regeneration provided by the mixing of metal-based fillers into the polymeric matrix. For instance, interactions between peptides and solid inorganic surfaces result in rearranged molecular structures, which then are adjusted for greater bioactivity with synthetic materials to enhance the regenerative performance of implanted device [38].

11.3.1.7 Bioinspired Metallic Nanoparticles

Metallic nanoparticles that are bioinspired are created through a green chemistry technique using a variety of plant extracts or cellular components. Utilizing extracellular or intracellular extracts, the metallic salt is bioreduced throughout the manufacturing process. For example, the surface plasmon resonance of silver and gold nanoparticles has numerous uses in the diagnosis of disease and its status. The optical characteristics of synthetic metallic nanoparticles can be altered. These metallic nanoparticles function as both an antibacterial agent and a catalyst for the breakdown of hazardous, nonbiodegradable dyes and contaminants. These compounds are gaining popularity as bioimaging and gene-silencing therapeutics [37].

11.3.1.8 Bioinspired Hydroxyapatite Nanocomposites

In bioinspired hydroxyapatite nanocomposites, nanohydroxyapatite (nHA) is integrated with bioinspired nanofiller. nHA is a natural mineral found in the hard tissue. Hydroxyapatite in its natural form contains fluoride and phosphide ions in traces, if consumed in bulk, can lead to metabolic poison in patients. The use of nHA is limited only to nonstressed brimming regions in clinical ortho-pedic and dental applications, due to its durability and lower fracture strength [37]. It is reported that bioinspired multifunctional mineralized nano-hydroxyapatite polyacrylic acid (nHa-PAA) hydrogel provides an alternate path to manage brittle bone disease (osteoporosis) [39].

11.3.1.9 Bioinspired Rosette Nanotube Composites

Bioinspired rosette nanotube composites are synthesized by combining synthetic DNA bases, in which hydrogen bonding arrays between guanine and cytosine are formed. The base resembles the structure of the collagen fibers of extracellular matrix (ECM). Bioactivity and cell adhesion can be improved by adding amino acids and peptide sequences within the scaffold. They are used in regenerative medicine [40,41].

Some more categories are emerging with time; as notified previously, it is a topic under research so everyday some or other papers on bionanocomposites are being written. Now let us discuss about the various components that make bionanocomposites.

11.3.2 COMPONENTS OF BIONANOCOMPOSITES

The basis of the formation of bionanocompsoite is somewhat same as that of polymer nanocompos-ite; the difference comes due to the addition of living/biomaterials of specific property. It has been discussed earlier that the matrix will decide the primary characteristic of the nanocomposite and will impart stability to nanocomposite while the reinforcement/filler induce some special proper-ties. By the selection of different reinforcements/fillers which are biologically active can enhance the properties of nanocomposite or use biologically active matrix material to induce biological properties in nanocomposites. So as discussed in the classification of nanocomposites in Section 11.2 of this chapter, matrix and reinforcement/filler are the main components to synthesize bionano-composites. The proper ratio of the components at specified conditions will provide us the desired bionanocomposite as depicted in Figure 11.4.

The materials obtained from biological sources like plants, animals and microorgan-isms are known as biomaterials. Generally, following are the biomaterials used for making of bionanocomposites:

- **Cellulose**

  Cellulose can be used as filler as well as matrix. It is a polysaccharide, mostly found in plants having significant number of hydroxyl groups. In composite synthesis, cellulose generally acts as a natural polymer and hydrophilic bulking agent. The advancement of nanotechnology led to the development of nanofibrillated cellulose, nanofibers or microfi-brillated cellulose. Cellulose provides biofunctionality and biodegradability and is gener-ally used in tissue engineering and regenerative medicines [42].

- **Chitosan**

  Chitosan is a type of linear natural polysaccharide that resembles cellulose in structure and composed of main amino ($-NH_2$) and secondary ($-OH$) groups. It can be made by deacetylating chitin at commercial scale [43]. It serves as a bacteriostatic adhesive for fabrics and an antimicrobial wound adhesive in bandages in medical applications due to its gel-forming ability [44].

- **Polylactic Acid (PLA)**

  Polylactic acid (PLA) or polylactide is a biodegradable and biologically active thermo-plastic aliphatic polyester derived from renewable resources such as sugar and starch. PLA is used as a polymeric scaffold for drug-delivery systems owing to its biocompatibility and biodegradability properties [45].

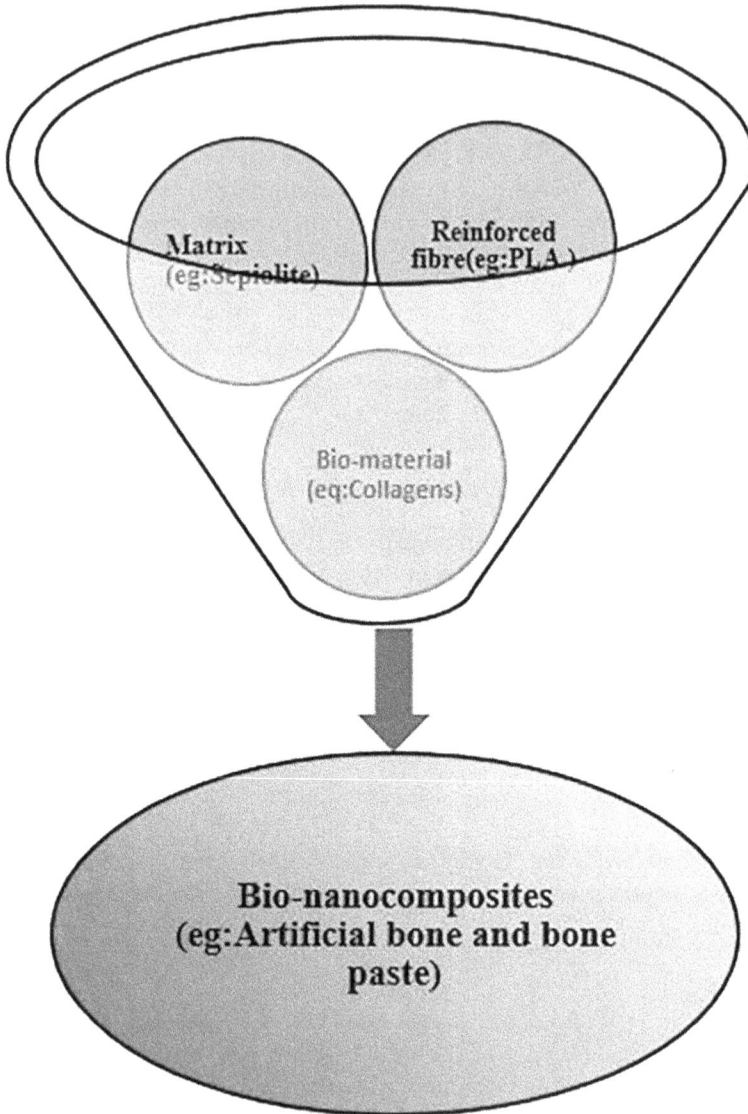

FIGURE 11.4   Components of bionanocomposites.

- **Starch**

     Amylose and amylopectin are the two components of starch, a polymeric carbohydrate. Amylopectin makes up 75%–80% of plant starch, while amylose makes up 20%–25%. It is mostly utilized in the pharmaceutical business to make paper and create different adhesives. The starch can be obtained from potatoes, maize, tapioca and wheat. The modified starch is used in textile printing applications as well as used in pharmaceutical preparations as an excipient [46].

- **Chitin**

     Chitin is a fibrous component, and its glucose derivative is obtained from a long-chain polymer of N-acetylgucosamine. Chitosan is created by acetylating chitin. It is made in various forms, such as gels, films, beads and sponges, and is mostly employed as an efficient dye and fabric binder [43].

- **Polyhydroxy Alkanonates (PHA)**

  Polyhydroxyalkanoates (PHA) are naturally available, biocompatible, water-resistant and biologically degradable polyesters. They are formed biochemically by microorganisms. They can be thermoplastic such as polyhydroxybutyrate or elastomeric materials such as polyhydroxyalvalerate having the melting point in the range from 40°C to 180°C [47].

11.4 METHODS OF PROCESSING OF BIONANOCOMPOSITES

As discussed earlier, bionanocomposites are an emerging field in nanoengineering and have different methods to prepare different kinds of bionanocomposites, but bionanocomposites can be viewed as a derivative of a nanocomposite itself, so some of the methods of preparation of nanocomposites can be altered to devise the method of preparation of bionanocomposites. With emerging science and technology, vast categories of method for synthesis of bionanocomposites can be classified as per Figure 11.5.

11.4.1 SOLUTION INTERCALATION

In this approach, dispersed nanoscale particles and biomaterial solutions are combined at first. Starch and protein, two biopolymers or bioprepolymers that are entirely soluble in solvent, are added. As a result of the components' interactions and their relationship, the mixing causes thickening, jellification or precipitation. The precipitation or separation into two phases indicates poor compatibility of the biomaterial with nanoparticles.

Silicate platelets (inorganic nanofillers) are swollen in solvents such as water, chloroform or toluene. When the biopolymer and the solution of swollen nanoparticles are mixed, the polymer chains insert between layers and displace the solvent within the interlayer of the silicate platelets. The solvent is removed, and the intercalated structure remains forming layered silicate bionanocomposite.

In layer-by-layer assembly method, the precipitation of interacting bioorganics and nano-sized inorganics is utilized to form nanocomposites [48,49]. It is also a solution technique in which firstly a layer of, for example, biopolymer is formed on a solid surface by adsorption and then a film is constructed by applying alternate assembling of the oppositely charged nanoparticles and biomacromolecules in the repeated sequential change of their solutions.

Bionanocomposite coatings, films, shells, and capsules, etc., can be synthesized using this method. Uniform distribution of nanoscale components in bulk bionanocomposites is required, which is not possible by means of simple mixing. So, vigorous stirring or ultrasonic treatment may be required to improve the homogeneity.

11.4.2 IN SITU INTERCALATIVE POLYMERIZATION

In this method, the nanoparticles are premixed with the liquid monomer or in monomer solution. Polymerization reaction is initiated by heat, radiation or by using suitable initiators. The polymer formation can occur between the intercalated sheets [50].

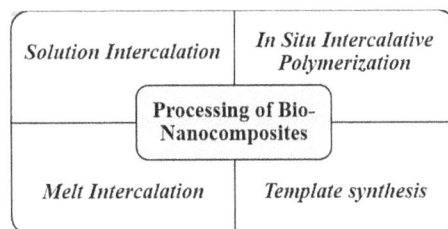

Solution Intercalation	In Situ Intercalative Polymerization
	Processing of Bio-Nanocomposites
Melt Intercalation	Template synthesis

FIGURE 11.5 Methods of processing of bionanocomposites..

Before in situ polymerization takes place, a monomer is combined with a layered silicate. It is significantly simpler to produce uniform mixing of particles in the monomer using a high shear mixer due to the low viscosity of the monomer (compared to melt viscosity) [1]. In addition, a faster rate of monomer diffusion into the interlayer region is produced by the low viscosity and high diffusivity. In this system, the reaction conditions and clay surface modification can be combined to regulate the nanocomposite shape. In situ polymerization is the only practically feasible way to create nanocomposites from the thermoset polymers.

11.4.3 MELT INTERCALATION

Solvent can be eliminated in this process, and it is compatible with industrial polymer extrusion and blending processes which are based on the melting point of polymers. As solvent is not used as the medium, nanoparticles are directly mixed with a molten polymer thereby offering an economically efficient route in producing polymer nanocomposites. A polymer is heated at an appropriate temperature to melt it and mixed directly with nanoparticles to distribute them properly in the polymer matrix. The composite is formed in the extruder [51]. The melt-mixing technique can be applied only for thermoplastics. Proteins, such as soy, wheat gluten, corn zein, caseinate, whey and gelatin, can also be processed. PHAs are also among the materials being worked by the extrusion [52,53].

This process for preparing clay-containing bionanocomposites is environmentally friendly as no organic solvents are used and biopolymers that were not suitable for in situ polymerization are used. The characteristic properties of bionanocomposites formed by this process depend upon the thermodynamic interaction between the polymer and the nanoparticles [54]. As the nanoscale particles, natural layered clay minerals are frequently used. The conditions of extrusion treatment make it possible to provide the good intercalation of macromolecules in the intergalleries of closely stacked silicate sheets and their exfoliation into individual nanoplatelets [55]. This method is convenient for the preparation of bionanocomposite on lab-to-commercial scales.

Thermal decomposition before melting is a parameter to consider. It will affect the biocompatibility and biodegradability of the nanocomposite. The exception is starch, but in its native form, the polysaccharide is not a true thermoplastic material. It is converted into a thermoplastic-like state by admixing plasticizers at 90°C–180°C under shear forces [53]. The sole problem in this method is that some biopolymers break down because of the mechanical shearing force or the processing temperature. This results in the degradation of polymer chains and, as a result, a reduction in molecular weight. So, to prevent the breakdown of biopolymer matrices, processing temperature and pressure adjustment are to be done carefully [4].

11.4.4 TEMPLATE SYNTHESIS

In this method, inorganic substances are created as a template from a precursor using biomolecules, fragments or whole cells, and microorganisms [56,57]. The bioorganics are trapped by templates as nanoparticles, a coating or shell, or a mesoporous matrix. This method is very adaptable and can be used to create a wide variety of bionanocomposite preparations. Sol-gel technique is frequently used to synthesize the inorganic component [58,59]. The biomineralization process of living cells is carried out in a highly controllable manner at ambient conditions and provides superior properties to structured biopolymer-inorganic nanocomposites [60–62]. The biomineralization processes follow "green" chemistry protocols, which call for low temperatures (0°C–30°C) and neutral media. The biosynthesis occurs in diluted or extremely diluted natural media, which is remarkable [63].

The research has documented the creation of many innovative functional materials that focus on supporting drug-delivery methods. Examples include conducting nanowires, photocatalysts and biocatalysts. The incompatibility of biomaterials with the sol-gel process restricts the opportunities to biomimic the biomineralization processes of living cells [64].

11.5 CHARACTERIZATION OF BIONANOCOMPOSITES

After synthesis, characterization of bionanocomposites is an important part as the properties such as biological, physical, mechanical, electrical, and thermal properties acquired by the bionanocomposites are justified/certified through various characterization methods. So, it becomes necessary to study surface characteristics and other properties like structural, morphological, and topological properties, because on these the physical property of bionanocomposites would depend. After synthesizing, the bionanocomposites study of characterization would later help us to devise the method for processing specific bionanocomposites for the desired application. The main characteristics along with their methods applied are discussed below.

11.5.1 PARTICLE SIZE

The particle size and their distribution along the surface and volume must be mainly determined because it gives an idea of organization of nanoparticles in bionanocomposites in 3D. It determines different structures possible for a given system of bionanocomposites. The different techniques such as dynamic light scattering (DLS), transmission electron microscopy (TEM), zeta potential, inductively coupled plasma mass spectrometry (ICP-MS) and matrix-assisted laser desorption/ionization mass spectrometry (MALDI-MS), etc. are generally used for particle size distribution determination [65].

11.5.2 SURFACE MORPHOLOGY

Surface morphology is an analytical method that uses high-tech microscopes to produce images of objects and samples that are invisible to the unaided eye. It is a sophisticated form of high-spatial resolution imaging. Images provide details on surface flaws and features. Field emission scanning electron microscopy (FE-SEM), high-resolution optical microscopy, focused ion beam-scanning electron microscopy (FIB-SEM), scanning transmission electron microscopy (STEM), ultrahigh-resolution-SEM (UHR-SEM) and other techniques can be utilized to perform surface morphology [65].

11.5.3 THERMAL PROPERTIES

In most cases, nanomaterial is employed to produce desirable bulk properties in materials. These novel materials are used in a variety of application areas, including building, healthcare, sensing, drug delivery, and diagnostic equipment. For thermal analysis, four primary methodologies have been reported. Differential scanning calorimetry (DSC), thermal gravimetric analysis (TGA), thermal mechanical analysis (TMA) and dynamical mechanical thermal analysis (DMTA) are the most widely used. The above-mentioned testing methods can be used to measure the properties of the original nanomaterial or the modified bulk material as a function of temperature or time over a wide temperature range from −150°C to 1600°C [65,66].

11.5.4 MECHANICAL PROPERTIES

The mechanical properties of the composite such as tensile, compressive, shear strength and modulus of elasticity of the biocomposite must be determined to evaluate its mechanical behavior. In bionanocomposites, mechanical properties for a period are usually required for bone fracture healing. The mechanical properties of the individual components are also important as they collectively determine the total strength of the composite. The dynamical mechanical thermal analysis and Fe-SEM are reported to be done for studying the same [67,68].

11.6 BIOLOGICAL ASPECTS OF BIONANOCOMPOSITES

The field of biomedical science and research has developed over the past few decades, and the research community is attempting to substitute native bodily tissues with synthetic or natural biomaterials that have the desired qualities to be directly in contact with living tissue. The research is also going on processes through which mimic biological phenomena can be employed. Appropriate biocompatibility, biodegradability and mechanical properties that do not trigger the immune system are the characteristics which have to be studied to evaluate the performance of bionanocomposites. Cell behavior and antibacterial tests are necessary to evaluate the growth and compatibility of biocomposites in the body [69]. The next section deals in detail about the biological properties of bionanocomposites.

11.6.1 BIOCOMPATIBILITY

Biocompatibility never means an inert material in general; in fact, the suitability of the host response is decision-making point. Earlier, the selection criteria for implantable biomaterials were set as a list of events that had to be avoided, mostly associated with the release of some corrosive products or degradation, or additives to or contaminants of the main constituents of the biomaterial, and their subsequent biological activity. Figure 11.6 depicts that the biocompatibility is property of a bionanocomposite which cure patient's body without harming it. Biocompatibility was defined as "the ability of a material to perform with an appropriate host response in a specific application" [69,70].

Nowadays, the medical sciences are concerned with this property; the implants used should show high degree of biocompatibility with patients' body else the implants or biomaterials that is being used can even take a person's life. So, medical science defines biocompatibility as

> The ability of a material to perform its desired function with respect to a medical therapy, without eliciting any undesirable local or systemic effects in the recipient or beneficiary of that therapy but generating the most appropriate beneficial cellular or tissue response in that specific situation and optimizing the clinically relevant performance of that therapy [70].

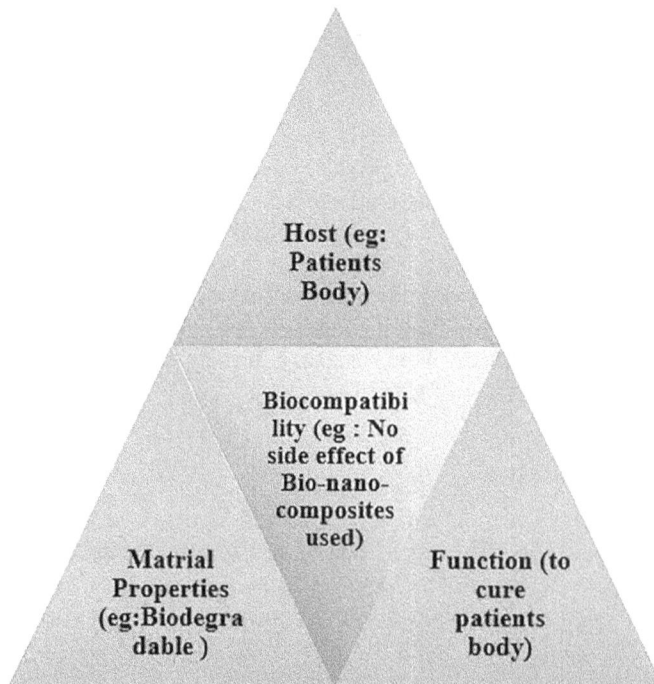

FIGURE 11.6 Factors that determine biocompatibility.

The mechanism of biocompatibility testing can be understood through Figure 11.7. By utilizing isolated cells in vitro, cell culture assays are evaluated for biocompatibility of a substance or extract. These methods evaluate the potential toxicity or irritation power of various substances and chemicals. They provide a right path to sort compounds before in vivo testing. Researchers can acquire information on how a cell reacts to a novel medicine in a regulated, controlled environment by using in vitro models. The in vivo test will involve further animal testing and level-wise clinical testing before approval. Animal testing also involves level-wise testing; firstly, bio-nanocomposite-based implants and medicines were tested on rats, if the test goes unsuccessful, the composition or properties of implant/drug are modified and retried till it shows the lowest fatality rate. Then, shift from rats to rabbit (a bigger mammal and repeat the above-explained process), similarly, further tests are conducted on dogs, goats and bulls before starting in vivo testing mechanism.

Let us understand the importance of biocompatibility via an example: Suppose we have injected a nanomaterial drug in the blood stream of a person without testing its biocompatibility and that nanomaterials have started coagulation of blood on its surface, then slowly, the blood will coagulate around the nanomaterial and can block the circulation and change the hemorheology resulting in life-threatening condition.

11.6.2 BIODEGRADABILITY

Biodegradability is the most wanted property of bionanocomposites and refers to the controlled decay of nanocomposites in in vitro or in vivo conditions [71,72]. Because they can decay inside the body because of normal biological processes, biodegradable polymeric materials have recently gained significant importance. As a result, there is no longer a need to remove a medical system once it has completed its purpose, such as once the active agent has been released from a medication delivery system. Biodegradable nanocomposites are of particular importance in bone tissue engineering. They can generate a place for a new tissue growth in the defect and can entirely be replaced by a new bone. Today's surgically implantable devices and drug-delivery systems come in different shapes and sizes and can be composed of polymers that can be cleared from the body through hydrolytic breakdown and subsequent metabolism after serving their intended purpose [72].

The degradation processes start with breaking of bonds such as through scissioning of bonds. Many forms of energy such as photo, thermal and mechanical can cause the scission of the bonds. The extent of deterioration can be stopped by the primary bond scission reactions, but they may lead

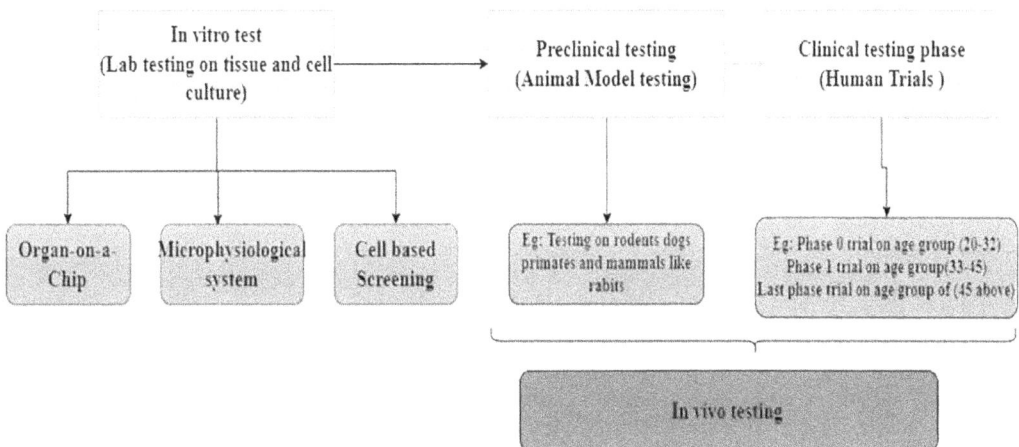

FIGURE 11.7   Level-wise testing process of biocompatibility of any medical implant/drug.

to secondary reactions, thereby generating more scissioning and recombination reactions. Polymers can be degraded by photo, thermal, mechanical, hydrolytic, oxidative, biodegradation, radiation, and chemical reactions [72].

One of the major difficulties that holds out the possibility of the development of unique and bioresponsive materials is the interaction between biomaterials and proteins or cells, which is relevant for tasks like medication transport or cell proliferation and differentiation. The innovation in the field of biomaterials is the design and synthesis of synthetic polymers that combine the information amassed from chemistry, material engineering and biological fields.

11.6.3 Antimicrobial Properties

Bionanocomposites exhibit antimicrobial potential and antimicrobial barriers through different mechanisms. The most often reported routes of antimicrobial action have been identified as (a) attachment of nanoparticles to microbial cells and reactive oxygen species, (b) their penetration inside the cells and (c) their ultimate infliction of bacterial death [73,74]. The usage of antimicrobial action of nanomaterial is taken not only in packaging and water treatment but also in medical sciences in drug delivery and cancer and tumor treatment. Before using any bionanocomposites, we have to check its antimicrobial properties. Suppose we are having the bone implants, these are one of the implants in which the retention time of nanomaterial in the human body is more, and these nanomaterials or bionanocomposites can become prone to microbial activity; the growth of host cells can lead to organ malfunction tumor and even cancer. So, it becomes necessary to study antimicrobial characteristics of bionanocomposite. A material used in the medical field shall possess high antimicrobial property. Different nanomaterials have different mechanisms to deal with microbial activity, and these can be summarized as given in Table 11.3 [75].

Using the mechanisms listed above, we can curb the effect of microbes on implants of bionanomaterials and bionanocomposites. Also, there are some materials that show some special biological properties like regenerative property and adaptation technology, which is beyond the scope of this chapter.

11.7 APPLICATION OF BIONANOCOMPOSITES

As we have discussed, bionanocomposites are a very important class of hybrid materials, composed of biopolymers and inorganic solids. Such biocompatible and biodegradable materials have proven to be a valuable gift to present and future generations. Environmental-friendly natural polymers, such

---

TABLE 11.3

Antimicrobial Action Mechanism of Nanomaterial in Bionanocomposites

Type of Nanomaterials	Antimicrobial Action Mechanism
Ag nanoparticle	It works through interaction with enzyme disulfide, causes the release of silver ions, disturbs metabolic processes, ruptures bacterial beoxy ribose nucleic acid (DNA) by impairing replication and stops the cell cycle.
ZnO nanoparticle	Causes bacterial membrane malfunction because of nanoparticle penetration and generation of reactive oxygen species (ROS) on the surface of the particles.
Al_2O_3 nanoparticle	Causes the bacterial membrane to develop pits and gaps, the bacterial cell wall to become disorganized, and the bacterial leakage of intracellular substances causes the bacteria to disappear.
MgO nanoparticle	Intracellular contents leaked out due to the rupture of cells and it leads to death of the bacterial cells.

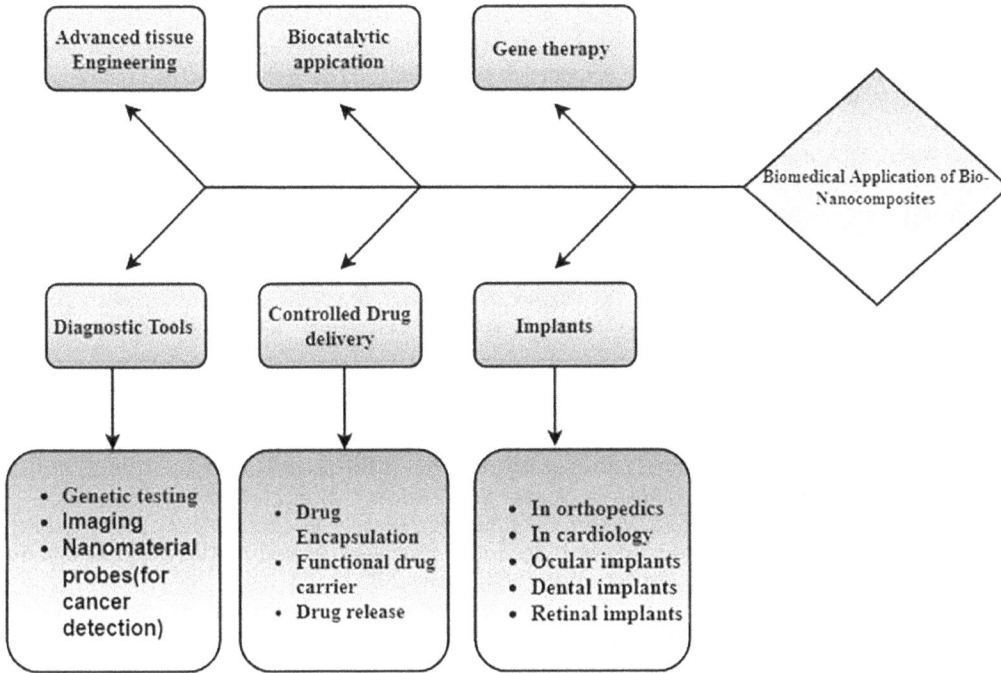

FIGURE 11.8   Application of bionanocomposites in the field of medical science.

as PLA, cellulose acetate, etc., have been widely used in the past few years. The bionanocomposites have large application in drug-delivery systems and tissue engineering in the food-packaging industry, etc. [76–80]. The food sector has used optically clear plasticized PLA-based bionanocomposite sheets for packaging. In vitro bone regeneration studies have been reported to use artificial bone tissue scaffolds based on natural hybrids of CA and nanohydroxyapatite (n-HA). Various medical science sector applications of bionanocomposites are depicted in Figure 11.8. However, the search for other new and economical synthesizing processes and materials for different requirements is a dynamic process.

11.8   CONCLUSION AND FUTURE SCOPE

The chapter covers introduction to bionanocomposites, their classification and characterization, and discussed in detail about the biological aspects of bionanocomposites. These materials have proven to be a milestone in the field of medical science due to their biological aspects. The bionanocomposites have proved to be highly potential for providing sustainable and ecofriendly solution of problems in the medical sector. With the use of nanomaterial technology, today's biomaterials possess improved biodegradability, biocompatibility and antimicrobial properties, without losing the bioadaptability. The graphene-based and other bionanocomposites discussed in this chapter are applied as biocatalyst, biomedical tools, bone repairing or replacement materials and food-packaging material, etc. This is the area that is quickly gaining attention of researchers and scientist community. However, there is a need on research and development for viable synthesis and processing methods, and for bionanocomposites, as it is still a challenge to gain a homogeneous dispersion of nano-sized reinforcement/filler in biopolymer matrix. The improper dispersion leads to inferior properties, and the cost factor is also to be considered and more analysis is required in this regard.

REFERENCES

1. R. K. Saini, A.K. Bajpai, E, Jain, Biodegradable and biocompatible polymer composites. In: *Biodegradable and Biocompatible Polymer Composites*. http://dx.doi.org/10.1016/B978-0-08-100970-3.00012-2.

2. M. Wu, Y. Wang, M. Wang, M. Ge, Effect of SiO_2 nanoparticles on the wear resistance of starch films (2008). *Fibres and Textile Eastern Europe*, 16, 96–99.

3. S. Ajmal, F. A. Hashmi, I. Imran, Recent progress in development and applications of biomaterials (2022), *International Conference on Advances in Materials and Mechanical Engineering*, 62(1). https://doi.org/10.1016/j.matpr.2022.04.233.

4. B. K. G. Theng, *Formation and Properties of Clay–Polymer Complexes*, Elsevier, New York, 1979. https://doi.org/10.1002/jpln.19801430317.

5. R. Rimašauskienė, R. Bhandari, A. K. Sharma, A. Aherwar, Materials used in composites: A comprehensive study (2019), *Journals of Materials Today: Proceedings*, 21(3), 1559–1562. http://dx.doi.org/10.1016/j.matpr.2019.11.086.

6. T. A. Nguyen, B. Han, S. Sharma, L. Longbiao, K. S., *Fiber-Reinforced Nanocomposites: An Introduction* (1st edition), 2020. eBook ISBN: 9780128199107.

7. A. Moorthy, D. N. Natarajan, R. Sivakumar, M. Manojkumar, M. Suresh, Dry sliding wear and mechanical behavior of aluminium/fly ash/graphite hybrid metal matrix composite using Taguchi method (2012), *International Journal of Modern Engineering Research*, 2(3), 1224–1230.

8. P. Ravindran, K. Manisekar, K. S. Vinoth, P. Rathika, Investigation of microstructure and mechanical properties of aluminum hybrid nanocomposites with the additions of solid lubricant (2013), *Material and Design*, 51, 448–456. http://dx.doi.org/10.1016/j.matdes.2013.04.015.

9. D. S. Prasad, C. Shoba, N. Ramanaiah, Investigations on mechanical properties of aluminum hybrid composites (2014), *Journal of Material Research and Technology*, 3(1), 79–85.

10. A. K. Kasar, N. Gupta, P. K. Rohatgi, P. L. Menezes, A brief review of fly ash as reinforcement for composites with improved mechanical and tribological properties (2020), *The Journal of Minerals, Metals & Material Society (TMS)*, 72. http://dx.doi.org/10.1007/s11837-020-04170-z.

11. H. D. Beyene, T. G. Ambaye, Application of sustainable nanocomposites for water purification process (2019). https://doi.org/10.1007/978-3-030-05399-4_14.

12. T. Hanemann, D.V. Szabó, Polymer-nanoparticle composites: From synthesis to modern applications (2010), *Materials*, 3(6), 3468–3517. https://doi.org/10.3390/ma3063468.

13. M. Baibarac, P. Gómez-Romero, Nanocomposites based on conducting polymers and carbon nanotubes: From fancy materials to functional applications (2006), *Journal of Nanoscience and Nanotechnology*, 6, 289–302. https://doi.org/ 10.1166/jnn.2006.903.

14. D. R. Paul, L. M. Robeson, Polymer nanotechnology: Nanocomposites (2008), *Polymer*, 49(15), 3187–3204. ISSN 0032-3861. https://doi.org/10.1016/j.polymer.2008.04.017.

15. K. C. Yung, W. M. Wu, M. P. Pierpoint, F. V. Kusmartsev, Introduction to graphene electronics - A new era of digital transistors and devices (2013), *Contemporary Physics*, 54, 233. https://doi.org/10.1080/00 107514.2013.833701.

16. S. Masoud, D. Ghanbari, Chapter 21: Polymeric nanocomposite materials. In *Advances in Diverse Industrial Applications of Nanocomposites*, 2011, pp. 501–518. https://doi.org/10.5772/1931.

17. M. Birkholz; U. Albers, T. Jung, Nanocomposite layers of ceramic oxides and metals prepared by reactive gas-flow sputtering (2004). *Surface and Coating Technology*, 179(2–3), 279–285. http://dx.doi.org/10.1016/S0257-8972(03)00865-X.

18. I. M. Low, Advances in ceramic matrix composites: Introduction. In I.M. Low (ed.) *Woodhead Publishing Series in Composites Science and Engineering, Advances in Ceramic Matrix Composites* (Second Edition), Woodhead Publishing, 2018, pp. 1–7. ISBN 9780081021668. https://doi.org/10.1016/B978-0-08-102166-8.00001-3.

19. A. Sommers, Q. Wang, X. Han, C. T'Joen, Y. Park, A. Jacobi, Ceramics and ceramic matrix composites for heat exchangers in advanced thermal systems – A review (2010), *Applied Thermal Engineering*, 30, 1277–1291. http://dx.doi.org/10.1016/j.applthermaleng.2010.02.018.

20. V. T. Le, N. S. Ha, N. S. Goo, Advanced sandwich structures for thermal protection systems in hypersonic vehicles: A review (2021), *Composites Part B: Engineering*, 226, 109301. https://doi.org/10.1016/j.compositesb.2021.109301.

21. G. H. Staab, Introduction to composite materials. In *Laminar Composites*, Butterworth-Heinemann, 1999, pp. 1–16. ISBN 9780750671248. https://doi.org/10.1016/B978-075067124-8/50001-1.

22. K. Palanikumar, Analyzing surface quality in machined composites. In eBook: *Machining Technology for Composite Materials*, 2012, pp. 154–182. https://doi.org/10.1533/9780857095145.1.154.

23. S. Aktas, E. Diler, Effect of ZrO$_2$ nanoparticles and mechanical milling on microstructure and mechanical properties of Al-ZrO$_2$ nanocomposites (2021), *Journal of Engineering Materials and Technology*, 143, 1–28. https://doi.org/10.1115/1.4050726.

24. J. Njuguna, K. Pielichowski, J. R. Alcock, Epoxy-based fibre reinforced nanocomposites (2007), *Advanced Engineering Materials*, 9(10), 835–847. https://doi.org/10.1002/adem.200700118.

25. M. K. Singh, T. Shokuhfar, J. J. D. A. Gracio, A. C. M. de Sousa, J. M. D. F. Fereira, H. Garmestani, S. Ahzi, Hydroxyapatite modified with carbon-nanotube-reinforced poly(methyl methacrylate): A nanocomposite material for biomedical applications (2008), *Advanced Functional Materials*, 18: 694–700. https://doi.org/10.1002/adfm.200700888.

26. C. L. Reichert, E. Bugnicourt, M.-B. Coltelli, P. Cinelli, A. Lazzeri, I. Canesi, F. Braca, B. M. Martínez, R. Alonso, L. Agostinis, S. Verstichel, L. Six, S. D. Mets, E. C. Gómez, C. Ißbrücker, R. Geerinck, D. F. Nettleton, I. Campos, E. Sauter, P. Pieczyk, M. Schmid, Bio-based packaging : Materials, modifications, industrial applications and sustainability (2020), *Polymers*, 12, 1558. https://doi.org/10.3390/polym12071558.

27. P. Gouma, R. Xue, C. P. Goldbeck, P. Perrotta, C. Balázsi, Nano-hydroxyapatite—Cellulose acetate composites for growing of bone cells (2012), *Materials Science and Engineering: C*, 32, 607–612. https://doi.org/10.1016/j.msec.2011.12.019.

28. L. Hsu, C. Weder, S. J. Rowan, Stimuli-responsive, mechanically-adaptive polymer nanocomposites (2011), *Journal of Materials Chemistry*, 21, 2812–2822. https://doi.org/10.1039/C0JMO2383C.

29. A. Pourjavadi, S. Rahemipoor, M. Kohestanian, Synthesis and characterization of multi stimuli responsive block copolymer-silica hybrid nanocomposite with core-shell structure via RAFT polymerization (2020), *Composites Science and Technology*. https://doi.org/10.1016/j.compscitech.2019.107951.

30. A. Bratek-Skicki, Towards a new class of stimuli-responsive polymer-based materials – Recent advances and challenges (2021), *Applied Surface Science Advances*, 4,100068. https://doi.org/10.1016/j.apsadv.2021.100068.

31. X. Qiu, S. Hu, "Smart" materials based on cellulose: A review of the preparations, properties, and applications (2013), *Materials (Basel)*, 6(3), 738–781. https://doi.org/10.3390/ma6030738.

32. A. Muhammad, S. Mohammed, A novel organoclay reinforced UHMWPE nanocomposite coating for tribological applications (2018), *Progress in Organic Coatings*, 118, 97–107. https://doi.org/10.1016/j.porgcoat.2018.01.028.

33. M. Erol, J. Hum, A. R. Boccaccini, Bioactive nanocomposites containing silicate phases for bone replacement and regeneration. In *Biomimetic Approaches for Biomaterials Development*, 2012, pp. 353–379. https://doi.org/10.1002/9783527652273.ch16.

34. Y. Xu, Q. Wu, Y. Sun, B. Hua, G. Shi, Three-dimensional self-assembly of graphene oxide and DNA into multifunctional hydrogels (2010), *Ameracian Chemical Society Nano*, 4, 7358–7362. https://doi.org/10.1021/nn1027104.

35. C. O. Crosby, B. Stern, N. Kalkunte, S. Pedahzur, S. Ramesh, J. Zoldan, Interpenetrating polymer network hydrogels as bioactive scaffolds for tissue engineering (2022), *Reviews in Chemical Engineering*, 38(3), 347–361. https://doi.org/10.1515/revce-2020-0039.

36. Y. Li, T. Xu, Z. Tu, W. Dai, Y. Xue, C. Tang, W. Gao, C. Mao, B. Lei, C. Lin. Bioactive antibacterial silica-based nanocomposites hydrogel scaffolds with high angiogenesis for promoting diabetic wound healing and skin repair (2020), *Theranostics*, 10(11), 4929–4943. https://doi.org/10.7150/thno.41839.

37. A. K. Gaharwar, S. A. Dammu, J. M. Canter, C.-J. Wu, G. Schmidt. Highly extensible, tough, and elastomeric nanocomposite hydrogels from poly(ethylene glycol) and hydroxyapatite nanoparticles (2011), *Biomacromolecules*, 12(5), 1641–50. https://doi.org/10.1021/bm200027z.

38. D. Khatayevich, M. Gungormus, H. Yazici, C. So, S. Cetinel, H. Ma, A. Jen, C. Tamerler, M. Sarikaya, Biofunctionalization of materials for implants using engineered peptides (2010), *Acta Biomaterialia*, 6(12), 4634–4641. https://doi.org/10.1016/j.actbio.2010.06.004.

39. Y. Zhao, Z. Li, Y. Jiang, H. Liu, Y. Feng, Z. Wang, H. Liu, J. Wang, B. Yang, Q. Lin, Bioinspired mineral hydrogels as nanocomposite scaffolds for the promotion of osteogenic marker expression and the induction of bone regeneration in osteoporosis (2020), *Acta Biomaterialia*, 113, 614–626. https://doi.org/10.1016/j.actbio.2020.06.024.

40. A. Childs, U. D. Hemraz, N. J. Castro, H. Fenniri, L. G. Zhang, Novel biologically-inspired rosette nanotube PLLA scaffolds for improving human mesenchymal stem cell chondrogenic differentiation (2013), *Biomedical Materials*, 8(6), 065003. https://doi.org/10.1089%2Ften.tea.2014.0138.

41. M.-H. You, M. K. Kwak, D.-H. Kim, Synergistically enhanced osteogenic differentiation of human mesenchymal stem cells by culture on nanostructured surfaces with induction media (2010), *Biomacromolecules*, 11(7), 1856–1862. https://doi.org/10.2144-0097.

42. F. W. Herrick, R. L. Casebier, J. K. Hamilton, K. R. Sandberg, Microfibrillated cellulose: Morphology and accessibility (1983), *Journal of Applied Polymer Science: Applied Polymer Symposium*, 37, 797–813. https://doi.org/10.12691/nnr-4-1-3.

43. D. Elieh-Ali-Komi, M. R. Hamblin, Chitin and chitosan: Production and application of versatile biomedical nanomaterials (2016), *International Journal of Advance Research (Indore)*, 4(3), 411–427.

44. C. P. Jiménez-Gómez, J. A. Cecilia, Chitosan: A natural biopolymer with a wide and varied range of applications (2020), *Molecules*, 25(17), 3981. https://doi.org/10.3390/molecules25173981.

45. S. Liu, S. Qin, M. He, D. Zhou, Q. Qin, H. Wang, Current applications of poly(lactic acid) composites in tissue engineering and drug delivery (2020), *Composites Part B: Engineering*, 199, 108238. https://doi.org/10.1016/j.compositesb.2020.108238.

46. R. V. Manek, P. F. Builders, W. M. Kolling, M. Emeje, O. O. Kunle, Physicochemical and binder properties of starch obtained from Cyperus esculentus (2012), *American Association of Pharmaceutical Scientiests Pharmaceutical Science and Technology*, 13(2), 379–88. https://doi.org/10.1208/s12249-012-9761-z.

47. B. S. Thorat Gadgil, N. Killi, G. V. N. Rathna, Polyhydroxyalkanoates as biomaterials (2017), *MedChemComm*, 8(9), 1774–1787. https://doi.org/10.1039/c7md00252a.

48. P. Gentile, I. Carmagnola, T. Nardo, V. Chiono, Layer-by-layer assembly for biomedical applications in the last decade (2015), *Nanotechnology*, 26, 422001. https://doi.org/10.1088/0957-4484/26/42/422001.

49. R. Nohria, R. Khillan, Y. Su, R. Dikshit, Y. Lvov, K. Varahramyan, Humidity sensor based on ultrathin polyaniline film deposited using layer by-layer nano-assembly (2006), *Sensors and Actuators B: Chemical*, 114, 218–222. https://doi.org/10.1016/j.snb.2005.04.034.

50. O. Güven, Radiation-assisted synthesis of polymer-based nanomaterials (2021), *Applied Sciences*, 11(17), 7913. https://doi.org/10.3390/app11177913.

51. S. Fu, Z. Sun, P. Huang, Y. Li, N. Hu, Some basic aspects of polymer nanocomposites: A critical review (2019), *Nano Materials Science*, 1(1), 2–30. https://doi.org/10.1016/j.nanoms.2019.02.006.

52. R. A. Vaia, E. P. Giannelis, Lattice model of polymer melt intercalation in organically-modified layered silicates (1997), *Macromolecules*, 30, 7990–7999. https://doi.org/10.1021/ma9514333.

53. M. Pommet, A. Redl, M.-H. Morel, S. Domenek, S. Guilbert, Thermoplastic processing of protein-based bioplastics: Chemical engineering aspects of mixing, extrusion and hot molding (2003), *Macromolecular Symposia*, 197, 207–218. https://doi.org/10.1002/masy.200350719.

54. Z. Shen, G.P. Simon, Y.-B. Cheng, Comparison of solution intercalation and melt intercalation of polymer–clay nanocomposites (2002), *Polymer*, 43, 4251–4260. https://doi.org/10.1016/S0032-3861(02)00230-6.

55. E. P. Rebitski, G. P. Souza, S. A.A. Santana, S. B.C. Pergher, A.C.S. Alcântara, Bionanocomposites based on cationic and anionic layered clays as controlled release devices of amoxicillin (2019), *Applied Clay Science*, 173, 35–45. https://doi.org/10.1016/j.clay.2019.02.024.

56. Y. Liu, J. Goebl, Y. Yin, Templated synthesis of nanostructured materials (2012), *Chemical Society Reviews*. https://doi.org/10.1039/c2cs35369e.

57. S. Murali, S. Kumar, J. Koh, S. Sinha, P. Singh, Bio-based chitosan/gelatin/Ag@ZnO bionanocomposites: Synthesis and mechanical and antibacterial properties (2019), *Cellulose*, 26, 5347–5361. https://doi.org/10.1007/s10570-019-02457-2.

58. A. D. Pomogailo, Polymer sol-gel synthesis of hybrid nanocomposites (2005), *Colloid Journal*, 67, 658–677. https://doi.org/10.1007/s10595-005-0148-7.

59. A. V. Vinogradov, V. V. Vinogradov, Low – temperature sol – gel synthesis of crystalline materials (2014), *Royal Society of Chemistry Advances*, 4, 1–17. https://doi.org/10.1039/C4RA04454A.

60. Y. Shchipunov, Bionanocomposites: Green sustainable materials for the near future (2012), *Pure and Applied Chemistry*, 84(12), 2579–2607. http://dx.doi.org/10.1351/PAC-CON-12-05-04.

61. A. W. Xu, Y. Ma, Biomimetic mineralization (2007), *Journal of Materials Chemistry*. http://dx.doi.org/10.1039/b611918m.

62. J. Xie, H. Ping, T. Tan, L. Lei, H. Xie, X.-Y. Yang, Z. Fu, Bioprocess-inspired fabrication of materials with new structures and functions (2019), *Progress in Materials Science*, 105, 100571. https://doi.org/10.1016/j.pmatsci.2019.05.004.

63. Y. Chen, Y. Feng, J. G. Deveaux, M. A. Masoud, F. S. Chandra, H. Chen, D. Zhang, L. Feng. Biomineralization forming process and bio-inspired nanomaterials for biomedical application: A review (2019), *Minerals*, 9(2), 68. https://doi.org/10.3390/min9020068.

64. P. Kohli, C. R. Martin, Template-synthesized nanotubes for biotechnology and biomedical applications (2005), *Journal of Drug Delivery Science and Technology*, 15(1), 49–57. https://doi.org/10.1016/S1773-2247(05)50006-6.

65. S. M. Alay-e-Abbas, K. Mahmood, A. Ali, M. I. Arshad, N. Amin, M. S. Hasan, Chapter-5: Characterization techniques for bionanocomposites. In eBook: *Bionanocomposites*, 2020. https://doi.org/10.1016/B978-0-12-816751-9.00005-2.

66. N. Pa'e, M. H. Salehudin, N. D. Hassan, A. M. Marsin, I. I. Muhamad, Thermal behavior of bacterial cellulose based hydrogels with other composites and related instrumental analysis. In M. Mondal (ed.) *Cellulose-Based Superabsorbent Hydrogels.* Polymers and Polymeric Composites: A Reference Series, Springer, Cham, 2018. https://doi.org/10.1007/978-3-319-76573-0_26-1.

67. S. Rana, M. Gupta, Variations in the mechanical properties of bionanocomposites by water absorption (2021), *Proceedings of the Institution of Mechanical Engineers, Part L: Journal of Materials: Design and Applications*, 235(7), 1655–1664. https://doi.org/10.1177/1464420721999694.

68. N. Baneshi, B. K. Moghadas, A. Adetunla, M. Yusmiaidil, P. M. Yusof, M. Dehghani, A. Khandan, S.-S, Saeed, D. Toghraie, Investigation the mechanical properties of a novel multicomponent scaffold coated with a new bio-nanocomposite for bone tissue engineering: Fabrication, simulation and characterization (2021), *Journal of Materials Research and Technology*, 15, 5526–5539. https://doi.org/10.1016/j.jmrt.2021.10.107.

69. A. Radhakrishnan, A. P. Saravana, Emerging strategies in bone tissue engineering. In *Tissue Engineering*, 2022. https://doi.org/10.1016/b978-0-12-824064-9.00013-7.

70. M. Kowalczuk, Intrinsically biocompatible polymer systems (2020), *Polymers (Basel)*, 12(2), 272. https://doi.org/10.3390/polym12020272.

71. Y. Tokiwa, B. P. Calabia, C. U. Ugwu, S. Aiba. Biodegradability of plastics (2009), *International Journal of Molecular Science*, 10(9), 3722–3742. https://doi.org/10.3390/ijms10093722.

72. G. Scott, Initiation processes in polymer degradation (1995), *Polymer Degradation and Stability*, 48(3), 315–324. https://doi.org/10.1016/0141-3910(95)00090-9.

73. R. S. Matche Baldevraj, R. S. Jagadish, Chapter 14: Incorporation of chemical antimicrobial agents into polymeric films for food packaging. In eBook: *Multifunctional and Nanoreinforced Polymers for Food Packaging*, Woodhead Publishing, 2011, pp. 368–420. ISBN 9781845697389. https://doi.org/10.1533/9780857092786.3.368.

74. A Staroń, O. Długosz, J. Pulit-Prociak, M. Banach, Analysis of the exposure of organisms to the action of nanomaterials (2019), *Materials (Basel)*, 5, 311–511. https://doi.org/10.3390/ma13020349.

75. L. Wang, C. Hu, L. Shao, The antimicrobial activity of nanoparticles: Present situation and prospects for the future (2017), *International Journal of Nanomedicine*, 12, 1227–1249. https://doi.org/10.2147/IJN.S121956.

76. K. R. Saravana, R. Vijayalakshmi, Nanotechnology in dentistry (2006). *Indian Journal of Dental Research*, 17(2), 62–65. https://doi.org/10.1016/j.sdentj.2012.12.002.

77. M. X. Tang, C. T. Redemann, F. C. Szoka, In vitro gene delivery by degraded polyamidoamine dendrimers (1996), *Bioconjugate Chemistry*, 7, 703–714. https://doi.org/10.1021/bc9600630.

78. M. Jorfi, M. N. Roberts, E. J. Foster, C. Weder, Physiologically responsive, mechanically adaptive bionanocomposites for biomedical applications (2013), *Americal Chemical Society: Applied Materials & Interfaces*, 5(4), 1517–1526. https://doi.org/10.1021/am303160j.

79. B. Joseph, V. K. Sagarika, C. Sabu, N. Kalarikkal, S. Thomas, Cellulose nanocomposites: Fabrication and biomedical applications (2020), *Journal of Bioresources and Bioproducts*, 5(4), 223–237. https://doi.org/10.1016/j.jobab.2020.10.001.

80. H. Hamedi, S. Moradi, S. M. Hudson, A.E. Tonelli, M. W. King, Chitosan based bioadhesives for biomedical applications: A review (2022), *Carbohydrate Polymers*, 282, 119100. https://doi.org/10.1016/j.carbpol.2022.119100.

12 Strategies to Improve the Micro-electrical Discharge Machining Performance of CFRP

*Hrishikesh Dutta, Kishore Debnath,
Deba Kumar Sarma, and J. Jayaramudu*

CONTENTS

12.1 INTRODUCTION

With the introduction of composite materials into the manufacturing industry with a motive to gain superior property characteristics by replacing the conventional counterparts, Carbon fiber reinforced polymer (CFRP) is also finding its place, specifically in the aerospace and automotive sector, because of its exhibition of better mechanical strength along with good fatigue and corrosion resistance [1,2]. There is evidence of high use of CFRP (up to 50% of the structural components) in the aerospace industry, including components like wing spars and fuselage [3]. The study also revealed that the overall turbulence in an airplane can be minimized by micro-perforating the

DOI: 10.1201/9781003343912-12

tail which causes the transition of the airflow around the plane from turbulent to laminar [4]. The prevalence of CFRP products can also be observed in the sports goods manufacturing sectors that produce racquet frames, golf clubs, fishing rods, and tripods [5]. Application of CFRP components having precisely fabricated micro-features can be seen in the products that are used for the measurement of vibration and temperature [6]. Normally, CFRP products are obtained by using conventional machining methods such as slotting, milling, drilling, etc. But, these methods come with some disadvantages like delamination, fiber swelling, burr formation, and fiber pullout [7,8]. Hence, nonconventional processes are being used for machining CFRP so that the damages induced by mechanical forces and thermal load during machining can be avoided. Manufacturers around the globe have started using nontraditional machining processes namely electrical discharge machining (EDM) to machine CFRP. However, it is again a tough task to machine CFRP by the EDM process due to some particular reasons. Carbon fiber possesses a high value thermal degradation temperature than that of polymers. Therefore, the sensitivity of polymer matrices toward heat is high and the average melting temperature of polymers is approximately 391 K [1]. Moreover, there is a huge dissimilarity in the values of electrical conductivity of epoxy resin (10^{-16} S/cm) [9] and carbon fiber ($7.14 \times 10^2 - 12.25 \times 10^2$ S/cm) [1]. A feasibility study was carried out by Lau et al. [7] for machining CFRP by the EDM process. In their study, it was revealed that the copper tool has better capability to machine CFRP than graphite tool considering tool wear. When a higher value of current was used, the fiber–resin debonding, matrix melting, and expansion of fibers due to intense thermal load were apparent during machining. A study was carried out by Lau and Lee [10] to compare machining performance of CFRP while using the laser cutting process and wire electrical discharge machining (WEDM) process. The WEDM process proved to be the better among the two methods in terms of the ability to produce features having fewer damages and better surface finish compared to laser cutting. Also, the extent of heat-affected zone in WEDM was less than that of laser cutting. Lodhi et al. [11] studied how the discharge current, pulse-on time, and gap voltage affect the roughness of the CFRP surface during EDM. The discharge current was found to be the most significant factor. A low value of surface roughness was attained by applying a pulse duration of 70 µs, current of 2 amp, and gap voltage of 50 volt. Gourgouletis et al. [12] carried out experiments to establish the feasibility of using EDM to make holes in CFRP. The best machining performance in terms of material removal rate (MRR) was achieved by applying straight polarity with pulse-on times in the range of 500–750 µs and a current of 1.5 amp. A comparison between dry EDM and vibration (ultrasonic)-assisted EDM of CFRP was carried out by Kurniawan et al. [13]. While assessing the burr-removing capability of the processes, it was revealed that copper tool could yield better results compared to that of brass tool. Yue et al. [14] studied the material removal mechanism involved in EDM of CFRP. It was revealed that during EDM of CFRP, material is removed through a combination of various physical processes such as melting and vaporization, sublimation, and oxidation. Sheikh-Ahmad and Shinde [15] fabricated holes in CFRP by EDM using tools made of graphite and copper. Graphite tool showed better performance in terms of MRR at higher values of pulse duration and current. Seikh-Ahmad [16], in another work, fabricated holes in CFRP by EDM process and the study revealed that the main region of occurrence of delamination was the hole inlet. It was observed that the extent of delamination was mostly affected by the gap current. Kumar et al. [17] used the micro-EDM process for fabricating through micro-holes in CFRP laminate. According to their study, drilling of CFRP by EDM could be done if only the discharge was targeted on the conductive carbon fiber. Dutta et al. [1] explained the material removal mechanism involved in the making of blind micro-holes in CFRP by EDM aided by assisting electrode and rotating tool. The effect of the parameters namely tool speed, voltage, and pulse duration on surface morphology and MRR were studied. The most significant factor that affected the MRR was voltage. Dutta et al. [18] studied the process of micro-electrical discharge machining (µEDM) on CFRP and found the optimum machining condition for minimum hole deviation through grey relational analysis.

12.2 PRESENT EXPERIMENTAL WORK ON MICRO-ELECTRICAL DISCHARGE MACHINING OF CFRP

In the work by Kumar et al. [17], conductive carbon fibers were uncovered from the polymer resin using mechanical methods such as rubbing and shaping so that the initial electrical connection between the tool and workpiece could be established. In the present work, during the preliminary investigation, experiments were carried out following the methods mentioned in the available literature such as Kumar et al. But the methods of exposing conductive carbon fibers to initiate the sparking process did not prove to be efficient.

In order to improve the µEDM performance of CFRP, an innovative method of µEDM assisted by conductive layers and a rotating tool was used in the present endeavor. The work on micro-hole fabrication was divided into three parts. In the first part, blind holes were machined in CFRP plates (Figure 12.1), and in the second part, through-micro-holes were fabricated. In the last part, powder-mixed µEDM (PMµEDM) of CFRP was carried out.

12.2.1 FABRICATION OF BLIND-MICRO-HOLES IN CFRP

Blind micro-holes were fabricated in CFRP by µEDM process. The µEDM of CFRP was assisted by the rotational motion of the tool and also by the presence of the conductive sheet.

12.2.1.1 Fabrication of CFRP Workpiece

The work material for the present study was polymer composite reinforced with carbon fiber or CFRP composite that was prepared by the process of hand lay-up. The carbon fiber used for the work was in the form of plain weave mats (3k). Araldite LY556 resin was used as the matrix and HY951 was used as the hardener. Both the hardener and the resin were mixed in a proportion of 1:10. The resin was poured over a steel plate and rectangular carbon fiber mats of the required size were stacked one after another in between the layers of resin. In the end, another steel plate was put over the stack, and the whole arrangement was allowed to cure for 20 hours at room temperature.

12.2.1.2 Assisting Conductive Sheet and Tool Materials

A 70 µm thick sheet layered with copper on its surface was used for assisting the sparking process during µEDM of CFRP. The cylindrical tool of 960 µm diameter was made from a copper block by WEDM. The image of the copper tool is shown in Figure 12.2.

FIGURE 12.1   Copper tool.

12.2.1.3 Machining Setup

A micro-machining setup (HYPER-15) was used for the fabrication of micro-features in CFRP. The photographic view of the machine labeling all the important parts is shown in Figure 12.3. The motions of the worktable along the X, Y, and Z axes are limited to 130, 75, and 80 mm, respectively. The machine comes with ±5 µm position accuracy and ±1 µm repeatability for all directions. For drive control of the machine, it has a multi-phase step control motor with 0.1 µm resolution. A brushless D.C. motor drives the electrically insulated spindle of the setup within a speed range of 200–3000 rpm. Hydrocarbon oil was used as the dielectric for the process. A schematic of the experimental setup along with the machining process is shown in Figure 12.4.

12.2.1.4 Parameters

For carrying out the experiments, a few parameters were chosen so that the machining performance could be enhanced by adjusting the values of these parameters. The input parameters selected for the study were pulse duration, voltage, and tool speed. Previous literature on similar works was

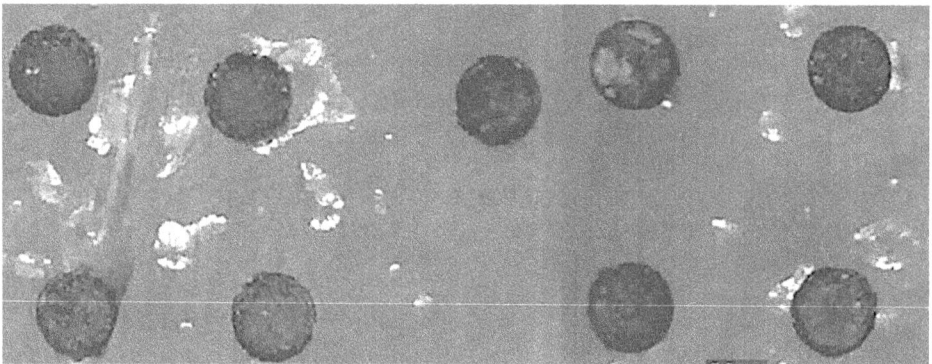

FIGURE 12.2 Blind-micro-holes.

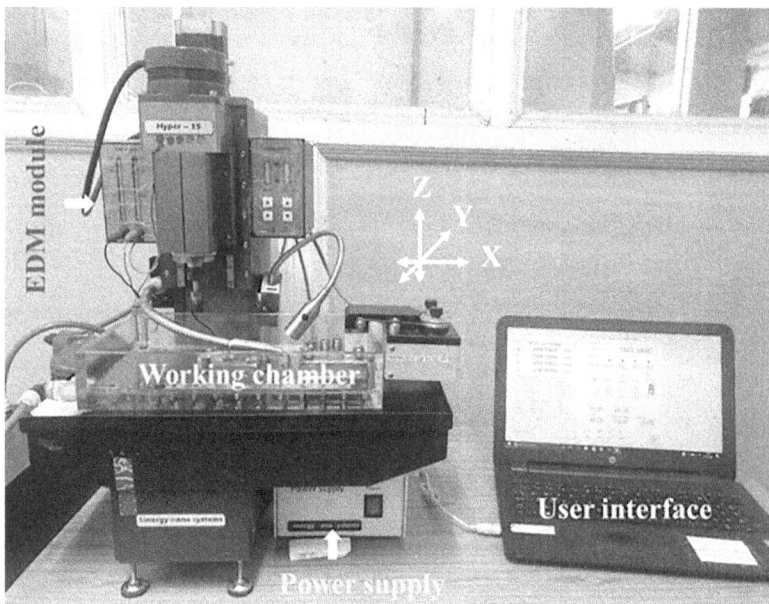

FIGURE 12.3 Micro-machining setup.

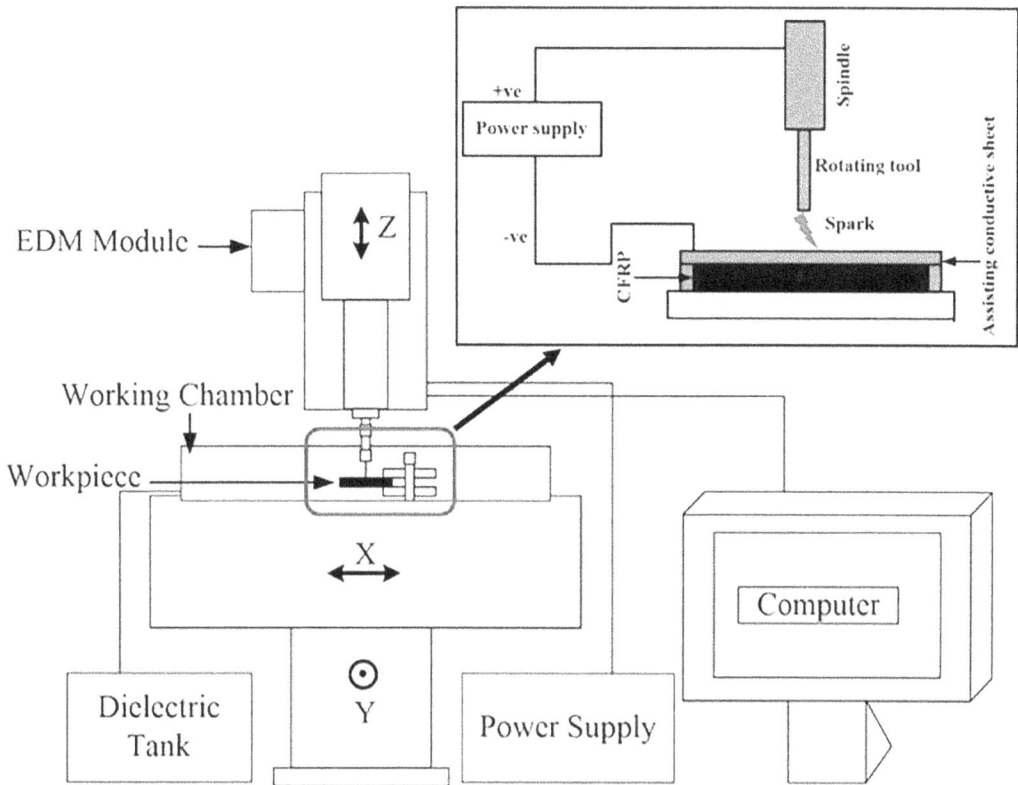

FIGURE 12.4 Schematic of experimental setup and the fabrication process.

followed to select the parameters and their levels [4,7]. The behavior of these parameters on the rate of material removal was studied by performing different analysis and surface characterization. The factors, their units, and their different levels are listed in Table 12.1.

12.2.1.5 Experimental Design

The design of experiment for the current study was an orthogonal array (L9) as per Taguchi's method. The "larger the better" approach was adopted for the current study. The L9 orthogonal array was chosen for the experimentation in order to minimize the number of experiments. Use of L9 orthogonal array also helped reduce the experimentation time and provided idea formulation. The output response selected for the analysis was MRR. The experiment was designed and analyzed using "MINITAB 17" software. Also, regression analysis was done to understand the behavior of the parameters and how their variation impacts the fluctuation in MRR.

12.2.1.6 Taguchi and Regression Analysis

The results of the Taguchi analysis are presented in Table 12.2. The "signal-to-noise" ratio for different experimental runs considering various combinations of input parameters was evaluated and is presented in the table. According to the "larger the better" criteria for MRR, a high value of the "signal-to-noise" ratio indicates a better machining condition. A higher MRR is expected to lead to better productivity of the machining process. Therefore, the motive of the analysis was to find out the optimum level parameters that yield the maximum MRR. The responses for "signal-to-noise" ratio for MRR are shown in Table 12.3. From the table, it can be concluded that the highest "signal-to-noise" ratio for voltage, tool speed, and pulse duration corresponds to the 3rd, 1st, and 3rd level

TABLE 12.1
Parameters and Their Levels

Parameters	Level 1	Level 2	Level 3
Voltage (volt)	115	135	155
Pulse Duration (µs)	20	40	60
Tool Speed (rpm)	250	350	450

of the mentioned parameters respectively. Therefore, it can be stated that the optimum values of input parameters for the best machining performance in this work are 155 volt, 20 µs, and 450 rpm. The results from the "analysis of variance" are presented in Table 12.4. It is evident from the table that voltage is the parameter having the highest influence on the variation of MRR as it has the highest F-value. The significance of the parameters was understood by looking at the p-values for them. As seen in Table 12.4, voltage and tool speed are significant parameters in this study as their p-values are less than 0.05 (95% confidence interval) while pulse duration proved to be an insignificant parameter as it has a p-value of 0.079 (>0.05). The graphs between data means and mean of "signal-to-noise" ratio are shown in Figure 12.5. The graphs confirm the fact that voltage affects MRR variation the most during the present study. An increase in the heat energy at a higher voltage led to faster MRR by melting more amount of workpiece material [1]. The tool speed is also fairly significant according to the current analysis. During µEDM of CFRP, the rotational movement of the tool helps flush out the debris from the vicinity of the working gap. Therefore, a variation in the tool speed certainly affects the MRR.

12.2.1.7   Morphological Study

The morphology of the workpiece surface after machining was investigated by field emission scanning electron microscope (FESEM) imaging. As observed in Figure 12.6a, breakage of fiber is prominent during µEDM of CFRP. The main reason for fiber breakage is the action of spalling. Spalling is a phenomenon where a material fails by breaking due to development of cracks in it. During µEDM of CFRP, excessive heat is generated and is experienced by both matrix and fibers. Due to the thermal load provided by the heat energy, numerous micro-carcks are developed on the surface of the carbon fibers. As the machining continues, these cracks propagate and finally result in breakage of the fibers by spalling. Some amount of overcut is also observed in Figure 12.6a. This overcut is the result of sparking at the interface of the tool's surface and inner surface of the machined hole. A centrifugal force is induced due to the tool's rotation in the dielectric around the tool–workpiece interphase, and with the action of this force, the debris are thrown away from the working zone, hence facilitating efficient machining [2]. When the tool speed is low, there may be some debris stuck to the inside surface of the hole that are unable to escape the machining zone due to low centrifugal force. Clusters of debris accumulated on the machined surface are visible in Figure 12.6b. Sometimes, these debris along with some protruded fibers (Figure 12.6a) can act as the spot for spark generation and this may lead to overcutting of the hole.

12.2.1.8   Assessment of Material Removal Mechanism

It is a very tough task to understand the mechanism of material removal in µEDM of CFRP, the reason being the anisotropic and inhomogeneous composition of CFRP that leads to a complex process of material removal. Although the primary mechanism of material removal during µEDM of CFRP is melting and vaporization, still it is not clearly understood how exactly the materials are removed while machining CFRP by µEDM. Previous studies on machining of materials with very little or no electrical conductivity suggested that continuous removal of material is possible during EDM

TABLE 12.2
"Signal-to-Noise" Ratio for Different Experimental Runs

Exp. No.	Voltage (volt)	Pulse Duration (μs)	Tool Speed (rpm)	MRR (μm³/s)	S/N Ratio (db)
1	115	20	250	169484.18	104.9095
2	115	40	350	151298.76	103.9621
3	115	60	450	198956.41	106.2544
4	135	20	350	267467.81	108.7540
5	135	40	450	273292.40	108.9367
6	135	60	250	212906.77	106.8250
7	155	20	450	334764.94	110.6618
8	155	40	250	262279.41	108.5879
9	155	60	350	267571.54	108.7573

TABLE 12.3
Response for "Signal-to-Noise" Ratio

Level	Voltage	Pulse Duration	Tool Speed
1	104.7	107.9	106.5
2	107.9	106.9	106.9
3	109.1	107.0	108.4
Delta	4.4	1.0	1.9
Rank	1	3	2

TABLE 12.4
Results from "Analysis of Variance"

Source	DF	Adj SS	Adj MS	F-Value	P-Value
Voltage	2	20663513839	10331756919	137.71	0.0.007
Pulse duration	2	1752244773	876122386	11.68	0.079
Tool speed	2	4739354195	2369677098	31.59	0.031
Error	2	150048497	75024248		
Total	8	27305161303			

of these kinds of materials with the assistance of a conductive layer [3,4]. The conductive layer is composed of carbon particles that are formed by pyrolysis of the hydrocarbon-based dielectric oil. This conductive carbon layer helps in the continuation of the material removal mechanism during μEDM of CFRP by assisting the sparking process.

In the present study, an attempt was made to understand the complex mechanism of material removal during μEDM of CFRP. The process of material removal can be divided into two phenomena:

1. Spark generation and its continuation.
2. Material removal by spalling.

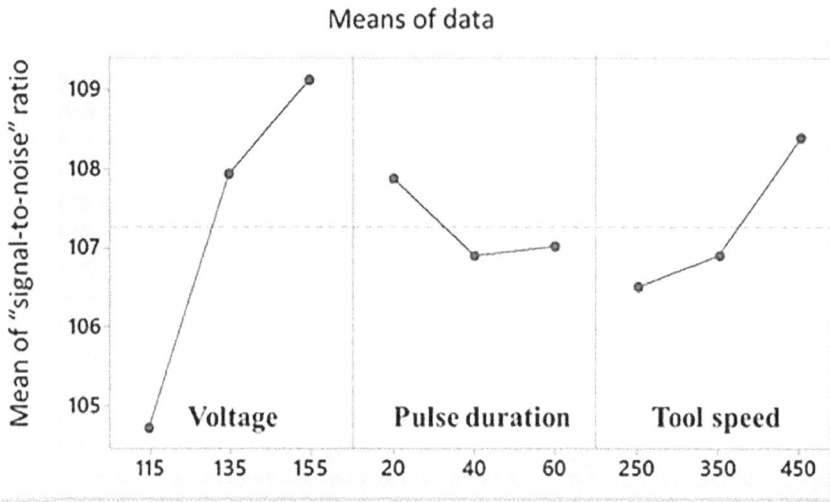

FIGURE 12.5   Main effect plots for "signal-noise-ratio."

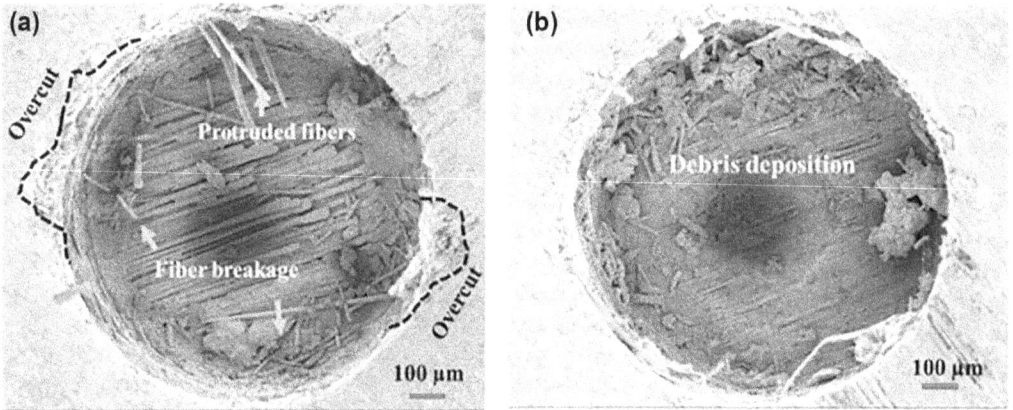

FIGURE 12.6   FESEM images of micro-holes.

12.2.1.8.1   Spark Generation and Its Continuation

First, when the tooltip is brought near the workpiece surface, an electrical spark generates between the tooltip and the conductive sheet present over the workpiece surface. This is the beginning of the machining process. Next, after the conductive sheet is melted out, sparking continues further due to the presence of the pyrolytic carbon layer.

12.2.1.8.2   Material Removal by Spalling

The generated heat, when propagates through the fiber and matrix, exerts an immense thermal load on them and micro-cracks are developed due to this thermal stress. These micro-cracks finally lead to failure of the materials by spalling. A previous study on EDM of nonconductive materials also confirmed that spalling is the phenomenon that is responsible for material removal [5].

12.2.2   Fabrication of Through-micro-holes in CFRP

In the next part of the current work, through-micro-holes were fabricated on CFRP composites. A copper tool of 600 μm in diameter was used to make 1500 μm deep through-micro-holes.

12.2.2.1 Parameters and Design of Experiment

The input parameters for the study were voltage, pulse duration, and tool speed. The effect of these parameters on machining time was assessed by performing the experiments as per Taguchi's L16 design of experiment. The input parameters and their level values are presented in Table 12.5.

12.2.2.2 Taguchi Analysis

Taguchi analysis was carried out using Minitab 17 for L16 algorithm. The results of the analysis showing the value of "signal-to-noise" ratio are shown in Table 12.6. The responses for "signal-to-noise" ratio are presented in Table 12.7. As per the "smaller-the-better" criteria, the optimum levels of input parameters for a minimum machining time are level 4 for all the three parameters. Hence, the optimum values of input parameters for the present study are a voltage of 160 volt, pulse duration of 70 μs, and tool speed of 600 rpm. This can also be confirmed in Figure 12.7 where the graphs for the input parameters are presented. The respective inclination of the graphs denotes the effectiveness of the parameters in the variation of machining time.

12.2.2.3 Analysis of Variance

Analysis of variance was carried out using MINITAB 17 in order to find out the significance of the input parameters and their relative level of impact on the variation of machining time. The results

TABLE 12.5

Input Parameters and Their Levels for Through-Micro-Hole Fabrication

Factors	Level 1	Level 2	Level 3	Level 4
Voltage (volt)	100	120	140	160
Pulse duration (μs)	10	30	50	70
Tool speed (rpm)	150	300	450	600

TABLE 12.6

Results of Experiments Showing Machining Time and "Signal-to-Noise" Ratio

Exp. No.	Voltage (volt)	Pulse Duration (μs)	Tool Speed (rpm)	Machining Time (s)	S-N Ratio (db)
1	100	10	150	2451	−67.7869
2	100	30	300	2260	−67.0822
3	100	50	450	2200	−66.8485
4	100	70	600	2050	−66.2351
5	120	10	300	2195	−66.8287
6	120	30	150	2240	−67.005
7	120	50	600	1986	−65.9596
8	120	70	450	2145	−66.6285
9	140	10	450	2005	−66.0423
10	140	30	600	1892	−65.5384
11	140	50	150	1950	−65.8007
12	140	70	300	1900	−65.5751
13	160	10	600	1653	−64.3655
14	160	30	450	1700	−64.609
15	160	50	300	1745	−64.8359
16	160	70	150	1780	−65.0084

TABLE 12.7

Response Table for "Signal-to-Noise" Ratio

Level	Voltage	Pulse Duration	Tool Speed
1	−66.99	66.26	66.40
2	66.61	−66.06	−66.08
3	−65.74	−65.86	−66.03
4	−64.70	−65.86	−65.52
Delta	2.28	0.39	0.88
Rank	1	3	2

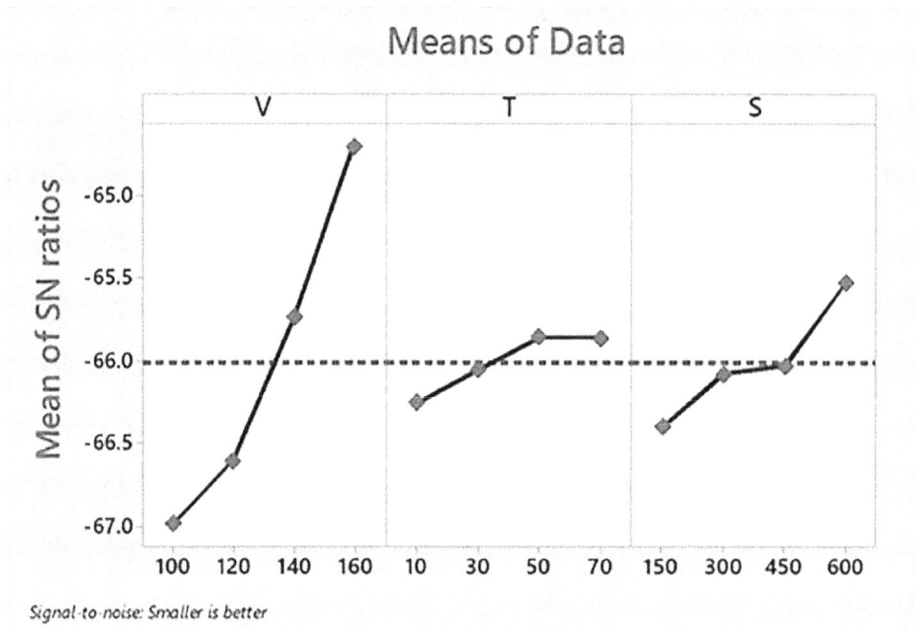

FIGURE 12.7 Graphs between means of "signal-to-noise" ratio and data means.

found from the analysis are shown in Table 12.8. For a confidence interval of 95%, the p-value for a factor must be less than 0.05 to be significant for the study. As seen in the table, the p-values for voltage and tool speed are less than 0.05, and hence, these factors can be considered significant in the context of the current study, whereas the p-value for pulse duration is 0.07 which is greater than 0.05, and it reveals that pulse duration is a relatively insignificant factor in this study.

12.2.2.4 Morphological Study

Morphological study of the through-micro-holes was done by analyzing them by FESEM. A hole with good circularity was obtained as seen in Figure 12.8. Some amount of peripheral surface damage resulted from the excessive heat produced during the μEDM of CFRP. Redeposition of material around the hole periphery can be seen in Figure 12.9. During μEDM of CFRP, the rotation of the tool electrode produces a centrifugal force around the tool–workpiece interface. This centrifugal force throws out the debris that are produced from removal of material from the CFRP workpiece. At a slower tool speed, the intensity of the centrifugal force is low, and hence, some debris cannot escape the interelectrode gap due to lack of assistance of the required throwing force. These debris get accumulated on the inside surface of the hole and around the periphery of the hole and are seen as redeposited layers of materials.

TABLE 12.8
Results of Analysis of Variance

Source	DF	Adj SS	Adj MS	*F*-Value	*P*-Value
Voltage	1	640248	213416	82.22	0.000
Pulse duration	1	31223	10408	4.01	0.070
Tool speed	1	89882	29961	11.54	0.007
Error	12	15574	2596		
Total	15	776926			

R-sq: 94.61% R-sq(adj): 93.27% R-sq(pred): 90.27%

FIGURE 12.8   FESEM image of through-micro-hole.

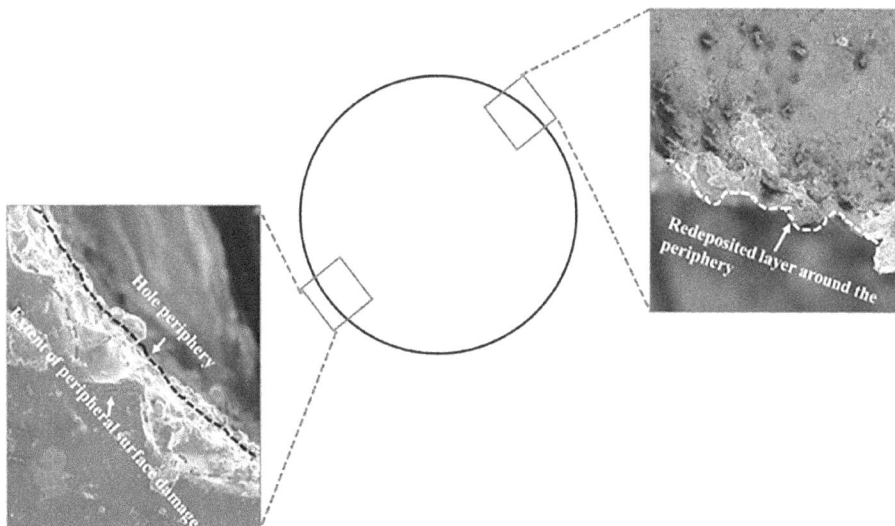

FIGURE 12.9   FESEM image showing peripheral damage and redeposition.

12.2.3 Powder-Mixed µEDM (PMµEDM) of CFRP

In this work, blind micro-holes were fabricated in CFRP composite by the PMµEDM method.

12.2.3.1 Parameters and Design of Experiment

The input parameters considered for the study were powder concentration (1, 3.5, and 6 g/L), capacitance (100, 1000, and 10,000 µF), and tool rotation (350, 450, and 550 rpm) (Table 12.9). A full factorial design of experiment was adopted for the experiments. In order to assess the significance of the input parameters, regression analysis was done.

12.2.3.2 Variation in Machining Time

The results of the experiment are shown in Table 12.10. The minimum machining time of 414 s was seen when the concentration of zinc powder was 6 g/L. The corresponding capacitance and tool speed for minimum machining time were 1000 pF and 550 rpm. It can be concluded from the results that the machining time decreased with an increase in the powder concentration. When the amount of powder particles is increased, early breakdown of the dielectric occurs due to the presence of charged particles which reduce the dielectric strength leading to the early breakdown of the dielectric [6]. Hence, machining time tends to decrease with the enhancement of powder concentration. It can be noted that capacitance also has a similar effect on machining time as that of powder concentration. With the increase in capacitance, the discharge energy increases, and with the increased

TABLE 12.9
Input Parameters and Their Levels for PMµEDM

Factors	Level 1	Level 2	Level 3
Powder concentration (mg/L)	1	3.5	6
Capacitance (µF)	100	500	1000
Tool rotation (rpm)	350	450	550

TABLE 12.10
Experimental Results for Machining Time

Exp. Run	Powder Concentration (g/L)	Capacitance (pF)	Tool Speed (RPM)	Machining Time (s)
1	1	100	350	1054
2	1	100	450	901
3	1	100	550	815
4	1	500	350	804
5	1	500	450	786
6	1	500	550	702
7	1	1000	350	676
8	1	1000	450	651
9	1	1000	550	635

(Continued)

TABLE 12.10 *(Continued)*
Experimental Results for Machining Time

Exp. Run	Powder Concentration (g/L)	Capacitance (pF)	Tool Speed (RPM)	Machining Time (s)
10	3.5	100	350	617
11	3.5	100	450	592
12	3.5	100	550	587
13	3.5	500	350	576
14	3.5	500	450	570
15	3.5	500	550	566
16	3.5	1000	350	560
17	3.5	1000	450	550
18	3.5	1000	550	541
19	6	100	350	539
20	6	100	450	478
21	6	100	550	466
22	6	500	350	449
23	6	500	450	441
24	6	500	550	436
25	6	1000	350	430
26	6	1000	450	421
27	6	1000	550	414

discharge energy, a higher amount of heat is generated at the workpiece–tool interface. Higher tool speed results in lower machining time as improved flushing condition helps in efficient machining leading to higher material removal [7].

12.2.3.3 Regression Analysis

To assess the influence of powder concentration, capacitance, and tool speed (S) on machining time, regression analysis was performed. The results of ANOVA for machining time are shown in Table 12.11. The low p-values (<0.05) of all three input parameters prove the significance of the parameters in the variation of machining time. The high R-sq. value of 87.45% (Table 12.12) confirms the existence of adequate fitness of the proposed statistic model.

12.2.3.4 Study of Machined Surface Morphology

The surfaces of the machined holes were studied by scanning electron microscope analysis. The debris accumulation is clearly visible on the surface of the hole as seen in Figure 12.10. The reason behind this accumulation of debris is inadequate flushing. Another way of formation of debris is the adherence of powder particles to the debris and getting settled on the machined surface. Damages such as spalling and matrix cracking are also visible in Figure 12.10. The tremendously high heat energy generated during PMµEDM leads to the generation of thermal stresses at the interface of the fiber and matrix. Micro-cracks are developed due to this thermal stress and these cracks propagate through the matrix and fibers. An overcut region was observed around the desired peripheral profile of the hole. The wobbling effect induced by the rotational motion of the tool and its eccentricity are the possible reasons for hole overcut.

TABLE 12.11
ANOVA for Machining Time

Source	D.F.	Adj. S.S.	Adj. M.S.	F-Value	P-Value
Regression	3	574445	191842	53.44	0.000
Powder concentration (PC)	1	483472	483472	134.94	0.000
Capacitance (C)	1	74952	74952	20.82	0.000
Tool speed (S)	1	16380	16380	4.57	0.043
Error	23	82406	3583		
Total	26	656851			

TABLE 12.12
Model Summary for Machining Time

R-Sq.	R-Sq. (Adj.)	R-Sq. (Pred.)
87.45%	85.82%	81.63%

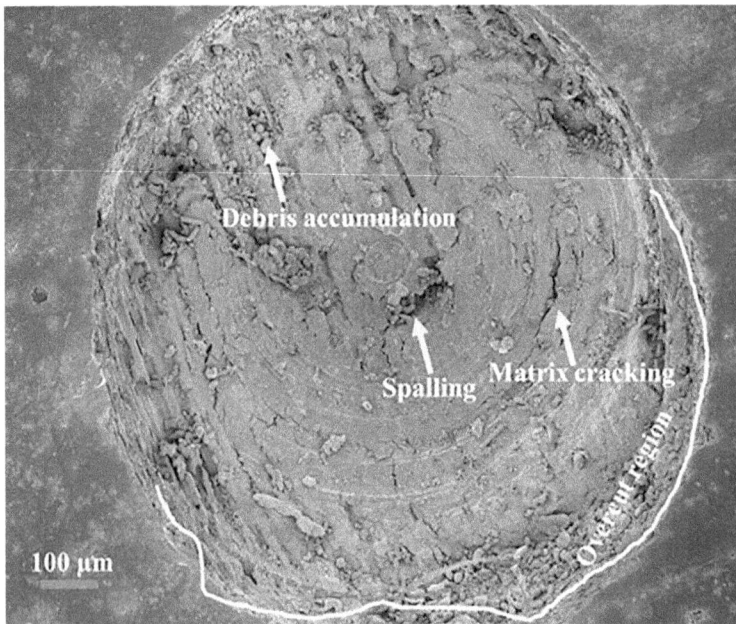

FIGURE 12.10 FESEM image of machined surface.

12.3 CONCLUSION AND FUTURE SCOPE

In the present study, µEDM of CFRP was performed using different strategies. The following salient points can be drawn from the current investigation:

1. Efficient machining of CFRP by µEDM is possible if we use an assisting electrode along with a rotating tool.
2. Voltage is the most important parameter that affects the MRR, machining time, and surface behavior of CFRP during µEDM of CFRP.

3. Statistical analysis shows that voltage, pulse duration, and tool speed are significant parameters.
4. PMµEDM of CFRP revealed that concentration of zinc powder plays a vital role in enhancing the machining performance.
5. Machining time is reduced at a higher concentration of zinc powder due to the bridging effect of the powder particles in the gap between the tool and workpiece.
6. Damages such as fiber breakage, matrix cracking, overcut, and spalling are encountered during µEDM of CFRP.

Although extensive investigation of µEDM of CFRP has been done in the current endeavor, further works can be carried out in the area of EDM of CFRP with the objectives of achieving micro-features with the highest surface quality and dimensional accuracy.

REFERENCES

1. Dutta, H., Debnath, K. and Sarma, D.K., 2019. A study of material removal and surface characteristics in micro-electrical discharge machining of carbon fiber-reinforced plastics. *Polym Compos*, 40(10), pp. 4033–4041.
2. Teimouri, R. and Baseri, H., 2012. Study of tool wear and overcut in EDM process with rotary tool and magnetic field. *Adv Tribol*, 2012, pp. 1–8.
3. Mohri, N., Fukuzawa, Y., Tani, T., Saito, N. and Furutani, K., 1996. Assisting electrode method for machining insulating ceramics. *CIRP Annals*, 45, pp. 201–204.
4. Mohri, N., Fukusima, Y., Fukuzawa, Y., Tani, T. and Saito, N., 2003. Layer generation process on work-piece in electrical discharge machining. *CIRP Annals*, 52(1), pp. 157–160.
5. Agarwal, N., Shukla, S. and Agarwal, V., 2015. Investigation of material removal method in EDM for non-conductive materials. *Eur J Adv Eng Technol*, 2(4), pp. 11–13.
6. Dutta, H., Debnath, K. and Sarma, D.K., 2022. Preliminary investigation of powder-mixed microelectrical discharge drilling of CFRP composite. *Surf Rev Lett*, 29(8), p. 2250111.
7. Lau, W.S., Wang, M. and Lee, W.B., 1990. Electrical discharge machining of carbon fiber composite materials. *Int J Mach Tools Manuf*, 30(2), pp. 297–308. https://doi.org/10.1016/0890-6955(90)90138-9.
8. Hocheng, H. and Tsao, C.C., 2005. The path towards delamination-free drilling of composite materials. *J Mater Process Technol*, 167(2–3), pp. 251–264. https://doi.org/10.1016/j.jmatprotec.2005.06.039.
9. Mun, S.Y., Lim, H.M. and Lee, S.H., 2018. Thermal and electrical properties of epoxy composite with expanded graphite-ceramic core-shell hybrids. *Mater Res Bull*, 97, pp. 19–23.
10. Lau, W.S. and Lee, W.B., 1991. A comparison between EDM wire-cut and laser cutting of carbon fibre composite materials. *Mater Manuf Process*, 6(2), pp. 331–342.
11. Lodhi, B.K., Verma, D. and Shukla, R., 2014. Optimization of machining parameters in EDM of CFRP composite using Taguchi technique. *Int J Mech Eng Technol (IJMET)*, 5(10), pp. 70–77
12. Gourgouletis, K., Vaxevanidis, N.M., Galanis, N.I. and Manolakos, D.E., 2011. Electrical discharge drilling of carbon fibre reinforced composite materials. *Int J Mach Mach Mater*, 10(3), pp. 187–201.
13. Kurniawan, R., Kumaran, S.T., Prabu, V.A., Zhen, Y., Park, K.M., Kwak, Y.I., Islam, M.M. and Ko, T.J., 2017. Measurement of burr removal rate and analysis of machining parameters in ultrasonic assisted dry EDM (US-EDM) for deburring drilled holes in CFRP composite. *Measurement*, 110, pp. 98–115.
14. Yue, X., Yang, X., Tian, J., He, Z. and Fan, Y., 2018. Thermal, mechanical and chemical material removal mechanism of carbon fiber reinforced polymers in electrical discharge machining. *Int J Mach Tools Manuf*, 133, pp. 4–17.
15. Sheikh-Ahmad, J.Y. and Shinde, S.R., 2016. Machinability of carbon/epoxy composites by electrical discharge machining. *Int J Mach Mach Mater*, 18(1–2), pp. 3–17. https://doi.org/10.1504/IJMMM.2016.075452.
16. Sheikh-Ahmad, J.Y., 2016. Hole quality and damage in drilling carbon/epoxy composites by electrical discharge machining. *Mater Manuf Processes*, 31(7), pp. 941–950.
17. Kumar, R., Agrawal, P.K. and Singh, I., 2018. Fabrication of micro holes in CFRP laminates using EDM. *J Manuf Processes*, 31, pp. 859–866. https://doi.org/10.1016/j.jmapro.2018.01.011.
18. Dutta, H., Debnath, K. and Sarma, D.K., 2020. Multi-objective optimization of hole dilation at inlet and outlet during machining of CFRP by µEDM using assisting-electrode and rotating tool. *Int J Adv Manuf Technol*, 110(9), pp. 2305–2322.

13 Polycarbonate (PC)-Based Material Design and Investigation of its Properties by Fused Deposition Modeling (FDM)

Umang Dubey, Unnikrishnan T. G., Mohan S., M. Ramesh, and K. Panneerselvam

CONTENTS

13.1 INTRODUCTION

Using a three-dimensional (3D) model as a starting point, an additive manufacturing technique builds products layer by layer. It's a rapid and flexible way to automate production. Half the production lead time is also reduced even with complicated items [1]. Fused deposition modeling (FDM) is a new type of fast manufacturing technique that generates aerospace prototypes and quick tooling with the greatest precision and ensures functionality without the need for laser to do so. Stereolithography (STL), a laser-based technology, demands more frequent and thorough maintenance than FDM does [2,3]. Unrolled plastic or metal wire filament coils are used to feed material into the extrusion nozzle. The filament is then heated and melted using the nozzle at appropriate temperatures. A numerically controlled mechanism and software package allow the nozzle to move

DOI: 10.1201/9781003343912-13

in both vertical and horizontal directions. The material is extruded once the nozzle has been heated to create layers and the beads begin to harden as soon as they are expelled [4].

The fill density (FD), layer thickness, and print speed (PS) are all factors that go into producing high-accuracy FDM items [5]. Products with flaws are the result of poor factor selection. According to Anitha et al. [6], layer thickness is an important consideration in the FDM process. Sood and Chaturvedi [7] examined FDM process characteristics such as layer thickness, orientation, raster width, raster angle, and air gap. According to Galantucci et al. [8], the construction direction of 3D-printed objects has no impact.

For example, Taguchi's experiments are used to reduce the number of variables and investigate how these variables interact with each other. The cost of conducting experiments is lowered when the number of experiments is reduced.

According to the prior review of the literature, there are just a few studies that discuss 3D printing using polycarbonate (PC) with FD as a consideration. PC filaments were employed in the current research to 3D print test components. The L_9 orthogonal array was used to design trials, which included Fill Density (FD), layer height (LH), and PS. Mechanical characterization, fractography, and other studies were completed. For studying the relevance of input factors, ANOVA was utilized in this study.

The purpose of this chapter is to investigate the influence of factors such as FD, layer thickness, and PS on the PC material. In order to examine the fracture of tensile specimens, researchers use optical microscopy in conjunction with flexural, hardness, impact, and tensile tests.

13.2 DETAILS OF EXPERIMENTS

13.2.1 FDM MACHINE

S Julia dual by factory works, India, provided the FDM machine utilized in this project. All three of the user-selected parameters can be processed by the 3D printing machine. PC threads were used in the 3D printing technique to create samples. SolidWorks was used to design pieces that will be manufactured. The 3D CAD model is then converted into a tessellation format STL file using SolidWorks. It assists in changing build settings and creates the G-code necessary to command an FDM machine's extrusion head [9,10]. In order to ensure that no adhesives interfere with the process, the construction plate was thoroughly cleaned before each step.

13.2.2 CHARACTERIZATION OF 3D-PRINTED PC

Following ASTM D638 standards, the 3D-printed PC was tested for tensile strength (Figure 13.1). It was decided to keep the crosshead speed at 2 mm/min. As shown in Figure 13.2, flexural testing on 3D-printed parts was performed at a crosshead speed of 2 mm/min in accordance with ASTM D790-10 standards. According to ASTM D256-10, the amount of energy absorbed by a 3D-printed PC during an impact test was measured (Figure 13.3) and 6.5 J of force was delivered. According to ASTM D2240-05, Shore D hardness was used to measure the hardness of 3D-printed PC samples. An optical microscope was used to examine the shattered 3D-printed PC sample microstructure.

13.2.3 SELECTION OF PARAMETERS AND DETAILS OF THE EXPERIMENT

Filtration density, layer thickness, and PS were determined after a series of trial-and-error tests. For the 3D printing of components, these three characteristics were set at three different levels. Table 13.1 lists the levels of 3D-printed components. Taguchi L_9 experiments were used to carry out the tests.

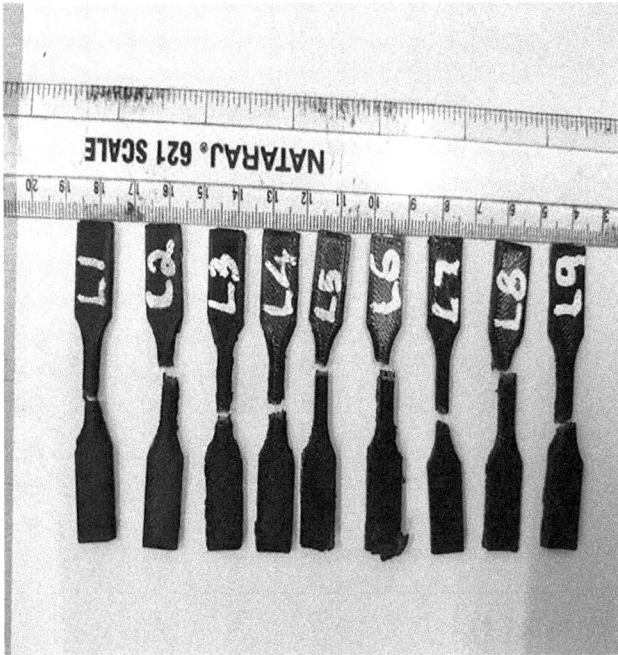

FIGURE 13.1 Tensile test specimens that have fractured.

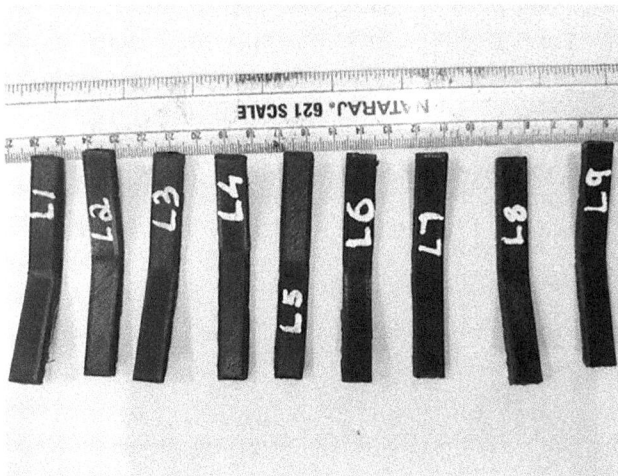

FIGURE 13.2 Test specimens that were flexed and fractured.

13.3 RESULTS AND DISCUSSION

Mechanical properties such as tensile strength, Shore D hardness, impact strength, flexural strength, and cracked surfaces were measured using optical microscopy on the 3D-printed PC materials. Data from these tests are included in Tables 13.1 and 13.2, which summarize the findings.

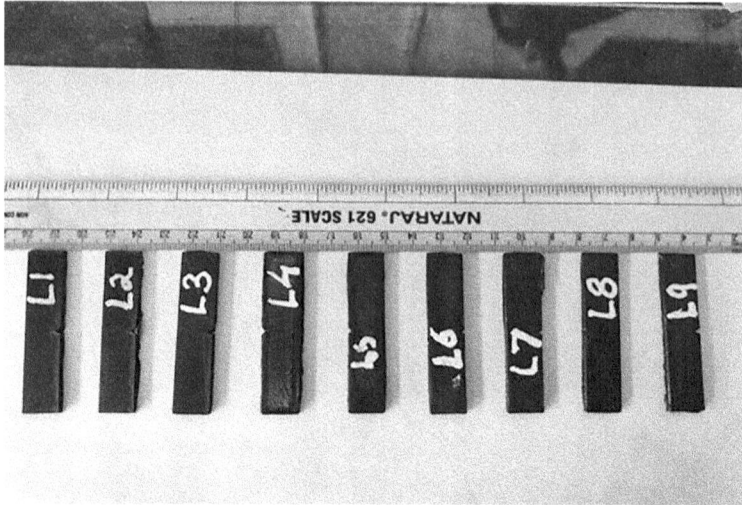

FIGURE 13.3 After an impact test, specimens were shattered.

TABLE 13.1
Planned Tests in Various Stages

Factors	PS (mm/s)	LH (mm)	FD (%)
Level 1	60	0.1	50
Level 2	65	0.2	75
Level 3	70	0.3	100

TABLE 13.2
L_9 Orthogonal Array

Factors	(LH) (mm)	(FD) %	(PS) (mm/s)
1	0.1	50	70
2	0.1	75	65
3	0.1	100	60
4	0.2	75	65
5	0.2	50	60
6	0.2	100	70
7	0.3	75	60
8	0.3	50	70
9	0.3	100	65

13.3.1 Tensile Strength of PC Material

Figure 13.4 depicts the main effect graphs of the tensile strength of a 3D-printed PC filament. At 0.2 LH, the tensile strength is showing the maximum value, whereas between 0.1 and 0.3 LH, the strength is below the mean. The maximum tensile strength of a PC material enhances with the addition of filler. The tensile strength lowers to 42.725 MPa when the PC is filled to 50% of its maximum capacity. To explain why the tensile strength of a PC material is stronger, while the layers are wrapped together and lower when they are unwrapped, one possibility is the formation of small holes between layers.

TABLE 13.3
Response of Experiments

S. No.	Flexural Strength (MPa)	Impact Strength (J)	Ultimate Tensile Strength (MPa)	Shore D Hardness
1	114.345	5.6254	44.986	63.5
2	114.730	5.9640	44.988	63.6
3	118.628	6.2290	48.596	68.5
4	121.650	5.1870	50.363	74.3
5	111.716	5.7480	42.725	58.2
6	116.938	6.4213	47.911	59.7
7	128.498	5.7337	50.767	75.5
8	116.158	6.6550	45.579	59.8
9	132.880	5.6254	53.098	78.5

Main Effects Plot for Means
Data Means

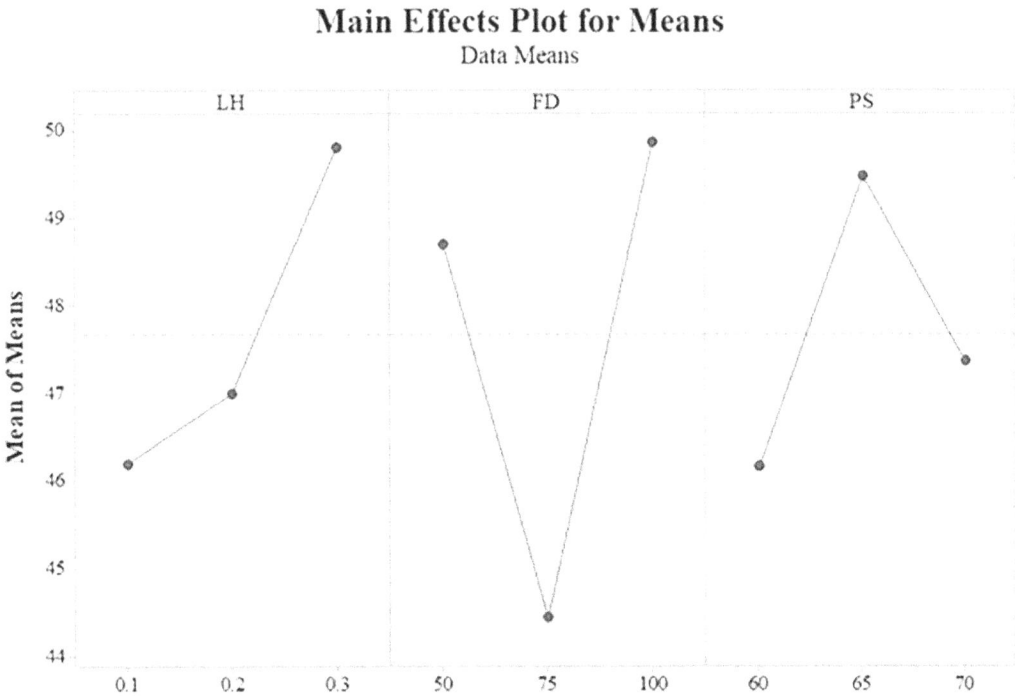

FIGURE 13.4 For ultimate tensile strength, these are the main effect charts.

13.3.2 FLEXURAL STRENGTH OF PC MATERIAL

It was demonstrated in Figure 13.5 that at 100% FD, the greatest flexural strength value was 132.88 MPa, while the lowest was 111.716 MPa at 50% of FD. At a FD of 75%, the flexural strength was higher than at a FD of 50%. Flexural strength is higher at a PS of 65 mm/s than at a speed of 60 mm/s, though. Flexural strength decreases at 70 mm/s; possibly layer-by-layer formations may have created a wraparound effect. With increasing PC FD concentration and layer thickness, the overall pattern exhibits a rising flexural strength, as can be seen in the figure.

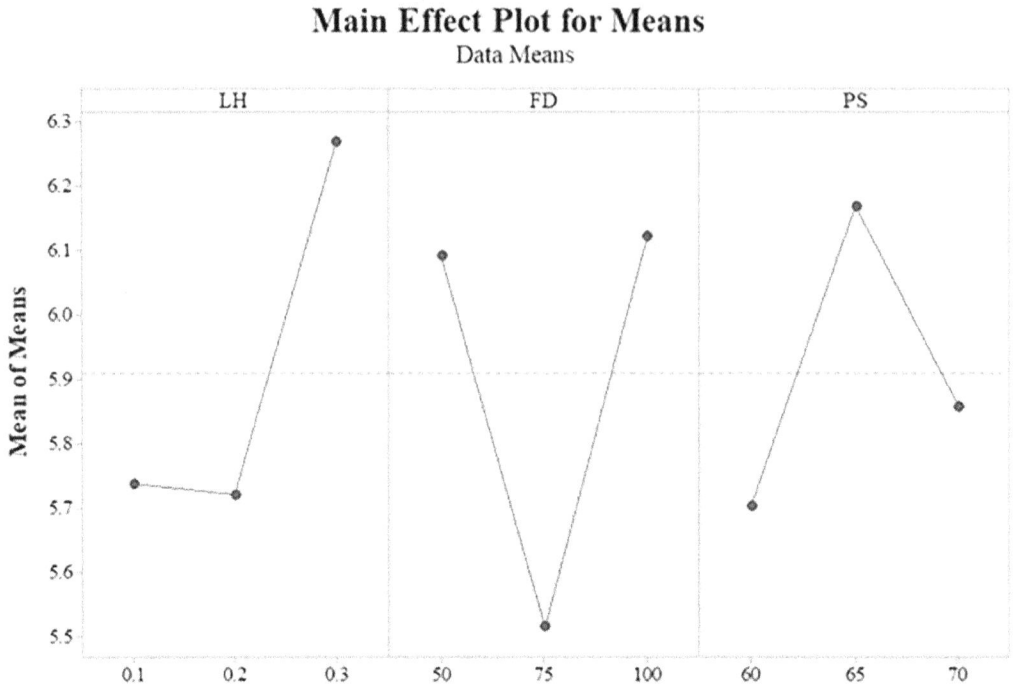

FIGURE 13.5   For flexural strength, these are the main effect charts.

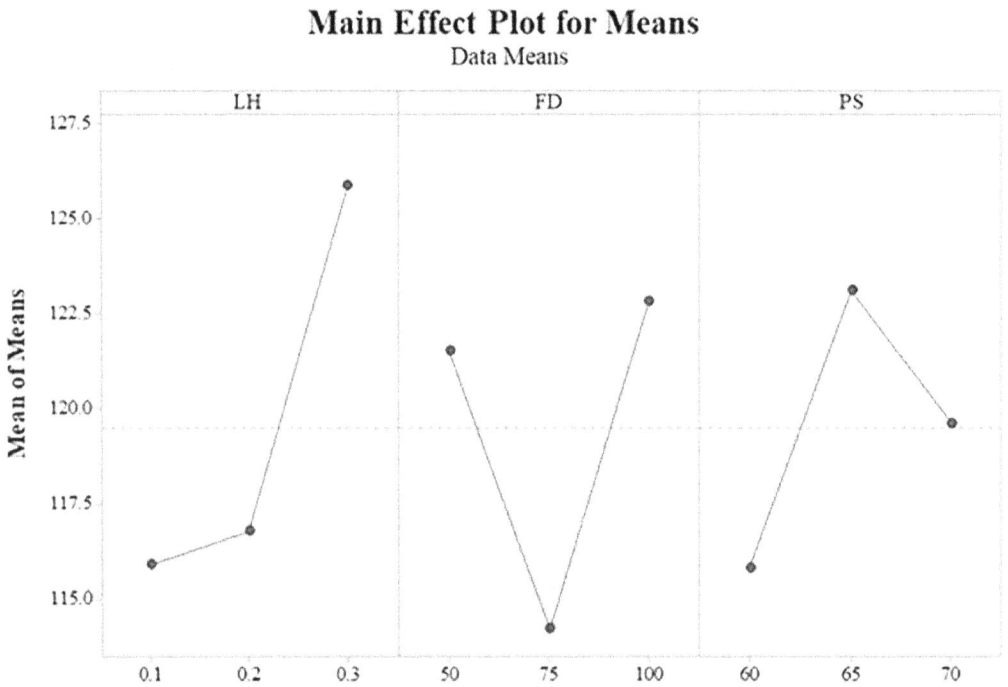

FIGURE 13.6   For impact strength, these are the main effect charts.

13.3.3 Impact Strength of PC Material

Figure 13.6 depicts the impact strength's principal effect. LH was at 0.3, which resulted in a higher impact strength than that of 0.2 and 0.1. At 0.2 LH, the strongest bonds are formed. It was shown that when the proportion of PC FD was only 75%, the impact strength was the lowest. PC FD of 50% results in a 6.6550 J impact strength, which is the highest. However, the strength decreases at 100% fill because of the stress energy absorbed. At 70 mm/s PS, the greatest impact strength recorded; it was somewhat higher than the 65 mm/s PS value. Because of the low speed of 60 mm/s, the layer-by-layer formation was not uniform and had some voids.

13.3.4 Shore D Hardness of PC

Figure 13.7 depicts the primary Shore D hardness effect plot. Shore D hardness was the lowest at 0.1 LH and rose to 0.2 LH when the layer thickness was increased. Shore D hardness decreased as a result of the uneven structure formed when the LH was set to 0.1. From 50% to 100%, the Shore D hardness value grew in value. Improved Shore D hardness values were achieved when pores were completely eliminated by setting the FD to 100%. When printing at 65 mm/s, a uniform layer was formed, and hardness levels were at their maximum. In the presence of pores and non-uniform layer development, the Shore D hardness value decreased.

13.3.5 Fractography Studies

In the presence of pores and non-uniform layer development, the Shore D hardness value decreased. Figures 13.8–13.10 show the fracture surface as a result of an ideal microscope analysis. When the FD was at 50%, a considerable number of pores were present, which resulted in a decrease in the material's strength. There are fewer holes on the broken surface of 75% FD than there are on the fractured surface of 50% FD. This is mostly due to the fact that 3D printing uses a higher proportion

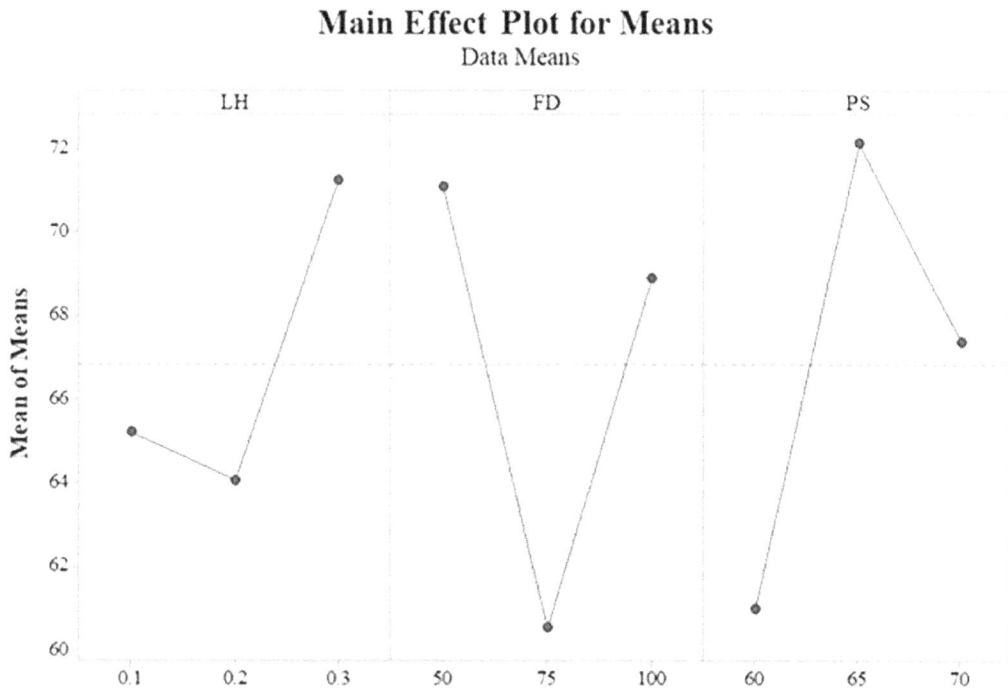

FIGURE 13.7 For Shore D hardness, these are the main effect charts.

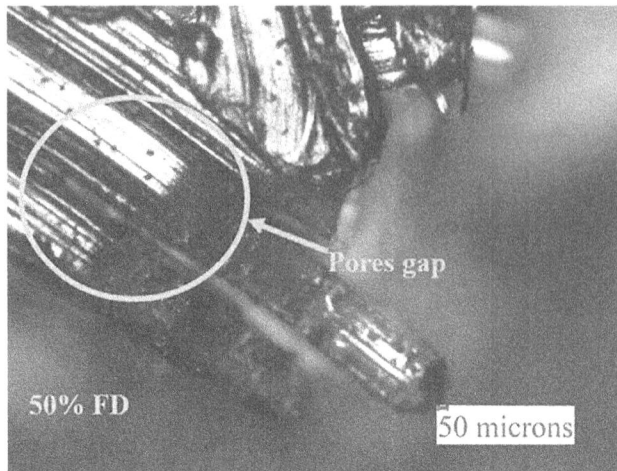

FIGURE 13.8   Image obtained with optical microscopy of 50% FD.

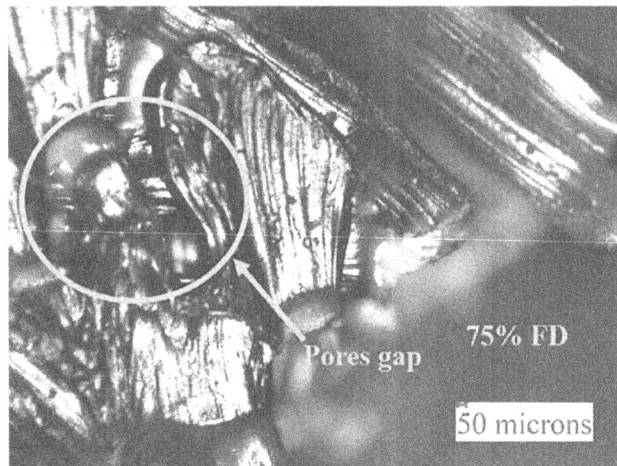

FIGURE 13.9   Image obtained with optical microscopy of 75% FD.

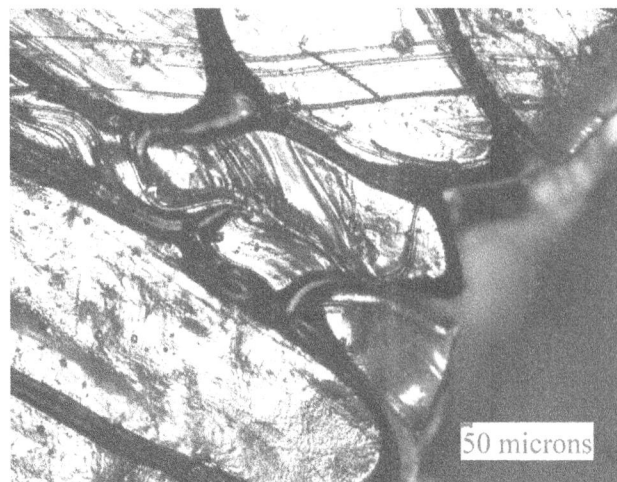

FIGURE 13.10   Image obtained with optical microscopy of 100% FD.

of PC material than traditional methods. Comparing the three materials, the FD at 100 showed the least amount of pores. The material was filled to 100% FD, resulting in improved performance. As a result, careful consideration must be given to the FD if one is to achieve optimal performance. Experiments and optical microscopic examinations had a high connection.

13.3.6 ANOVA ANALYSIS

ANOVA was used to examine the relationship between output and input parameters. Tables 13.4–13.7 shows the ANOVA results for Shore D hardness, impact strength, flexural strength, and tensile

TABLE 13.4
ANOVA for Tensile Strength

Source	DF	Seq SS	Contribution	Adj SS	Adj MS	F-Value	P-Value
LH	2	21.718	24.01%	21.718	10.859	8.56	0.105
FD	2	49.194	54.39%	49.194	24.597	19.39	0.049
PS	2	16.997	18.79%	16.997	8.498	6.70	0.130
Error	2	2.536	2.80%	2.536	1.268		
Total	8	90.445	100.00%				

TABLE 13.5
ANOVA for Flexural Strength

Source	DF	Seq SS	Contribution	Adj SS	Adj MS	F-Value	P-Value
LH	2	182.039	46.32%	182.039	91.020	75.54	0.013
FD	2	129.174	32.87%	129.174	64.587	53.60	0.018
PS	2	79.398	20.20%	79.398	39.699	32.95	0.029
Error	2	2.410	0.61%	2.410	1.205		
Total	8	393.021	100.00%				

TABLE 13.6
ANOVA for Impact Strength

Source	DF	Seq SS	Contribution	Adj SS	Adj MS	F-Value	P-Value
LH	2	0.58420	35.12%	0.58420	0.29210	15.64	0.060
FD	2	0.70158	42.18%	0.70158	0.35079	18.78	0.051
PS	2	0.34028	20.46%	0.34028	0.17014	9.11	0.099
Error	2	0.03735	2.25%	0.03735	0.01867		
Total	8	1.66341	100.00%				

TABLE 13.7
ANOVA for Shore D Hardness

Source	DF	Seq SS	Contribution	Adj SS	Adj MS	F-Value	P-Value
LH	2	89.929	19.29%	89.929	44.964	36.52	0.027
FD	2	186.496	40.00%	186.496	93.248	75.74	0.013
PS	2	187.316	40.18%	187.316	93.658	76.08	0.013
Error	2	2.462	0.53%	2.462	1.231		
Total	8	466.202	100.00%				

strength. For 3D-printed PC components, it can be deduced from table that FD is an important element to consider.

LH was the least significant component in Shore D hardness, impact strength, and flexural strength, whereas PS was the least significant element in tensile strength.

13.3.7 PREDICTED MEAN

The mean response (μ) at the optimal condition may be anticipated from the ANOVA table's optimal value determination. Using ANOVA and Equation 13.1, the response characteristic's optimum value may be determined by the parameters that are statistically significant.

$$\mu_{\text{Pred}} = \bar{A} + \bar{B} + \bar{C} - 2\bar{Y} \tag{13.1}$$

$\bar{Y}$ =arithmetic average of responses;
$\bar{A}, \bar{B}, \bar{C}$ =values of the average response at relevant parameters levels;

13.3.7.1 Determination of Confidence Intervals

In order to get an accurate estimate of the mean (μ), average of the experiment's results must be taken into account. When using statistical approaches, it is expected that parameter values will be anticipated with a range of values within which they are most likely to fall. This is standard practice.

$$\text{CI}_p = \sqrt{\frac{F_\alpha(v_1, v_2)\text{MS}_e}{\eta_{\text{eff}}}} \tag{13.2}$$

$F_\alpha(v_1, v_2)$ =the confidence interval's F-ratio;
v_1 = the average's amount of degrees of freedom;
v_2 = is the error's degree of freedom;
N=Quantitative sum total;
MS_e =variance of errors;
η_{eff} the sample size that is most useful.

$$\eta_{\text{eff}} = \frac{N}{1 + \text{DOF}_{\text{opt}}} \tag{13.3}$$

To calculate the projected confidence interval and optimal confidence interval for the expected mean, we used Equations 13.1–13.3 and significant ANOVA parameters as described before. The results are shown in Table 13.8.

TABLE 13.8
Mean and Confidence Interval (CI) Estimates for Several Tests

Tests	Confidence Interval of Predicted Mean		Predicted Optimum Response
	Upper Bound	Lower Bound	
Shore D Hardness	76.136	71.897	74.72
Impact	6.314	5.898	6.156
Flexural	138.929	126.841	130.83
Tensile	52.2442	47.1325	49.55

13.3.8 CONFIDENCE INTERVAL

A parameter's prevalence may be estimated using the confidence interval (CI), with the significant parameter being the most accurate estimate within the CI's range. It is possible to estimate the frequency of observed intervals by calculating their confidence level or coefficient of assurance. CI width represents the degree of uncertainty surrounding an unknown quantity. There may be more data needed to narrow down a parameter with such a broad range if that is the case.

13.4 CONCLUSIONS

Various characterization methods were used to assess the 3D printing of PC components using the FDM technology. The following conclusions may be derived from the results of PC 3D printing tests using a variety of settings, including FD, LH, and PS. The Shore D hardness, impact strength, flexural strength, and tensile strength were tested. FD of 100% has the greatest Shore D hardness, tensile strength, flexibility, and flexural strength, whereas 75% of FD has greater impact strength than 50% of FD. The PC material's mechanical properties are most strongly influenced by its FD when it is 3D printed. However, the flexural strength has been enhanced by 0.3 mm of LH, which affects the mechanical properties such as tensile strength, Shore D hardness, and impact strength. The mechanical characteristics of PS were not influenced by the parameters that were selected, unless greater speeds result in the fault of wrapping and lower speeds result in heat-affected zones. CI of 95% suggested a tensile strength of 49.65 ± 1.78 MPa as the ideal value. According to the projected range, the optimal condition was 48.77 MPa, which is within the range. The values obtained from the optimal parameter were likewise within the projected range for the other hardness, impact, and flexural tests. As a result, this study contributes to the growing body of information about 3D printing using a PC filament in Indian precision casting companies. Fillers can be added to the PC matrix, and the re-related characteristics can be examined and compared to pure PC.

REFERENCES

1. Chua, C. K., Leong, K. F. *Rapid Prototyping: Principles and Applications in Manufacturing*, Wiley, Singapore, 1997.
2. Jain, P., Kuthe, A. M. Feasibility study of manufacturing using rapid prototyping: FDM approach. *Procedia Engineering*. 2013; 63: 4–11.
3. Karapatis, N. P., Van Griethuysen, J. P. S., Glardon, R. Direct rapid tooling: A review of current research. *Rapid Prototyping Journal*. 1998; 4(2): 77–89.
4. Gregorian, A., Elliot, B., Navarro, R., Ochoa, F., Singh, H., Monge, E., Foyos, J., Noorani, R., Fritz, B., Jayanthi, S. Accuracy improvement in rapid prototyping machine (FDM-1650). *2001 International Solid Freeform Fabrication* Symposium, University of Texus, Austin, Texus, 2001; pp. 77–84.
5. Zhou, J., Herscovici, D., Calvin, C. C. Parametric process optimization to improve the accuracy of rapid prototyped stereolithography parts. *International Journal of Machine Tools & Manufacture*. 2000; 40(3): 363–379.
6. Anitha, R., Arunachalam, S., Radhakrishnan, P. Critical parameters influencing the quality of prototypes in fused deposition modelling. *Journal of Materials Processing Technology*. 2001; 118: 385–388.
7. Sood, A. K., Chaturvedi, V., Datta, S., Mahapatra, S. S. Optimization of process parameter infused deposition modelling using weighted principle component analysis. *Journal of Advanced Manufacturing Systems*. 2011; 10(02): 2241–2250.
8. Galantucci, L. M., Lavecchia, F., Percoco, G. Experimental study aiming to enhance thesurface finish of fused deposition modeled parts. *Manufacturing Technology*. 2009; 58: 189–192.
9. Marcos, M., Wendt, C., Fernandez-Vidal, S., Gomez-Parra, A., Batista, M. *Advances in Material Science Engineering*. 2016: Article ID 5780693. http://dx.doi.org/10.1155/2016/5780693.
10. Groover, M. P. *Modern Manufacturing*, Fifth edition, John Wiley & Sons, New Jersey, 2013.

14 Mechanics of Composite Material Machining
An Overview

Murugabalaji V., Bikash Chandra Behera,
and Matruprasad Rout

CONTENTS

14.1 INTRODUCTION

Composite materials find their applications in many fields due to their supreme specific strength and stiffness properties compared with monolithic metals [1]. These materials are constituted of two or more materials, out of which one material is called as a matrix and other constituents are called reinforcements (particles or fibre). The matrix part of the composite material serves as a protective element by aiding in stress distribution to the reinforcement material(s) and also imparts the form to the composite material part. The reinforcements aid in attaining superior properties and strengthen the matrix in favourable directions [2]. The conventional machining methods like milling [3,4], drilling [5] and turning [6] of these composite materials are complex due to the non-homogeneity, anisotropy and abrasive reinforcements [7]. The tool wear and surface integrity studies are of major concern while machining these materials. Tool wear occurs as the material possesses high hardness and fracture toughness, which leads to increased cutting forces and brittle fracture of the workpiece [8]. The values of hardness and fracture toughness, as depicted by Gavalda Diaz et al. [8], clearly show (Figure 14.1) the regime of difficult-to-machine composites for various matrix materials. This chapter focuses on the aspects of machining some common types of composites and the difficulties encountered in terms of surface integrity, tool life and some practical techniques suggested by researchers for the effective machining of composite materials.

DOI: 10.1201/9781003343912-14

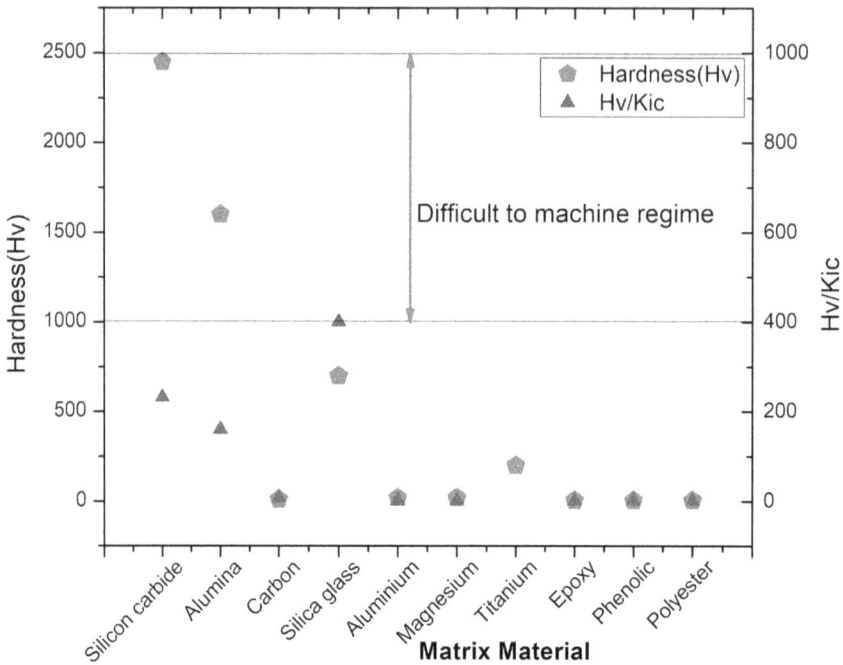

FIGURE 14.1 Approximate values of H_v and H_v/K_{ic} for different materials used in manufacturing of composites [8].

14.2 TYPES OF COMPOSITES

Composites are divided into two general categories based on the size of the matrix material and reinforcements used. Based on the matrix material, the composite materials are classified as polymer matrix composites (PMCs), metal matrix composites (MMCs) and ceramic matrix composites (CMCs). Similarly, based on the size of the reinforcement, the composite materials are named either conventional (reinforcement particle size of few microns in all dimensions) or nanocomposites (reinforcement particle size at nanoscale in at least one dimension) [2]. Further, depending upon the shape of the reinforcement, composite materials are classified as fibre-reinforced composites, particle-reinforced composites and structural composites (laminated sheet and sandwich composites) [9]. The properties of the composite material so formed depend upon the following factors [10]:

   i. The nature and properties of both constituent phases, i.e., the matrix and the reinforcement.
  ii. The physical form of the reinforcement, i.e., fibre or particle.
 iii. The proportion of matrix and reinforcement present in the composite material, i.e., the volume fraction of the portion of the matrix (V_m) and volume fraction of the portion of the reinforcement (V_f) [2]

$$V_m = \frac{\text{Volume of matrix}}{\text{Volume of composite}}$$

$$V_f = \frac{\text{Volume of reinforcement}}{\text{Volume of composite}}$$

where $V_m + V_f = 1$.

14.2.1 POLYMER MATRIX COMPOSITES

The polymer matrix composites (PMCs) are reinforced with brittle fibres of high strength embedded in ductile polymer matrix of soft nature. They are called fibre-reinforced plastic (FRP) composites. The fibres may be glass, carbon or aramid and are named as GFRP (glass fibre-reinforced plastic), CFRP (carbon fibre-reinforced plastic) and AFRP (aramid fibre-reinforced plastic), respectively. Aramid fibre is usually sold under the trade name 'Kevlar' [11]. The fibres may be long or short and continuous or discontinuous depending upon the application. The long fibres can be oriented either unidirectional (orientation in which fibres are arranged parallel) or woven (orientation in which fibres are woven in the form of a cloth or fabric). Unidirectional fibres are found to possess superior mechanical properties in a composite [12]. The primary mechanical properties possessed by FRP composites [1] are given in Table 14.1. The most common matrix materials that are used in FRP composites include thermoset polymers (commonly used thermosets are epoxy and polyester) or thermoplastic polymers (Polyamide and Polyetheretherketone). The epoxy resin is costlier and stronger compared with polyester resin, and it offers good fabrication accuracy, thereby finding its application as CFRP and AFRP composites in military, aerospace and medical applications [2]. Polyester resin, on the other hand, is cheaper and has low strength. They are used in boat hull fabrication, structural parts like panels of automobiles and aircraft parts and electrical appliances [2].

14.2.2 METAL MATRIX COMPOSITES

The metal matrix composites (MMCs) are applicable for usage in higher temperature than PMCs [1]. The most common applications of MMCs are in the aerospace and automotive industries. Continuous or discontinuous fibres as well as particle reinforcements are used in MMCs. Particle-reinforced MMCs termed as PMMCs are found to possess superior properties of strength, wear resistance and ductility with low anisotropy compared with fibre-reinforced MMCs [2]. Some of the commonly reinforced fibres and matrix materials in MMCs, with their corresponding mechanical properties [10], are listed in Table 14.2.

TABLE 14.1
Mechanical Properties of FRP Composites

FRP Material	Tensile Strength (MPa)	Elastic Modulus (MPa)	Strain to Failure (%)	Density (g/cm3)
GFRP				
Unidirectional $V_f = 60\%$	1000	45,000	203	2.1
Woven cloth $V_f = 20\%-50\%$	100–300	10,000–20,000	-	1.5–2.1
Chopped roving $V_f = 20\%-50\%$				
50–200	6000–12,000	-	1.3–2.1	
Sheet moulding compound $V_f = 20\%-50\%$	10–20	500–2000	-	1.3–1.9
CFRP				
Unidirectional $V_f = 60\%$ (high strength)	1200	145,000	0.9	1.6
Unidirectional $V_f = 60\%$ (high modulus)	800	220,000	0.3	1.6
AFRP				
Unidirectional $V_f = 60\%$	1000	75,000	1.6	1.4

Source: Adapted from Teti [1].

TABLE 14.2
Mechanical Properties of MMCs

Fibre	Matrix	Fibre Content (vol%)	Density (g/cm3)	Longitudinal Tensile Modulus (GPa)	Longitudinal Tensile Strength (MPa)
Carbon	Al 6061	41	2.44	320	620
Boron	Al 6061	48	—	207	1515
Silicon carbide	Al 6061	50	2.93	230	1480
Alumina	Al 380.0	24	—	120	340
Carbon	Mg AZ31	38	1.83	300	510
Borsic	Ti	45	3.68	220	1270

Source: Adapted from Callister and Rethwisch [10].

14.2.3 CERAMIC MATRIX COMPOSITES

Ceramic matrix composites (CMCs) are recently developed advanced engineering materials to impart improved fracture toughness to non-reinforced ceramics, which exhibits superior specific stiffness and high-temperature resistance properties [1]. The reinforcements used in CMCs can be carbonaceous (carbon fibre, carbon nanotubes, graphene, carbon black, etc.) [13] or non-carbonaceous like SiC, Si_3N_4 and Al_2O_3 [2]. The CMCs find their applications in the manufacturing of high-value parts for the aerospace, nuclear and automotive industries. The room temperature mechanical properties of CMCs reinforced with whiskers of SiC [1] are given in Table 14.3.

14.3 MACHINING OF COMPOSITE MATERIALS

The schematic diagram representing the machining of monolithic metal and composite material is shown in Figure 14.2a and b, respectively. Orthogonal machining involves the cutting of materials with an alignment in which cutting edge of the tool is always perpendicular to the direction of tool motion. During orthogonal machining of monolithic metals, material removal occurs by shearing along the shear plane due to the slip mechanism. Below the shear plane, the material remains undeformed, and above the shear plane, chip formation occurs [14]. Similarly, in the case of composite materials, the relative motion between tool and the workpiece surfaces causes the chip formation during the machining process. On the shear plane, the material is sheared and shaped into chips. The actual uncut chip thickness varies when the cutting tool has a nose radius. The effective chip thickness, in this case, is $t - \Delta t$ (Figure 14.2b). The material under the zone of the uncut chip thickness undergoes both elastic and plastic deformation [15]. The components

TABLE 14.3
The Room Temperature Mechanical Properties of CMCs Reinforced with Whiskers of SiC

Matrix Material	SiC Whisker Volume Fraction, V_f (%)	Flexural Strength (MPa)	Fracture Toughness (MPa)
Si_3N_4	0	400–650	30–45
	10	400–550	40–60
	30	350–500	45–65
Al_2O_3	0	-	30
	10	400–500	45

Source: Adapted from Teti [1].

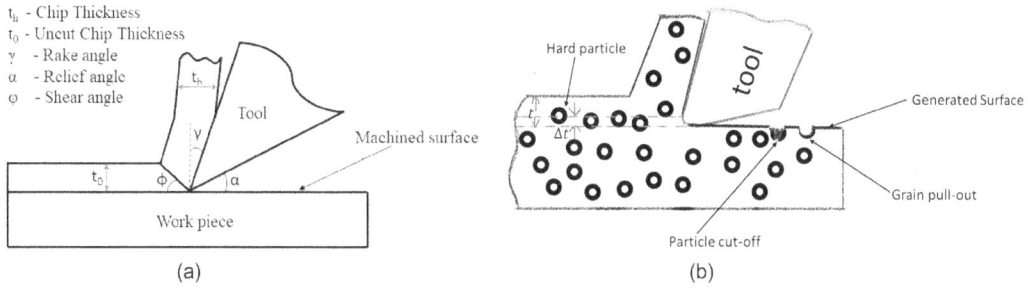

FIGURE 14.2 Schematic of machining of (a) monolithic material and (b) composite material.

of the matrix may be removed or crushed when a compressive load is placed between a tool and the work. The method of material removal in composite machining is influenced by the location of the reinforcement. The reinforcement pushed in the zone of the chip when it was above the cutting line and within the matrix when it was below. If reinforcement is present along the cutting, it will either pull out or be cut off, as seen in Figure 14.2b. This results in the formation of voids or cavities on the machined surface. It also creates more variation in the dynamic force components. Good surface finish was observed when particles/reinforcement sheared instead of pull out [16].

14.3.1 MACHINING OF PMCs

The FRP composites are difficult to machine because of the oscillating cutting forces caused by the fragmentary fracture of the fibres [17]. The physical and mechanical characteristics of the matrix and the fibre, the volume percentage of fibre and the fibre orientation are the typical parameters that affect the machinability of FRPs [17,18]. During machining, the glass and carbon fibres fracture in brittle manner before the cutting edge of the tool. On the other hand, the more durable aramid fibres avoid shearing and exhibits the tendency to bend before the cutting edge. Thus, the type of fibre reinforcement and its orientation has a significant impact on the surface integrity of the work material's machined edge [19]. Orthogonal cutting of unidirectional FRP composite was carried out for the first time to study the chip formation mechanism and explore the different types of chips by Koplev et al. [20]. Wang et al. [21] studied the impact of fibre orientation on the cutting mechanism and concluded that all the aspects concerned with material removal are dependent upon fibre orientation. They categorized the different modes of chip formation during the orthogonal machining of FRP composites based on fibre orientation of the composite material and the cutting-edge rake angle of the tool as shown in Figure 14.3. Arola et al. [22] and Nayak et al. [23] conducted tests using a charge-coupled device (CCD) camera to determine the mechanism of chip formation during the orthogonal machining of unidirectional CFRP and GFRP composites, respectively. They concluded that cutting-edge rake angle and fibre orientation are crucial factors in chip creation.

The different modes of chip formation, as described by Wang et al. [21], are as follows (refer Figure 14.3):

i. *When the rake angle is positive and the fibre orientation (θ) is zero degrees*, the delamination type of chip formation takes place. In this type, the tool penetration on the work material initiates a crack at the tool point that spreads along the matrix–fibre interface. This led to bending-induced fracture in front of the cutting edge.

ii. *When the rake angle is negative and the fibre orientation (θ) is zero degree*, the fibre buckling type of chip formation takes place. The fibres are subjected to a compressive load, which results in buckling, inplane shearing and failure at the matrix–fibre interface.

Failure due to repeated buckling occurs in a direction normal to the fibre length, resulting in discontinuous chips.

iii. Fibre shearing type of chip formation occurs *when the fibre orientation (θ) is greater than 0° and less than or equal to 90°.* This is the same for all rake angle values. The compressive force and inplane shearing cause the failure, as the tool advances into workpiece material. The continuity of the chip depends upon the intensity of the inplane shear. As the fibre orientation increases from 0° to 90°, there is a gradual increase in the interlaminar shear and fracture of fibres along the interface of matrix and fibre takes place. The icroscopic view of the machined surface shows transverse cracks at fibre ends on the machined surface.

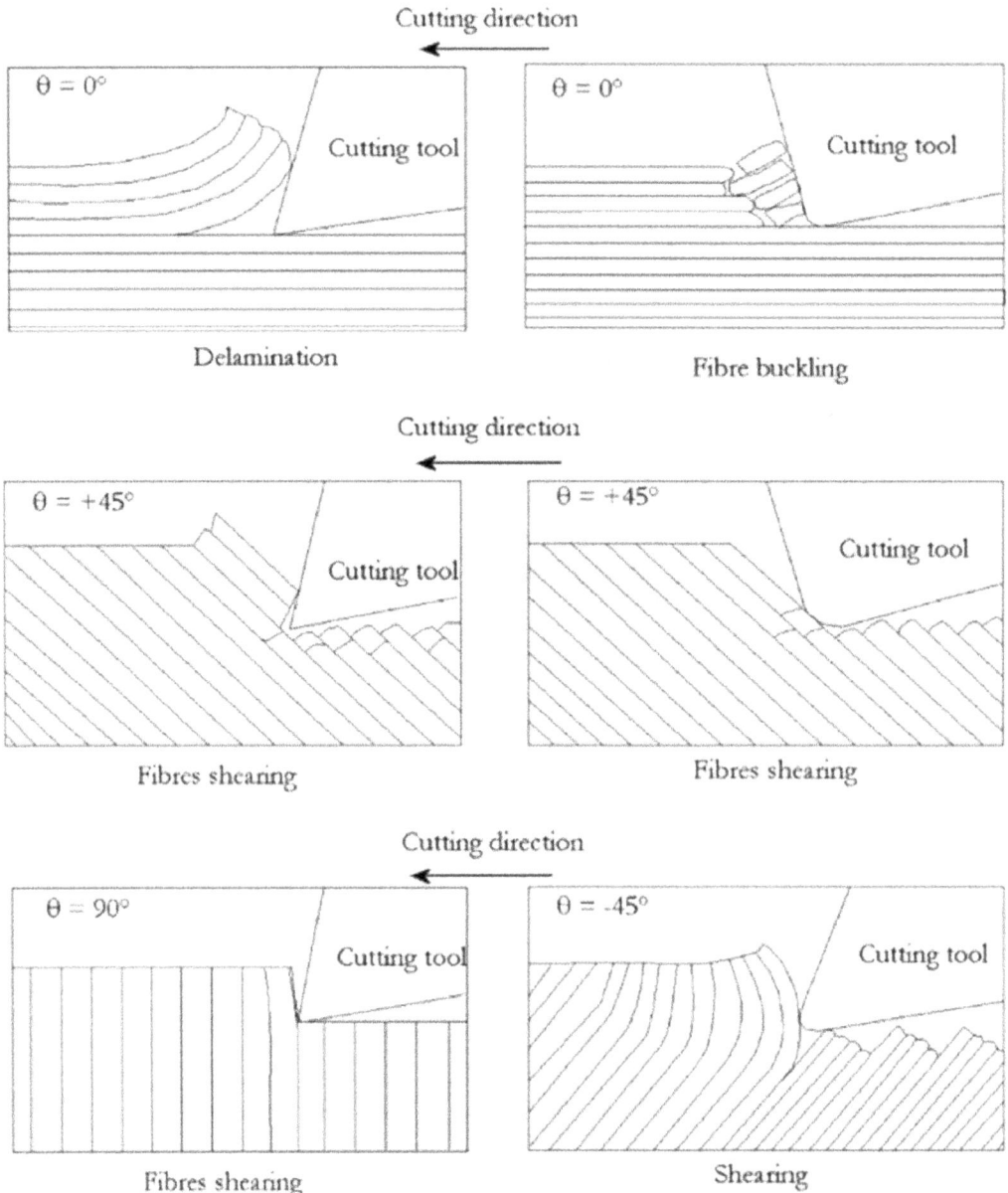

FIGURE 14.3 Schematic of chip formation mechanism in FRP composites. (Adapted from Wang et al. [21] modified by Ghidossi et al. [24].)

iv. *When the fibre orientation is greater than 90°,* fibre shearing takes place by edge trimming causing severe deformation of fibres as the tool advances. The massive elastic bending of fibres causes delamination, interlaminar shear and out-of-plane displacement leading to discontinuous chips. Figure 14.4 depicts the machined surface of various types of chips. It clearly shows the subsurface damages caused during the machining and its dependence on the different orientations of the fibre at specified rake angle and depth of cut values. The damages are severe in the fibre orientation of 120° compared with other orientations [18].

14.3.2 Machining of MMCs

Machining of MMCs is difficult as the ceramic reinforcements are generally very harder than the cutting tools used. Manna and Bhattacharyya [25] studied the influence of different tooling systems on tool wear while machining Aluminium-based MMCs reinforced with SiC. The different tooling systems considered are fixed circular tooling (FCT), fixed rhombic tooling (FRT), fixed square tooling (FST) and rotary circular tooling (RCT). They found that in RCT system tool wear was minimum compared with other tooling systems and was found to be effective. Researchers discovered the emergence of built-up edge (BUE) during the low-speed machining of MMCs, which is why the cutting force becomes low at these lower cutting speeds [26–28]. Lin and Bhattacharyya [29] and Hung et al. [30] analysed the chip development mechanism during the machining of MMCs and compared it with monolithic metals. They noticed that when cutting the aluminium matrix composite material reinforced with SiC, semi-continuous chips are produced. During machining, the particles severely impede the matrix's plastic deformation and slow down the chip's crack propagation. Thus, the reinforcement provides strength as well as better chip control while machining the composite [30].

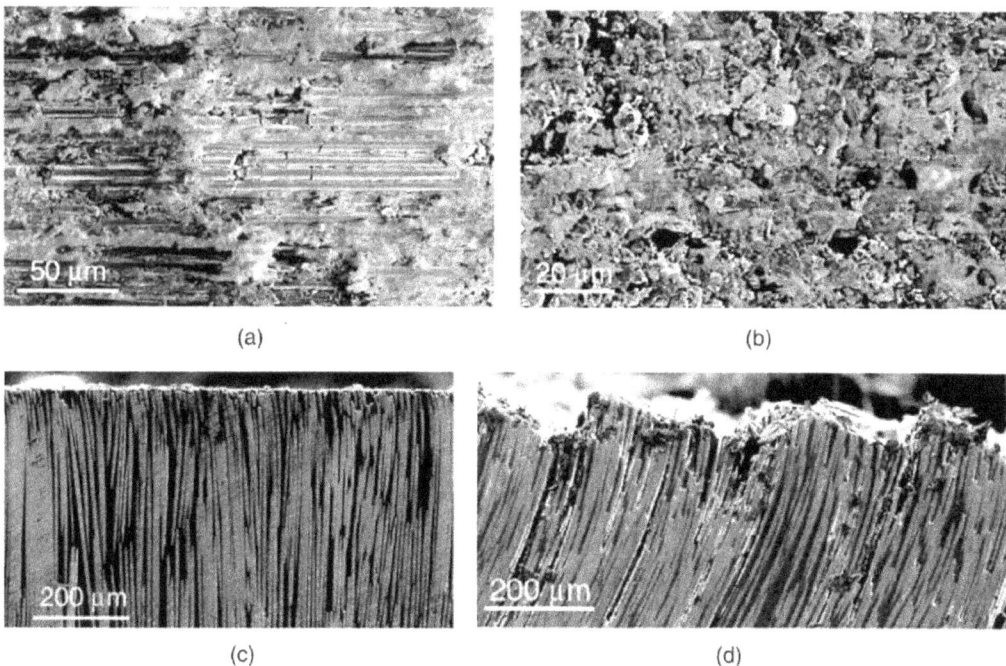

(a)

(b)

(c)

(d)

FIGURE 14.4 Microstructures of machined FRP composites, Tool rake angle $(\gamma)=20°$ and depth of cut $(d)=0.1$ mm, (a) $\theta=0°$, (b) $\theta=90°$, (c) $\theta=90°$, (d) $\theta=120°$ [18].

14.3.3 Machining of CMCs

The CMCs are commonly manufactured with long fibre reinforcements with three mutually perpendicular planes of symmetry, which implies nine independent elastic constants are required to define the material properties of CMCs. They can be referred to as orthotropic as they are not entirely anisotropic [31]. The orthogonal cutting of CMC composite in three different orientations *viz.* longitudinal, across and transverse was studied by Gavalda Diaz and Axinte [32]. They observed the following modes of chip formation (see Figure 14.5):

i. **Composite shearing when cutting along longitudinal fibre orientation**: When the application of shear stress while cutting exceeds the ultimate shear strength of the entire CMC material, this kind of shearing takes place. Due to the varied and delicate nature of the chips, discontinuous chips emerge as shown in Figure 14.5a.

ii. **Matrix shearing when cutting along longitudinal fibre orientation**: This type of shearing happens when the cutting shear stress is greater than the matrix material's shear strength but lower than the ultimate shear strength. Due to the shearing and buckling of the fibres, matrix debonding is caused as the tool advances in the cutting direction as shown in Figure 14.5b.

iii. **Matrix fracture when cutting along longitudinal fibre orientation**: When the application of shear stress during cutting is less than the matrix's shear strength, shearing cannot be accomplished. The fracture toughness is lowest at the matrix–fibre interface, where the crack propagates in the cutting direction as shown in Figure 14.5c.

iv. **Matrix shearing when cutting along across fibre orientation**: When the application of shear stress is higher than the matrix's ultimate shear strength, this mode of shearing manifests. Linear shear mode is used for material removal, just like in monolithic ceramic materials as shown in Figure 14.5d.

v. **Matrix fracture when cutting along across fibre orientation**: When the application of shear stress during cutting is less than the matrix's shear strength, shearing cannot be accomplished. According to the explanation for longitudinal fibres, the fracture first spreads in the matrix–fibre interface in the cutting direction before tending to cause rotation to the surface and discrete offcuts as shown in Figure 14.5e.

vi. **Matrix shearing when cutting transverse fibre orientation**: When the application of shear stress is higher than the matrix's ultimate shear strength, this mode of shearing manifests. The fibres are cut by tool indentation, and the composite functions as a monolithic ceramic as it does so across the fibre, as shown in Figure 14.5f.

vii. **Matrix fracture when cutting along transverse fibre orientation**: When the shear stress applied during cutting is less than the matrix's shear strength, shearing cannot be accomplished. As the matrix–fibre interface is seen along the vertical direction, crack propagation happens in that direction. Bending along the vertical axis in brittle manner removes material as shown in Figure 14.5g.

14.4 TOOL WEAR

The other important aspect of conventional machining of composites is the tool wear that result from the combined effects of the physical, chemical and thermomechanical aspects of the orthogonal cutting process. The contact as well as the relative sliding motion of the tool with the surface of the workpiece and the chip flow on the surface of the tool, all contribute to the tool wear [33]. The diagrammatic representation of common characteristics in the evaluation of tool wear is depicted in Figure 14.6. Tool wear is generally characterized by flank wear caused by the friction between flank of the tool and the newly generated surface of the workpiece and crater wear occurring on the rake surface. Flank wear is more dangerous than crater wear as

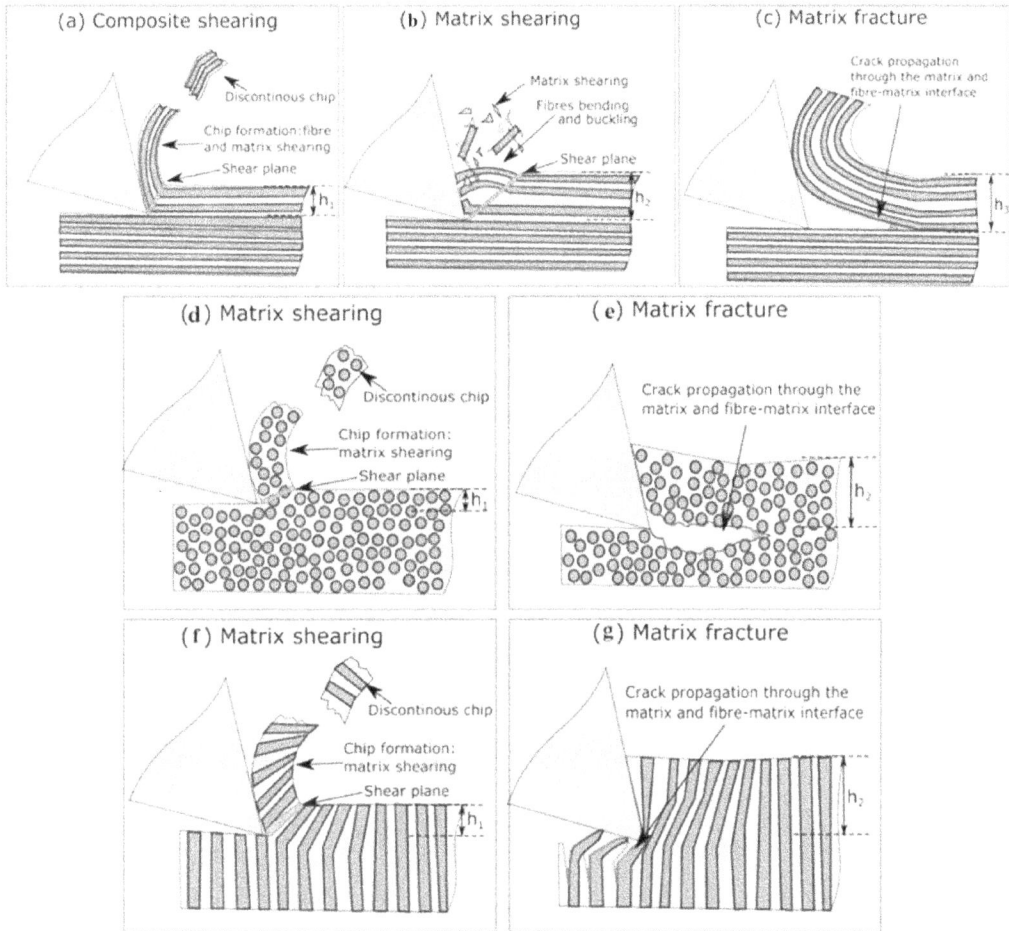

FIGURE 14.5 Mechanism of chip formation during orthogonal machining of CMC composite in different fibre orientations (a), (b), (c) for longitudinal orientation and increasing uncut chip thickness values $(h_3 > h_2 > h_1)$, (d) and (e) for across orientation and increasing uncut chip thickness values $(h_2 > h_1)$ and (f) and (g) for transverse orientation and increasing uncut chip thickness values $(h_2 > h_1)$ [32].

it affects surface integrity and causes material loss [34]. Increasing the clearance angle will help in minimizing flank wear, but it will weaken the tool. So, the optimum clearance angle for every tool has to be determined [35]. To the best of the knowledge of the authors, not much detailed work on tool wear during conventional machining of CMCs is available in the open literature. So, a brief review of the tool wear during the machining of PMCs and MMCs is presented here.

14.4.1 Tool Wear during Machining of PMCs

Wang et al. [36] studied the wear behaviour of cutting tool inserts made up of carbide and polycrystalline diamond (PCD) materials during the machining of Graphite/Epoxy composite. The PCD tool exhibited good life compared to the carbide tool due to its excellent hardness, abrasive resistance and thermal conductivity. They observed severe tapering, grooving and cutting-edge rounding in the flank of the carbide inserts, whereas PCD inserts were not grooved but cutting-edge rounding was observed. Ramulu et al. [37] found out that the microstructure of FRP composites strongly

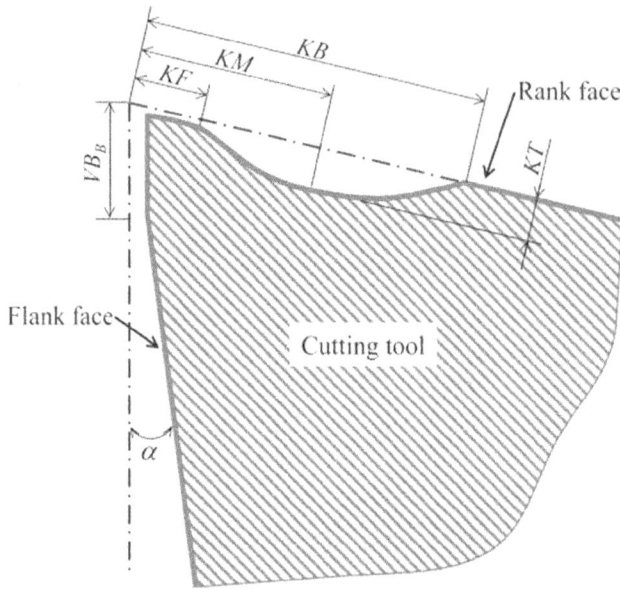

FIGURE 14.6   Characteristic features of tool wear while machining FRP composites [33].

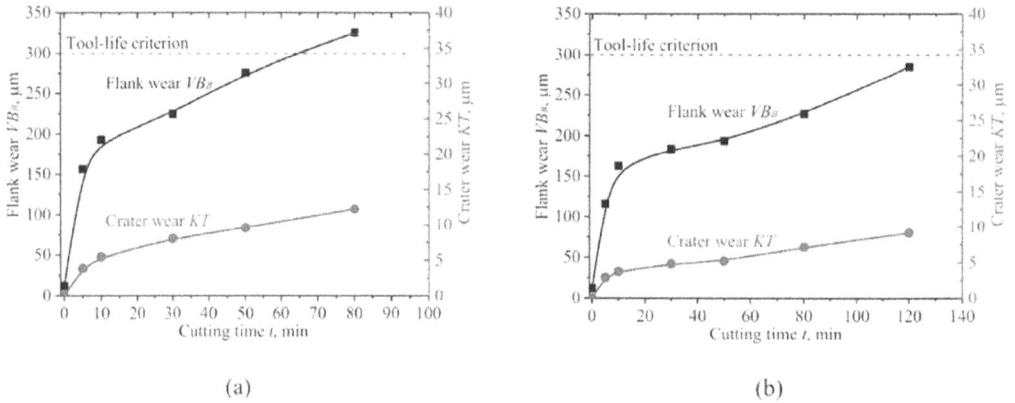

(a)   (b)

FIGURE 14.7   Flank wear and crater wear in (a) traditional machining (v =10 m/min, d = 50 μm) and (b) EVA machining of FRP composites (v= 10 m/min, ap=50 μm, f=27.32 kHz, a=2.85 μm, b=2.11 μm) [33].

influences the wear rate of the tool (PCD). Coarser grains have better wear resistance than finer ones. Xu and Zhang [33] compared the elliptic vibration assisted (EVA) machining of FRP composites and conventional machining and found that EVA machining has much-reduced tool wear compared with traditional machining. The variation in flank wear and crater wear over a period of cutting time, as obtained by Xu and Zhang [33], is shown in Figure 14.7. They also found lower cutting forces in EVA machining, compared with the traditional machining, as more localized material fracture and high-frequency recurrent material removal motion occurs in EVA machining.

14.4.2   Tool Wear during Machining of MMCs

Some of the common tool wear mechanisms reported while machining MMCs are as follows: crater wear, flank wear, notch wear, abrasion wear and edge chipping. El-Gallab and Sklad [38] found that the grooves present in the tool became wider as a result of abrasive action of harder particles of SiC and led to the formation of crater wear. However, the experimental findings of Sahoo et al. [39] clearly indicate that crater wear is not a serious concern during machining of MMCs. In general, the flank wear is considered while assessing the tool life. Manna and Bhattacharyya [25] noticed that flank wear was more while machining Al-SiC$_p$ MMC at low cutting speed. They claimed that strong cutting forces and the development of BUE were primarily responsible for flank wear. The increased cutting speed and the abrading action of high-energy chip particles result in an increase in flank wear. Similarly, due to the increased micro-cutting on the flank face, the flank wear increases as the depth of cut increases. Abrasive wear of tools is caused by two mechanisms, *viz.* two-body abrasive wear mechanism and three-body abrasive wear mechanism. Two-body abrasive wear occurs due to the interaction between a soft surface and a rough hard surface, whereas three-body abrasive wear is a result of hard particles dislodged and entrapped between the sliding surfaces [40]. It was found that abrasive wear due to the three-body mechanism was the main reason for the damage to the tool cutting edge [41–43]. Ding et al. [44] discovered that the machined surface had developed a sequence of wavy ridges, which caused notch wear on the flank face. This notch wear was more prevalent in wet cutting conditions under low cutting speeds, as the low cutting temperature increased the hardness of the workpiece matrix. High cutting speeds did not result in an increase in the matrix material's hardness, as effective dissipation of heat from the machined surface could not be achieved by the coolant. Chipping of cutting edge is also reported by some researchers [38,44,45]. The reported work shows an increase in edge chipping with the increase in cutting speed.

14.5   SURFACE INTEGRITY OF COMPOSITES

Surface integrity is a measure of the impact of machining on the surface topography (surface characteristics such as roughness and waviness), surface metallurgy (metallurgical alterations in the surface and subsurface) and mechanical properties (e.g. hardness) of the composite material [4]. Geometrical defects such as voids, pits, microcracks and craters are formed in composite machining due to breakage and pull-out of the particles (Figure 14.2) and interfacial debonding of matrix/reinforcement with particles present in the matrix. Such flaws are caused by the presence of reinforcing particles and the selection of process parameters. The interfacial bonding of reinforcement and matrix also affects the surface integrity. Particles in the bonding are separated and create voids because of weaker interfacial bonding. The cutting tool drags the removed particle from the matrix over the machined surface, thereby affecting the surface integrity [46]. During machining, high strain hardened layer results from low cutting speed, high depth of cut and smaller particle volume fraction [47]. There are three basic factors [16] that affect mechanical deformation and, thus, residual stresses. These are (a) limitations of matrix flow caused by particle presence, (b) particle indentation on machined surface and (c) matrix compression between the tool face and the particles. Low tool feed rates make these issues more prominent.

The occurrence of tool wear during the machining of FRP composites has a negative impact on the surface integrity of the machined part because of the increased contact length, contact friction and fibre fracture caused by bending of the fibres beneath the flank surface [33]. The EVA cutting of FRP composites has been discovered to be a promising method in which the time of contact between the flank face and the surface of the workpiece is reduced due to vibration, resulting in minor tool wear and good surface integrity [48]. The studies on particle tool interaction during the machining of MMCs by Ghandehariun et al. [49] indicated that surface integrity is greatly dependent on the position of the reinforcement particle relative to the cutting edge of the tool. If the particle is above the cutting plane, pull-out of the particle causes cavities and develops residual stresses. When the particle is at the bottom of the machining plane, it is elongated and when at the centre it is sheared by the tool flank. Therefore, the surface roughness and residual stress of the machined surface in MMCs are influenced by the particle and tool's relative positions. Micro cracking and micro plasticity are the major dominating factors while machining CMCs with small uncut chip thickness to obtain better surface roughness. The importance of reinforcement particles in the surface topography, after machining, is due to the fact that the mode of failure is affected by the surface integrity of the workpiece [8].

14.6 CONCLUDING REMARKS

This chapter presents an overview of the machining of composite materials with a focus on machining difficulties that arise during orthogonal cutting. These have been discussed with reference to tool wear and surface integrity of the workpiece material. The following key ideas are evident from the literature:

- The fibre orientation and the tool rake angle are important factors in defining the chip formation mechanism while machining FRP composites. The tool wear is highly dependent on the grain size of the FRP composites, coarser grains produce lesser tool wear compared with fine grains. The EVA cutting of FRP composites is found to be a promising technique to obtain good surface integrity and reduced tool wear.
- During machining of MMCs, flank wear and abrasive wear becomes significant due to large cutting forces and abrasive action of dislodged particles between two surfaces.
- The formation of chips during CMCs machining is determined by the magnitude of applied shear stress and the ultimate shear strength of the matrix material, when cutting along the longitudinal, cross and transverse fibre orientations.
- The presence of particle phase in the workpiece not only increases tool wear but also favours a number of surface flaws including, but not limited to, particle pull-out, grooves, pits, microcracks and craters. The reinforcing particles affect the material's susceptibility to strain hardening by acting as barriers.

14.7 FUTURE SCOPE

This chapter presented the machining of composite materials by conventional machining techniques. Non-conventional and hybrid machining techniques are explored by researchers to a certain extent in recent years [50–53]. The future scope pertaining to the traditional machining of composites will be on the application of unconventional sources like laser-assisted traditional machining of composites and ultrasonic-assisted machining. The cryogenic machining of composites is also one of the recent thrust areas of research. The numerical modelling, simulation and optimization of process variables can be done for these above said machining techniques to attain better quality and better productivity.

REFERENCES

1. Teti, R. (2002), Machining of composite materials. *CIRP Annals*, 51(2), 611–639.
2. Shyha, I., Huo, D. (2021), *Advances in Machining of Composite Materials Conventional and Non-conventional Processes.* Springer International Publishing Cham, Switzerland.
3. Voss, R., Seeholzer, L., Kuster, F., Wegener, K. (2017), Influence of fibre orientation, tool geometry and process parameters on surface quality in milling of CFRP. *CIRP Journal of Manufacturing Science and Technology*, 18, 75–91. doi: 10.1016/j.cirpj.2016.10.002.
4. Gao, T., Zhang, Y., Li, C., Wang, Y., Chen, Y., An, Q., Zhang, S., Li, H. N., Cao, H., Ali, H. M., Zhou, Z., Sharma, S. (2022), Fiber-reinforced composites in milling and grinding: Machining bottlenecks and advanced strategies. *Frontiers of Mechanical Engineering*, 17(2), 24.

5. Geier, N., Davim, J. P., Szalay, T. (2019), Advanced cutting tools and technologies for drilling carbon fibre reinforced polymer (CFRP) composites: A review. *Composites Part A: Applied Science and Manufacturing*, 125, 105552. doi: 10.1016/j.compositesa.2019.105552.

6. Rajasekaran, T., Palanikumar, K., Vinayagam, B. K. (2012), Experimental investigation and analysis in turning of CFRP composites. *Journal of Composite Materials*, 46, 809–821. doi: 10.1177/0021998311410500.

7. Abrate, S., Walton, D. A. (1992), Machining of composite materials. Part I: Traditional methods. *Composites Manufacturing*, 3, 75–83. doi: 10.1016/0956-7143(92)90119-F.

8. Gavalda Diaz, O., Garcia Luna, G., Liao, Z., Axinte, D. (2019), The new challenges of machining Ceramic Matrix Composites (CMCs): Review of surface integrity. *International Journal of Machine Tools and Manufacture*, 139, 24–36. doi: 10.1016/j.ijmachtools.2019.01.003.

9. Rajak, D. K., Pagar, D. D., Kumar, R., Pruncu, C. I. (2019), Recent progress of reinforcement materials: A comprehensive overview of composite materials. *Journal of Materials Research and Technology*, 8, 6354–6374. doi: 10.1016/j.jmrt.2019.09.068.

10. Callister, W. D., Rethwisch, D. G. (2007), *Materials Science and Engineering*. John Wiley and sons New york.

11. Ertekin, M. (2017), *Aramid fibers. In Fiber Technology for Fiber-Reinforced Composites* (pp. 153–167). Woodhead Publishing Philadelphia.

12. Aboudi, J., Arnold, S., Bednarcyk, B. (2006), *Micromechanics of Composite Materials*. CRC Press Boca Raton.

13. Morales-Flórez, V., Domínguez-Rodríguez, A. (2022), Mechanical properties of ceramics reinforced with allotropic forms of carbon. *Progress in Materials Science*, 128. doi: 10.1016/j.pmatsci.2022.100966.

14. Jain, S. K., Schmid, S. R. (2010), *Manufacturing Engineering and Technology*, Sixth Edition. Prentice Hall, New Jersey.

15. Pramanik, A., Arsecularatne, J. A., Zhang, L. C. (2008), Machining of particulate-reinforced metal matrix composites. *Machining: Fundamentals and Recent Advances*, 127–166. doi: 10.1007/978-1-84800-213-5_5.

16. Pramanik, A., Zhang, L. C., Arsecularatne, J. A. (2008), Machining of metal matrix composites: Effect of ceramic particles on residual stress, surface roughness and chip formation. *International Journal of Machine Tools and Manufacture*, 48, 1613–1625. doi: 10.1016/j.ijmachtools.2008.07.008.

17. Sheikh-Ahmad, J. Y. (2009), *The Machining of Polymer Composites*. Springer US, New York.

18. Wang, X. M., Zhang, L. C. (2003), An experimental investigation into the orthogonal cutting of unidirectional fibre reinforced plastics. *International Journal of Machine Tools and Manufacture*, 43, 1015–1022. doi: 10.1016/S0890-6955(03)00090-7.

19. Ramulu, M. (1997), Machining and surface integrity of fibre-reinforced plastic composites. *Sadhana*, 22, 449–472. doi: 10.1007/BF02744483.

20. Koplev, A., Lystrup, A., Vorm, T. (1983), The cutting process, chips, and cutting forces in machining CFRP. *Composites*, 14, 371–376. doi: 10.1016/0010-4361(83)90157-X.

21. Wang, D. H., Ramulu, M., Arola, D. (1995), Orthogonal cutting mechanisms of graphite/epoxy composite. Part I: unidirectional laminate. *International Journal of Machine Tools and Manufacture*, 35, 1623–1638. doi: 10.1016/0890-6955(95)00014-O.

22. Arola, D., Ramulu, M., Wang, D. H. (1996), Chip formation in orthogonal trimming of graphite/epoxy composite. *Composites Part A: Applied Science and Manufacturing*, 27, 121–133. doi: 10.1016/1359-835X(95)00013-R.

23. Nayak, D., Bhatnagar, N., Mahajan, P. (2005), Machining studies of Uni-Directional Glass Fiber Reinforced Plastic (UD-GFRP) composites part 1: Effect of geometrical and process parameters. *Machining Science and Technology*, 9, 481–501. doi: 10.1080/10910340500398167.

24. Ghidossi, P., El Mansori, M., Pierron, F. (2006), Influence of specimen preparation by machining on the failure of polymer matrix off-axis tensile coupons. *Composites Science and Technology*, 66, 1857–1872. doi: 10.1016/j.compscitech.2005.10.009.

25. Manna, A., Bhattacharyya, B. (2002), A study on different tooling systems during machining of Al/SiC-MMC. *Journal of Materials Processing Technology*, 123, 476–482. doi: 10.1016/S0924-0136(02)00127-9.

26. Chen, P., Hoshi, T. (1992), High-performance machining of SiC Whisker-reinforced aluminium composite by self-propelled rotary tools. *CIRP Annals – Manufacturing Technology*, 41, 59–62. doi: 10.1016/S0007-8506(07)61152-4.

27. Lin, J. T., Bhattacharyya, D., Lane, C. (1995), Case study "Machinability of a silicon carbide reinforced aluminium metal matrix." *Wear*, 189, 144. doi: 10.1016/0043-1648(95)06720-5.

28. Sahin, Y., Kok, M., Celik, H. (2002), Tool wear and surface roughness of Al_2O_3 particle-reinforced aluminium alloy composites. *Journal of Materials Processing Technology*, 128, 280–291. doi: 10.1016/S0924-0136(02)00467-3.

29. Lin, J. T., Bhattacharyya, D. (1998), Chip formation in the machining of SiC-Particle-reinforced Aluminium-matrix composites. *Composites Science and Technology*, 3538, 285–291.

30. Hung, N. P., Yeo, S. H., Lee, K. K., Ng, K. J. (1998), Chip formation in machining particle-reinforced metal matrix composites. *Materials and Manufacturing Processes*, 13, 85–100. doi: 10.1080/10426919808935221.

31. Kaw, A. K. (2006), *Mechanics of Composite Materials*, Second Edition. CRC Press, Taylor & Francis Group Boca Raton.

32. Gavalda Diaz, O., Axinte, D. A. (2017), Towards understanding the cutting and fracture mechanism in ceramic matrix composites. *International Journal of Machine Tools and Manufacture*, 118–119, 12–25. doi: 10.1016/j.ijmachtools.2017.03.008.

33. Xu, W., Zhang, L. (2018), Tool wear and its e ff ect on the surface integrity in the machining of fi bre- reinforced polymer composites. *Composite Structures*, 188, 257–265. doi: 10.1016/j.compstruct.2018.01.018.

34. Siddhpura, A., Paurobally, R. (2013), A review of flank wear prediction methods for tool condition monitoring in a turning process. 371–393. doi: 10.1007/s00170-012-4177-1.

35. Wang, D. H, Ramulu, M, Arola, D. (1995), Orrthogonal cutting mechanisms of graphite/epoxy composite. Part II: Multidirectional laminate. *International Journal of Machine Tools and Manufacture*, 35, 1639–1648.

36. Wang, D. H., Wern, C. W., Ramulu, M., Rogers, E. (1995), Cutting edge wear of tungsten carbide tool in continuous and interrupted cutting of a polymer composite. *Materials and Manufacturing Process*, 10, 493–508. doi: 10.1080/10426919508935040.

37. Ramulu, M., Faridnia, M., Garbini, J. L., Jorgensen, J. E. (1991), Machining of graphite/epoxy composite materials with polycrystailine diamond (PCD) tools. *Journal of Engineering Materials and Technology*, Transactions of the ASME, 113, 430–436. doi: 10.1115/1.2904122.

38. El-Gallab, M., Sklad, M. (1998), Machining of Al/SiC particulate metal-matrix composites Part I: Tool performance. *Journal of Materials Processing Technology*, 83, 151–158. doi: 10.1016/S0924-0136(98)00054-5.

39. Sahoo, A. K., Pradhan, S., Rout, A. K. (2013), Development and machinability assessment in turning Al/SiCp-metal matrix composite with multilayer coated carbide insert using Taguchi and statistical techniques. *Archives of Civil and Mechanical Engineering*, 13, 27–35. doi: 10.1016/j.acme.2012.11.005.

40. Li, X., Seah, W. K. H. (1999), Tool wear acceleration in relation to the reinforcement percentage in cutting of metal matrix composites. *ASME International Mechanical Engineering Congress and Exposition, Proceedings (IMECE)*, 1999-U, 379–388. doi: 10.1115/IMECE1999-0695.

41. Lin, C. B., Hung, Y. W., Liu, W. C., Kang, S. W. (2001), Machining and fluidity of 356Al/SiC(p) composites. *Journal of Materials Processing Technology*, 110, 152–159. doi: 10.1016/S0924-0136(00)00857-8.

42. Jaspers, S. P. F. C., Dautzenberg, J. H. (2002), Material behaviour in metal cutting: Strains, strain rates and temperatures in chip formation. *Journal of Materials Processing Technology*, 121, 123–135. doi: 10.1016/S0924-0136(01)01227-4.

43. Davim, J. P. (2002), Diamond tool performance in machining metal-matrix composites. *Journal of Materials Processing Technology*, 128, 100–105. doi: 10.1016/S0924-0136(02)00431-4.

44. Ding, X., Liew, W. Y. H., Liu, X. D. (2005), Evaluation of machining performance of MMC with PCBN and PCD tools. *Wear*, 259, 1225–1234. doi: 10.1016/j.wear.2005.02.094.

45. Ciftci, I., Turker, M., Seker, U. (2004), Evaluation of tool wear when machining SiCp-reinforced Al-2014 alloy matrix composites. *Materials and Design*, 25, 251–255. doi: 10.1016/j.matdes.2003.09.019.

46. Liao, Z., Abdelhafeez, A., Li, H., Yang, Y., Diaz, O. G., Axinte, D. (2019), State-of-the-art of surface integrity in machining of metal matrix composites. *International Journal of Machine Tools and Manufacture*, 143, 63–91. doi: 10.1016/j.ijmachtools.2019.05.006.

47. Pramanik, A., Zhang, L. C. (2017), Particle fracture and debonding during orthogonal machining of metal matrix composites. *Advances in Manufacturing*, 5, 77–82. doi: 10.1007/s40436-017-0170-0.

48. Xu, W., Zhang, L. C., Wu, Y. (2014), Elliptic vibration-assisted cutting of fibre-reinforced polymer composites: Understanding the material removal mechanisms. *Composites Science and Technology*, 92, 103–111. doi: 10.1016/j.compscitech.2013.12.011.

49. Ghandehariun, A., Kishawy, H. A., Umer, U., Hussein, H. M. (2016), Analysis of tool-particle interactions during cutting process of metal matrix composites. *International Journal of Advanced Manufacturing Technology*, 82, 143–152. doi: 10.1007/s00170-015-7346-1.

50. Rafighi, M. (2022), Effects of shallow cryogenic treatment on surface characteristics and machinability factors in hard turning of AISI 4140 steel. *Proceedings of the Institution of Mechanical Engineers, Part E: Journal of Process Mechanical Engineering*, 236, 2118–2130. doi: 10.1177/09544089221083467.

51. Zhai, C., Xu, J., Hou, Y., Sun, G., Zhao, B., Yu, H. (2022), Effect of fiber orientation on surface characteristics of C/SiC composites by laser-assisted machining. *Ceramics International*, 48, 6402–6413. doi: 10.1016/j.ceramint.2021.11.183.

52. Kim, J., Zani, L., Abdul-Kadir, A., Jones, L., Roy, A., Zhao, L., Silberschmidt, V. V. (2022), Hybrid-hybrid machining of SiC-reinforced aluminium metal matrix composite. *Manufacturing Letters*, 32, 63–66. doi: 10.1016/j.mfglet.2022.04.002.

53. Abedinzadeh, R., Norouzi, E., Toghraie, D. (2022), Study on machining characteristics of SiC–Al$_2$O$_3$ reinforced aluminum hybrid nanocomposite in conventional and laser-assisted turning. *Ceramics International*, 48, 29205–29216. doi: 10.1016/j.ceramint.2022.05.196.

15 Fabrication and Machinability (Drilling) Properties of Fiber Metal Laminate (FML) Composites (CARALL and GLARE)

Ergün Ekici and Ali Riza Motorcu

CONTENTS

15.1 INTRODUCTION

Fiber-reinforced polymer (FRP) composites are used in a variety of industries, including aerospace, ships, civil structures, and automobiles, due to their exceptional properties, which include superior fatigue strength, corrosion resistance, and high hardness-to-weight [1]. The development of fiber metal laminates (FML) over the past few decades has been fueled by the weight restrictions placed on metal, particularly in the aerospace industry, the limited applications of conventional composites, and the growing demand for high-performance, lightweight structures (FMLs). FMLs, which emerged by adding thin metal (aluminum, magnesium, or titanium) layers (0.2–0.5 mm thick) to traditional FRP, are mixed composite materials that combine the hardness and strength of FRP composites with the toughness and strength of metallic alloys (Figure 15.1). Fiber types and metals

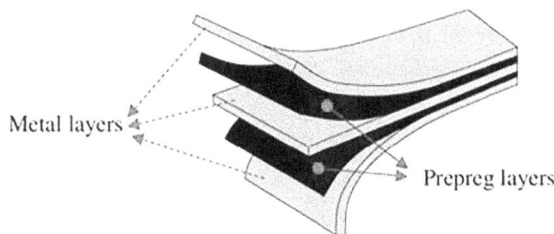

FIGURE 15.1 Fiber metal laminate.

DOI: 10.1201/9781003343912-15

used in FML are presented in Figure 15.2, and aluminum-based FMLs with different fiber types are presented in Figure 15.3.

Although aluminum (Al) is widely used in aerospace applications [2,3], its weakness against fatigue was already seen as a problem to be solved [4,5]. Aramid-reinforced aluminum laminates (ARALL) bonded as adhesives using the fatigue resistance of high-strength aramid fibers was first introduced in 1978. Four classes of ARALL composites (Figure 15.3) differ in base material quality [6]. Despite having high weight-saving potential and superior fatigue behavior, ARALL had

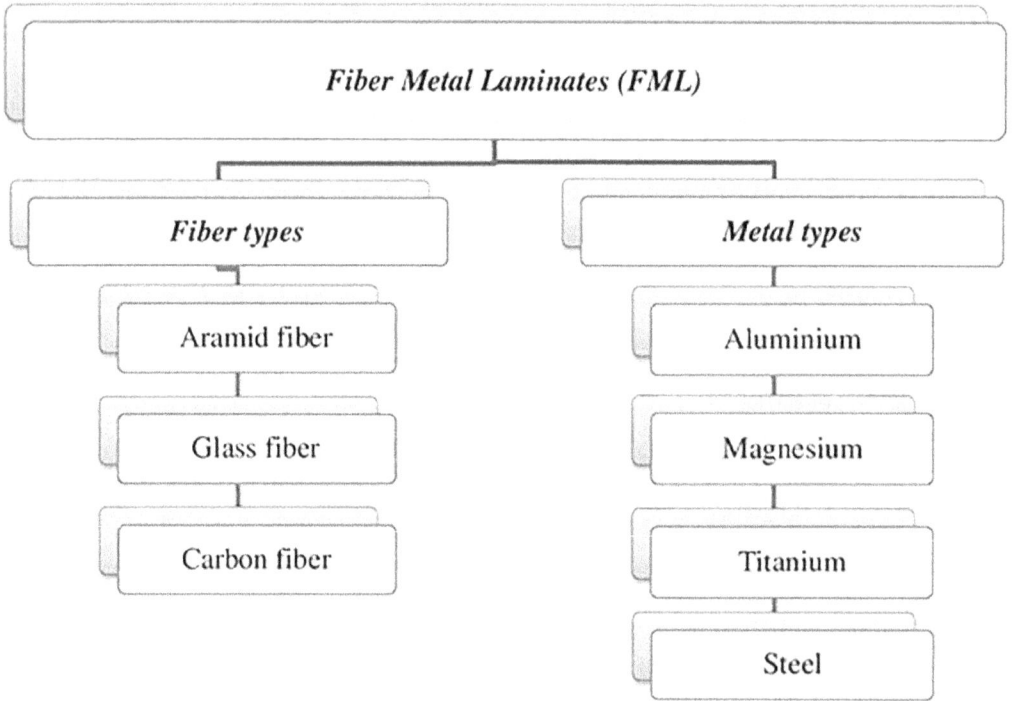

FIGURE 15.2 Types of fibers and metals used in FML.

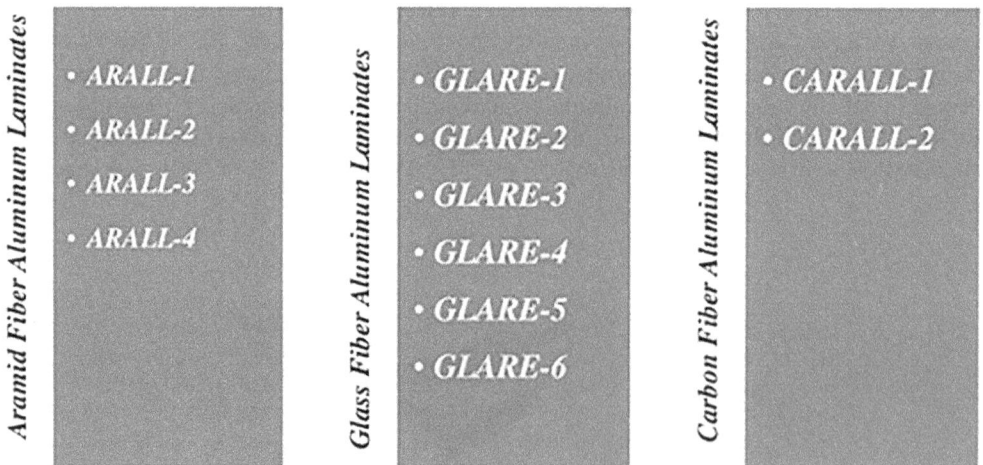

FIGURE 15.3 Aluminum-based FMLs with different fiber types.

problems such as low strength of the fiber-matrix connection, low strength at the metal-composite interface, and relatively poor notch strength behavior, especially compared with monolithic Al alloy [7,8]. With these problems mentioned above, the anisotropic properties of unidirectional fibers prevented their application in the trunk skin where biaxial stresses occurred. By 1990, ARALL's disadvantages were tried to be eliminated with GLARE, developed by using glass fiber instead of aramid fibers in FMLs [8]. GLARE impact resistance is attributed to Al and glass/epoxy properties, but it is significantly higher than monolithic Al impact resistance [9,10]. The need for lighter structures to enhance the strength of FMLs due to lower emissions and fuel consumption is driving the development of CARALL. Corrosion resistance, improved fatigue tolerance, strength-to-weight ratio, low thermal expansion, and formability are just a few of the benefits of carbon fiber composites combined with thin metal layers. However, there are still significant issues that need to be resolved for CARALL, such as galvanic corrosion between carbon and Al alloys and large thermal residual stresses in Al alloy brought on by a significant mismatch in the thermal expansion coefficients between CFRP and Al alloy [11]. For this purpose, titanium (Ti)-based FMLs have been created [12,13].

15.2 FABRICATION OF FMLs

The fabrication of the FML consists of two important steps: the stacking and bonding of the metal and FRP layers. The fabrication of FRP layers is similar to the preparation of conventional FRP laminate, while pretreatment is required for metal layers. The resin bond is usually sufficient for bonding metal and FRP sheets depending on the fiber type and the expected mechanical properties. In some cases, the adhesive film is used for interfaces.

15.2.1 INTERFACE ANALYSIS OF FMLs

High-strength and low-density Al and Ti alloys are used as metal components of FML composites. Al is a highly favored material in FML due to its exceptional qualities, including high thermal conductivity, non-magnetic properties, high strength, corrosion resistance, low density, low cost, good manufacturing properties, and many different classes of mechanical performance. During manufacture, the metal must be combined with different fiber components, which can vary in orientation and number of layers. The contact between the matrix, fiber surface, and metal layers significantly impacts the strength of FML [14]. Due to the discontinuity in the material's properties and sudden changes in the stress distribution, the interfaces are the weakest link in the load transfer chain in joined connections [15]. The most common damages in FML composites due to internal stresses between fiber and matrix or between composite layers are breakage of the matrix or fibers and separation (delaminations) between two adjacent layers [16]. Although adhesively bonded structural joints frequently transmit shear stresses, they are also subject to normal stresses, which frequently cause tearing or peeling [17]. Since unbonded areas, pores and inclusions at the metal–fiber interface will create a basis for delamination during the use of FML, and metal components are specially pretreated before bonding with fiber layers [18,19].

How the polymer interacts with the metal determines the strength of the bond between the metal and fiber layers [20]. Metals absorb oils and other pollutants from the atmosphere due to their high-energy surfaces. As a result, without properly cleaning the metal surface, obtaining a quality adhesive bond is impossible. Cleaning the surface of a metallic sample and providing the roughness that creates an increased contact area between the two adhered layers are accomplished in a series of different steps [21]. Mechanical (i.e., sanding [22] and grit blasting [23]), chemical (i.e., acid etching [24]), electrochemical (i.e., anodizing [25]), coupling agent (i.e., sol–gel and silane [26]), and novel dry surface treatments (i.e., laser ablation [27], plasma-sprayed [28]) are a variety of commonly used methods in the pretreatment of metal surfaces in FML [29].

Chemical etching and anodizing methods are commonly used to increase the surface activity of Al alloys. Etching can improve adhesion between adhesives and Al alloys (degreasing in organic solvents or alkaline solutions followed by etching in acid solutions). The Al surface must be anodized before bonding to decrease the susceptibility to corrosion after processing [30]. The chromic acid anodizing (CAA) treatment improves adhesion and produces an anodized layer with high corrosion resistance. However, its use for environmental protection is not recommended in practice [31,32]. The phosphoric acid anodizing (PAA) process is a more environmentally friendly pretreatment process that creates a more open pore, allowing high molecular primers/adhesives to penetrate. However, PAA treatment is a time-consuming pretreatment method that is not widely utilized due to a significant corrosion concern [25]. However, sulfuric acid anodizing (SAA) can produce a thicker oxide layer in less time than PAA. Phosphoric-sulpfric acid (PSA) anodizing is a promising alternative to CAA. The method chosen should be applied without skipping the steps meticulously, and the FML structure should be created immediately to preserve the surface properties.

Titanium (Ti) is a good alternative to Al alloys because it is completely resistant to galvanic corrosion when combined with carbon composites. However, the pure surface of Ti alloys is inefficient for charge transfer between Ti layers and CFRP due to poor adhesive properties. Surface preparation of Ti alloys is required prior to bonding to increase mechanical coupling and chemical bonding between CFRP and Ti alloys [33,34]. Sanding, grinding, and abrasive blasting are methods for increasing bond strength by roughening the surface of Ti and removing surface contaminants. It should only be used on thick-section components, and the substrate damage should be considered [35]. Another method is the cleaning and etching of Ti surfaces in alkaline mixtures [36]. However, these methods are sensitive to Ti alloy chemical composition, and hydrogen adsorption on Ti surfaces can cause embrittlement [37]. Plasma and laser surface treatments, on the other hand, are excellent alternatives to chemical and mechanical methods.

The bond strength of composite adhesive joints is increased by using additional adhesive (adhesive film) layers between the prepreg layers and metal. Trzepiecinski et al. [17] found that using adhesive film increased the peel strength of FML by 289.4%. However, they also stated that it increased the material cost of the double-layer adhesive film by about 10% in the weight of the composite, 30% in thickness, and 20% per prepreg layer. Dos Santos et al. [38] reported that the best failure mode and failure load results were obtained by adding adhesive to the Ti–CFRP interfaces, which significantly increased the CFRP connection's peel strength.

15.2.2 EFFECTS OF NANOPARTICLES ON FML

Nano-reinforcement elements are now being used to improve FML mechanical characteristics and to strengthen the interfacial bond between polymer composites and metal sheets. Graphene nanoplatelets (GNPs) [39], carbon nanotubes (CNTs) [39], and graphene nanoparticles [40,41] have been introduced as suitable reinforcement materials to enhance the mechanical properties of polymeric composites at very low contents.

While incorporating nanoparticles into polymeric composites on mechanical properties is well known, researchers are currently investigating the impacts of nanoparticles on the mechanical and interfacial behavior of FMLs. Yu et al. [42] examined the use of epoxy reinforced with carbon nanotubes in concentrations of 0%, 0.50%, 1%, 1.50%, 2%, 3.50%, and 5% by weight as an adhesive to join Al plates. The thermal degradation temperature of the adhesive containing 1% by weight of CNT was found to be approximately 14°C higher than that of pure epoxy, and it had the highest initial and final fracture toughness of all tested joints. According to Khoramishad et al. [43], adding multi-walled carbon nanotubes (MWCNTs) to FMLs improves resin-fiber adhesion and, as a result, reduces matrix cracking and composite delamination. However, it has been reported that MWCNTs weaken the bond between Al and composite layers and allow greater plastic deformation of the Al layer. According to Khurrama et al. [44], incorporating nanomaterials such as MWCNTs into the epoxy matrix enhances the bond strength between the Al metal and fiber-reinforced composite.

Jin et al. [45] reported that adding MWCNTs to the polyimide (PI) resin matrix improves the strength of the Ti/CFRP interface. By adding nanoclay (NC) to the resin, the delamination and interfacial resistance of FMLs can be improved [46]. The increased fatigue life of FMLs by embedding nanofilms in epoxy is due to interfacial strengthening, crack bridging, and shrinkage of nano alumina [47,48]. Megahed et al. [49] tried to improve the strength of GLARE by adding nanofillers such as copper (Cu), Al, aluminum oxide (Al_2O_3), silica (SiO_2), Ti oxide (TiO_2), and NC. According to the researchers, increasing the weight percent of nanofillers to 3% or higher by weight can result in agglomeration. These aggregations cause defects in nanocomposite laminates and harm mechanical properties by acting as stress concentrations, resulting in cracks and premature failure.

Most studies that found an increase in strength after incorporating CNTs into polymers attributed this to the nanotube-pulling mechanism and bridging cracks in the matrix. While the selection of nanoparticle type in FML is an issue to be considered, it is important to correctly determine the weight percentage due to improper dispersion and agglomerate formation and to follow the appropriate steps without skipping in order to obtain a homogeneous mixture.

15.2.3 PRODUCTION TECHNOLOGIES OF FMLs

After pretreated metal sheets and FRP layers are stacked alternately in the desired sequence, they are adhered to by applying heat and pressure. This step allows the resin to cure and bond within the FRP layers and at the interface with the metal plates. In addition to selecting the appropriate metal according to the expectations regarding the FML properties, the selected fiber and resin type are two other important parameters for the production conditions.

The thermoset-based epoxy resin used in FRP composites in aircraft manufacturing is also widely preferred for FML due to its low production and processing costs and good mechanical properties. One of the major disadvantages of epoxy-based FMLs is the significantly longer processing period needed to cure the polymeric resin matrix [50–52].

When used in composite materials, thermoplastics have higher operating temperatures than thermosets, higher fracture toughness, better impact damage resistance, and the capability to reshape them after manufacturing processes and repair them after a damaging event [51,53]. Furthermore, using a thermoplastic-based composite in FML enables the production of components that can be molded and shaped in a single step [54]. However, due to current production costs, the use of thermoplastics in aviation, whether as a matrix phase in FRP or as a structural adhesive, is very low compared with thermosets [51].

While the aerospace industry is focused on the next generation of FML, advanced thermoplastic fiber metal laminate (TFML) Ti, which can withstand high temperatures, is more attractive. The high-temperature capabilities of Ti and carbon fiber-reinforced polyetheretherketone (PEEK), which provide excellent toughness and thermal performance, draw attention to the new-generation FML [55,56].

The most difficult challenge in mass-producing FMLs is developing a good production process for small complex laminate parts [57]. The oldest and most fundamental manual technique for producing laminates is lay-up. While the first alternative is to obtain the FML shape by first performing each thin layer in production, the production of direct FML is the second alternative, which is more effortless and cheaper [58]. The press brake bending (PBB) method, used in producing FML components with a single curvature, is based on the plastic deformation of the material placed between the punch and mold on a conventional press brake. While thin FMLs do not interfere with bending, the degree of material deformation is constrained by the characteristics of the composite [59]. The autoclave is considered one of the most popular approaches in FML production due to its quality and reliability in producing large and expensive laminate parts that require high performance where high-volume production is not required [60]. Hot pressing (HP) of FML, which offers relatively short cycle times, is another alternative to autoclaving [60]. Although time and temperature are two main parameters in the HP process that must be properly monitored, preheating is used before the curing phase to reduce resin viscosity and ease stacked laminate deformation [60–62]. Depending on the type of resin used, both methods involve heating the FML to a high temperature and then cooling it to room temperature. This applied heat treatment creates residual stresses that reduce the fatigue strength of FML components due to the residual stress originating from metal and FRP with different coefficients of thermal expansion.

The post-stretch technique was developed to alter this stress system [63]. Shot peen forming (SPF) [64], which is based on the production of local deformation as a result of the interaction of the plate with the hard balls hitting the surface of the FML at incredible speed, and laser forming (LF) [65], which is used in the production of FML components are other methods. These two methods, however, are insufficient for producing FML components with geometrically complex and formability. However, it can create complex-shaped FML components with incremental sheet forming (ISF) [66]. In mold forming techniques (stamping, hydroforming, and electromagnetic forming) that focus on Al-based composites to produce complex and smaller FML parts, the energy required for deformation can be applied from various sources (hydro-mechanical, mechanical, electromagnetic). FML layer sequences and process parameters must be carefully selected to reduce material defects [67]. Mennecart et al. [68] present two novel processes for hybridizing FML components in a single processing step using conventional metal layers and woven fibers. Wet compression molding is combined with deep drawing in the first process. It is combined with deep drawing resin transfer molding (RTM) in the second process. Because of the ability to provide a more homogeneous part thickness and inject the resin until the end of the forming process, more complex shapes can be produced. Some component characteristics must be considered before forming reinforced thermoplastics and metallic in a single drawing process. While it is impossible to shape thermoplastic at low temperatures, it should be remembered that it will decompose at very high temperatures. Temperature balance is a prerequisite for a good forming process. While the metallic component requires relatively high forces, the FRP can only transmit a limited number of forces [69].

15.3 THE MACHINING (DRILLING) OF FMLs

While 25% of an Airbus 380 body consists of composite materials, carbon or glass-based FRP composites make up 22% and GLARE 3% [70]. Even though near-net-shape composite components are produced, holes must still be drilled to allow the mechanical joining of the parts (screws/rivets) to one another [71]. The total number of holes required can reach 300,000 for small jets and 1.5–3 million for commercial airplanes [72]. Drilling is the most commonly used method of creating holes due to its speed, simplicity, and efficiency. However, drilling is also one of the most complicated cutting operations [73]. The hole drilling process in FRP composites is a big problem due to the anisotropic and inhomogeneous nature of the fiber, its high abrasiveness and low thermal conductivity. At high temperatures, metal alloys (like Al and Ti) have a strong chemical affinity for tool materials. In this respect, there are more serious challenges regarding hole quality and integrity in FML, where a monolithic metal sheet with different machinability properties and anisotropic fiber-reinforced composite are considered. Burr formation in metal alloys in FML, fiber/matrix detachment in FRP, interlaminar cracks, thermal damage, and delamination are important problems [74,75]. In the aerospace sector, it is estimated that 60% of the parts are rejected during the assembly stage because of flawed or subpar holes after composite component drilling. Delamination occurs due to the perpendicular application of feed forces to the laminate layers during the drilling process [76]. In addition, during the drilling of metal and FRP composite layers, chips evacuated along with the hole rub against the inner surface wall of the hole, which causes delamination [72]. In the literature, numerous studies have been done on delamination that reduces the structural integrity of composite structures (FRP, CFRP-metallic stacks, GLARE, and CARALL), tool geometry [77–79], cutting parameters [80–82], different coating properties [83–85], and addition of nanoparticles [86,87] to reduce damage. In most studies, it has been observed that cutting forces [88–90] have been examined, and the feed rate, which is one of the drilling parameters, has the main effect on the cutting forces. The layered nature of the FML causes the thrust forces to varying continuously throughout the hole. Metal layers generate a much higher cutting force compared to FRP. It should be acknowledged that the tool geometry, material, type of drilling operation, tool wear, and FML adhesion method will all impact the cutting force. Dagger drill, which offers suitable hole quality and minimum delamination in drilling traditional CFRP composites, will cause delamination due to high thrust forces in the face of the high built-up edge (BUE) effect originating from metallic layers. The dimensions and mounting position of the FML parts will limit the type of drilling operation selected. Tool wear can be improved by coating applications on carbide tools or by choosing polycrystalline diamond (PCD) tools. Thirukumaran et al. [79] promoted the selection of marginless drill bits for drilling in FML assembly processes.

Another factor that determines hole quality is dimensional accuracy. Since the elastic modulus differences between metal and composites will cause different elastic deformations in each layer along the hole, the dimensional tolerance changes continuously throughout the drilled hole [91,92]. When drilling thin FMLs, increased workpiece bending due to the absence of a support plate is a likely cause of the higher hole perpendicularity error [93]. A large part of the energy consumed in the drilling process is converted into heat. Increasing temperatures in the cutting zone cause thermal distortions at the metal/fiber and fiber/resin interface. While this causes different dimensional changes in the metal and FRP layers, it increases the workpiece temperature and the hole perpendicularity error [93]. Giasin et al. [81] reported that increased vibrations and workpiece instability after increasing feed rate increased the hole size at the hole exit compared with the hole entry. Fiber orientation also affects hole size, and increasing workpiece thickness increases the circularity error [72]. Park et al. [75] revealed that cutting parameters affect hole size tolerance, and hole roundness error varies between 15 and 192 μm. Giasin et al. [94] stated that the coating of the drill is the only parameter that affects the hole cylindricality, and the spindle speed is the only factor that affects the hole perpendicularity. For FML, surface roughness is a complex process influenced by the ductile nature of the metal and the brittle nature of the fiber and FRP layer orientations. However, mean roughness (Roughness Average, Ra) in drilling is also affected by drilling parameters and tool geometry due to the continuous vibration of the cutting tool. Due to the nature of the components, matrix degradation in FRP and voids caused by fiber-matrix separation are common in drilling FML, while surface scratches occur in Al layers. In FML, chips formed after the plastic deformation of the ductile metallic component are pushed upward through the hole. Due to the fragile nature of the fiber, the cutting mechanism takes place, and brittle and dusty chips are formed. After the friction of metallic chips on the inner surface of the hole along the hole in the layered structure, much higher surface roughness values appear in FRP [85]. According to Giasin et al. [81,94], Ra surface roughness values between 1 and 4.56 m were primarily obtained in GLARE drilling, and Ra increased as feed rate and cutting speed increased. Kumar and Verma [95] stated that the graphene oxide (GO) additive is a processing limitation that affects surface roughness. Ra values increased with increasing GO ratio.

Burr formation, which creates safety problems in parts and affects part performance, can be defined as the accumulation of plastically deformed material at the edges and corners of the part by plastic flow from cutting operations. The burr formation produced in the entry and exit regions of the workpiece depends on the machining conditions, tool geometry and material, part geometry, and workpiece materials [96,97]. The workpiece material forced into plastic flow by the drill during the drilling process causes burr formation on the hole entry and exit surfaces [98]. A costly second process, deburring and reaming, is required to remove burr formation, which is harder than the workpiece material due to the strain hardening effect. This secondary process costs 30% of the manufacturing costs for high-precision components, while it is between 15% and 20% for medium-complexity components [99]. While drilling GLARE with TiAlN-coated carbide tools, burr formations of different heights and thicknesses occurred at feed and spindle speeds, while minimum feed rate provided minimum burr formation. Burr formation cannot be eliminated, but it can be minimized by choosing appropriate cutting parameters [72]. Exit burrs increasing with a higher feed rate are more than inlet burrs [80]. The cutting tool coating properties are another important parameter affecting the burr height [100]. Cryogenic treatment and minimum quantity lubrication (MQL) are important in reducing exit burr formation [101].

FMLs can be cut using non-traditional manufacturing processes such as laser and water jet cutting in addition to traditional machining processes. However, these two methods cannot be considered finishing because both methods provide insufficient edge quality due to the formation of a rough sandy edge that can cause fatigue cracks in water jet cutting and a small heat-affected zone in laser jet cutting [76].

15.4   CONCLUSIONS

The increasing need for high-performance and lightweight composite structures in the aerospace industry increases the demand for FML composite materials. This situation makes it necessary to improve existing FML materials' mechanical properties and develop production and assembly conditions. The pre-production conditions of CARALL and GLARE composites, which are currently under development, are evaluated. The effects of the metal group and fiber type on mechanical properties are presented. In addition to traditional thermoset composite materials, the effects of new-generation thermoplastic composite materials on the forming abilities and mechanical characteristics of FML composites are discussed. The influence of nanoparticles on the mechanical characteristics of FML was evaluated. FML production technologies are presented in detail.

In FML composites, drilling is a common post-production joining method, as is the assembly of FRP composite and metallic stacks. For this reason, the manufacturing processes required for assembling FML composites have been examined, and the drilling process has been comprehensively evaluated. In the drilling process of FML composites, the effects of drilling parameters, tool geometry, and cutting tool coating conditions are important. They affect thrust force, torque, and the examined hole quality, such as surface roughness and delamination formation. In addition, the circularity and ovality of the holes after drilling the composite structures, which are important in ensuring precise aerospace tolerances, are comprehensively evaluated.

REFERENCES

1. Xu J, Li C, Mi S, An Q, Chen M. Study of drilling-induced defects for CFRP composites using new criteria. *Compos Struct* 2018; 201: 1076–1087.
2. Straznicky PV, Laliberté JF, Poon C, Fahr A. Applications of fiber-metal laminates. *Polym Compos* 2000; 21(4): 558–567.
3. Vermeeren CAJR. An historic overview of the development of fibre metal laminates. *Appl Compos Mater* 2003; 10(4): 189–205.
4. Chandrasekar M, Ishak MR, Jawaid M, et al. An experimental review on the mechanical properties and hygrothermal behaviour of fibre metal laminates. *J Reinf Plast Compos* 2017; 36(1): 72–82.
5. Sinmazçelik T, Avcu E, Bora MÖ, Çoban O. A review: Fibre metal laminates, background, bonding types and applied test methods. *Mater Des* 2011; 32(7): 3671–3685.
6. Asundi A, Choi AY. Fiber metal laminates: An advanced material for future aircraft. *J Mater Process Technol* 1997; 63(1–3): 384–394.
7. Vogelesang LB, Gunnink JW. ARALL: A materials challenge for the next generation of aircraft. *Mater Des* 1986; 7(6): 287–300.
8. Roebroeks GH. Towards GLARE: The development of a fatigue insensitive and damage tolerant aircraft material. Doctoral dissertation, Technische Universiteit Delft, 1991.
9. Alderliesten RC, Vlot A. Fatigue Crack Growth Mechanism of Glare. In *Proceedings of the 22nd International SAMPE Europe Conference*, Paris, France, 1991, pp. 41–52.
10. Alderliesten RC, Hagenbee M, Homan JJ, et al. Fatigue and damage tolerance of glare. *Appl Compos Mater* 2003; 10(4): 223–242.

11. Wang WX, Takao Y, Matsubara T. Galvanic corrosion-resistant carbon fiber metal laminates. In *Proceedings of the 16th international Conference on Composite Materials*, Kyoto, Japan, 8–13 July 2007, pp. 8–13.
12. Li X, Zhang X, Zhang H, et al. Mechanical behaviors of Ti/CFRP/Ti laminates with different surface treatments of titanium sheets. *Compos Struct* 2017; 163: 21–31.
13. Salve A, Kulkarni R, Mache A. A review: Fiber metal laminates (FML's)-manufacturing, test methods and numerical modeling. *Int J Eng Technol Sci* 2016; 3(2): 71–84.
14. Özgür Bora M, Çoban O, Sinmazçelik T, Cürgül İ. The influence of different circular hole perforations on interlaminar shear strength of a novel fiber metal laminates. *Polym Compos* 2016; 37(3): 963–973.
15. Liu J, Sawa T. Stress analysis and strength evaluation of single-lap adhesive joints combined with rivets under external bending moments. *J Adhes Sci Technol* 2001; 15(1): 43–61.
16. Costa ML, Botelho EC, Rezende MC. Monitoring of cure kinetic prepreg and cure cycle modeling. *J Mater Sci* 2006; 41(13): 4349–4356.
17. Trzepiecinski T, Kubit A, Kudelski R, et al. Strength properties of aluminium/glass-fiber-reinforced laminate with additional epoxy adhesive film interlayer. *Int J Adhes Adhes* 2018; 85: 29–36.
18. Valenza A, Fiore V, Fratini L. Mechanical behaviour and failure modes of metal to composite adhesive joints for nautical applications. *Int J Adv Manuf Technol* 2011; 53(5): 593–600.
19. Park SY, Choi WJ, Choi HS. The effects of void contents on the long-term hygrothermal behaviors of glass/epoxy and GLARE laminates. *Compos Struct* 2010; 92(1): 18–24.
20. Dou X, Tunggal D. Finite element modeling of stamp forming process on fiber metal laminates. *World J Eng Technol* 2015; 3(03): 247.
21. Ebnesajjad S, Ebnesajjad C. *Surface Treatment of Materials for Adhesive Bonding*. William Andrew Publishing, New York, pp. 96–100, 2013.
22. Mohamad M, Marzuki HFA, Ubaidillah EAE, et al. Effect of surface roughness on mechanical properties of aluminium-carbon laminates composites. *Adv Mat Res* 2014; 879: 51–57.
23. Rudawska A, Danczak I, Müller M, Valasek P. The effect of sandblasting on surface properties for adhesion. *Int J Adhes Adhes* 2016; 70: 176–190.
24. Osman E, Warikh ARM, Yahaya SH, et al. Effects of different pre-treatments on the performance of kenaf fiber reinforced aluminum laminates sandwich composite. *J Adv Manuf Technol* 2019; 13(2): 97–111.
25. Bjørgum A, Lapique F, Walmsley J, Redford K. Anodising as pre-treatment for structural bonding. *Int J Adhes Adhes* 2003; 23(5): 401–412.
26. Rider AN. Factors influencing the durability of epoxy adhesion to silane pretreated aluminium. *Int J Adhes Adhes* 2006; 26(1–2): 67–78.
27. Moroni F, Musiari F, Pirondi A. Influence of laser ablation-induced surface topology on the mechanical behaviour of aluminium bonded joints. *Proc Inst Mech Eng. L: J Mat* 2019; 233(3): 505–520.
28. Mandolfino C, Lertora E, Genna S, et al. Effect of laser and plasma surface cleaning on mechanical properties of adhesive bonded joints. *Procedia CIRP* 2015; 33: 458–463.
29. Park SY, Choi WJ, Choi HS, et al. Recent trends in surface treatment technologies for airframe adhesive bonding processing: A review (1995–2008). *J Adhes* 2010; 86(2): 192–221.
30. Xu Y, Li H, Shen Y, et al. Improvement of adhesion performance between aluminum alloy sheet and epoxy based on anodizing technique. *Int J Adhes Adhes* 2016; 70: 74–80.
31. Thrall EW, Shannon RW. Adhesive bonding of aluminum alloys. *Met Finish* 2007; 05(9): 49–56.
32. Almeida RS, Damato CA, Botelho EC, et al. Effect of surface treatment on fatigue behavior of metal/carbon fiber laminates. *J Mater Sci* 2008; 43(9): 3173–3179.
33. Hu Y, Zhang J, Wang L, et al. A simple and effective resin pre-coating treatment on grinded, acid pickled and anodised substrates for stronger adhesive bonding between Ti-6Al-4V titanium alloy and CFRP. *Surf Coat Technol* 2022; 432: 128072.
34. Ye J, Wang H, Dong J, et al. Metal surface nanopatterning for enhanced interfacial adhesion in fiber metal laminates. *Compos Sci Technol* 2021; 205: 108651.
35. Khan AA, Al Kheraif AA, Alhijji SM, Matinlinna JP. Effect of grit-blasting air pressure on adhesion strength of resin to titanium. *Int J Adhes Adhes* 2016; 65: 41–46.
36. Molitor P, Barron V, Young T. Surface treatment of titanium for adhesive bonding to polymer composites: A review. *Int J Adhes Adhes* 2001; 21(2): 129–136.
37. Snogren RC. *Handbook on Surface Preparation*. Palmerton Publishing Co., New York, 1974.
38. Dos Santos DG, Carbas RJC, Marques EAS, da Silva LFM. Reinforcement of CFRP joints with fibre metal laminates and additional adhesive layers. *Compos B Eng* 2019; 165: 386–396.

39. Eslami-Farsani R, Aghamohammadi H, Khalili SMR, et al. Recent trend in developing advanced fiber metal laminates reinforced with nanoparticles: A review study. *J Ind Text* 2020; 1528083720947106.

40. Kumar J, Verma RK, Khare P. Graphene-functionalized carbon/glass fiber reinforced polymer nanocomposites: fabrication and characterization for manufacturing applications. In: Hussain C and Kumar V (eds.) *Handbook of Functionalized Nanomaterials. Environmental Health and Safety. Micro and Nano Technologies*, pp. 57–78, Elsevier, Oxford, UK, 2021.

41. Kesarwani S, Verma RK. A critical review on synthesis, characterization and multifunctional applications of reduced graphene oxide (rGO) / composites. *Nano* 2021; 16(09): 2130008.

42. Yu S, Tong MN, Critchlow G. Use of carbon nanotubes reinforced epoxy as adhesives to join aluminum plates. *Mater Des* 2010; 31: 126–129.

43. Khoramishad H, Alikhani H, Dariushi S. An experimental study on the effect of adding multi-walled carbon nanotubes on high-velocity impact behavior of fiber metal laminates. *Compos Struct* 2018; 201: 561–569.

44. Khurram AA, Hussain R, Afzal H, et al. Carbon nanotubes for enhanced interface of fiber metal laminate. *Int J Adhes Adhes* 2018; 86: 29–34.

45. Jin K, Wang H, Tao J, Zhang X. Interface strengthening mechanisms of Ti/CFRP fiber metal laminate after adding MWCNTs to resin matrix. *Compos B Eng* 2019; 171: 254–263.

46. Zakaria AZ, Shelesh-nezhad K. Introduction of nanoclay-modified fiber metal laminates. *Eng Fract Mech* 2017; 186: 436–448.

47. Prasad EV, Sivateja C, Sahu SK. Effect of nanoalumina on fatigue characteristics of fiber metal laminates. *Polym Test* 2020; 85: 106441.

48. Rauf OU, Khurram AA, Hussain R, et al. Nanoparticles enhanced interfaces of glass fiber laminate aluminum reinforced epoxy (GLARE) fiber metal laminates. *Polym Compos* 2021; 42(8): 3954–3968.

49. Megahed M, Abd El-baky MA, Alsaeedy AM, Alshorbagy AE. An experimental investigation on the effect of incorporation of different nanofillers on the mechanical characterization of fiber metal laminate. *Compos B Eng* 2019; 176: 107277.

50. Soutis C. Carbon fibre reinforced plastic in aircraft construction. *Mater Sci Eng A* 2005; 412: 171–176.

51. Cortés P. The fracture properties of a fiber metal laminate based on a self-reinforced thermoplastic composite material. *Polym Compos* 2014; 35(3): 427–434.

52. Cortés P, Cantwell WJ. The fracture properties of a fibre–metal laminate based on magnesium alloy. *Compos B Eng* 2005; 37(2–3): 163–170.

53. Balakrishnan P, John MJ, Pothen L, et al. Natural fibre and polymer matrix composites and their applications in aerospace engineering. *Adv Compos Mater Aerospace Eng* 2016; 365–383.

54. Reyes G, Kang H. Mechanical behavior of lightweight thermoplastic fiber–metal laminates. *J Mater Process Technol* 2007; 186(1–3): 284–290.

55. Cortes P, Cantwell WJ. The tensile and fatigue properties of carbon fiber-reinforced PEEK-titanium fiber-metal laminates. *J Reinf Plast Compos* 2004; 23(15): 1615–1623.

56. Shanmuga L, Kazemi ME, Qiu C, et al. Influence of UHMWPE fiber and Ti6Al4V metal surface treatments on the low-velocity impact behavior of thermoplastic fiber metal laminates. *Adv Compos Hybrid Mater* 2020; 3(4): 508–521.

57. Blala H, Lang L, Li L, Alexandrov S. Deep drawing of fiber metal laminates using an innovative material design and manufacturing process. *Compos Commun* 2021; 23: 100590.

58. Hu Y, Zhang W, Jiang W, et al. Effects of exposure time and intensity on the shot peen forming characteristics of Ti/CFRP laminates. *Compos Part A Appl Sci* 2016; 91: 96–104.

59. Sinke J. Forming technology for composite/metal hybrids. In Long AC (ed.) *Composites Forming Technologies*; Woodhead Publishing Series in Textiles. Woodhead Publishing, Sawston, UK, 2007, pp. 197–219.

60. Sherkatghanad E, Lang L, Liu S, Wang Y. Innovative approach to mass production of fiber metal laminate sheets. *Mater Manuf Process* 2018; 33(5): 552–563.

61. Liu S, Sinke J, Dransfeld C. An inter-ply friction model for thermoset based fibre metal laminate in a hot-pressing process. *Compos B Eng* 2021; 227: 109400.

62. Botelho EC, Silva RA, Pardini LC, Rezende MC. A review on the development and properties of continuous fiber/epoxy/aluminum hybrid composites for aircraft structures. *Mater Res* 2006; 9: 247–256.

63. Alderliesten R. On the development of hybrid material concepts for aircraft structures. *Recent Pat Eng* 2009; 3(1): 25–38.

64. Li H, Lu Y, Han Z, et al. The shot peen forming of fiber metal laminates based on the aluminum-lithium alloy: Deformation characteristics. *Compos B Eng* 2019; 158: 279–285.

65. Kant R, Joshi SN, Dixit US. *Research issues in the laser sheet bending process. In: Davim J. P. (eds.) Materials Forming and Machining: Research and Development.* Woodhead Publishing, Elsevier, Sawston, Cambridge, pp. 73–97, 2016.
66. Ding Z, Wang H, Luo J, Li N. A review on forming technologies of fibre metal laminates. *Int J Lightweight Mater Manuf* 2021; 4(1): 110–126.
67. Trzepieciński T, Najm SM, Sbayti M, et al. New advances and future possibilities in forming technology of hybrid metal–polymer composites used in aerospace applications. *J Compos Sci* 2021; 5(8): 217.
68. Mennecart T, Werner H, Ben Khalifa N, Weidenmann KA. Developments and analyzes of alternative processes for the manufacturing of fiber metal laminates. In *Proceedings of the MSEC 2018*, College Station, TX, USA, 18–22 June 2018.
69. Behrens BA, Hübner S, Neumann A. Forming sheets of metal and fibre-reinforced plastics to hybrid parts in one deep drawing process. *Procedia Eng* 2014; 81: 1608–1613.
70. Graham-Cumming J. *The Geek Atlas:128 Places Where Science and Technology Come Alive.* O'Reilly Media Inc., California, pp. 25, 2009 .
71. Eneyew ED, Ramulu M. Experimental study of surface quality and damage when drilling unidirectional CFRP composites. *J Mater Res Technol* 2014; 3(4): 354–362.
72. Giasin K, Ayvar-Soberanis S. An Investigation of burrs, chip formation, hole size, circularity and delamination during drilling operation of GLARE using ANOVA. *Compos Struct* 2017; 159: 745–760.
73. Rivero A, Aramendi G, Herranz S, López de Lacalle LN. An experimental investigation of the effect of coatings and cutting parameters on the dry drilling performance of aluminium alloys. *Int J Adv Manuf Technol* 2006; 28(1): 1–11.
74. Bonhin EP, David-Müzel S, de Sampaio Alves MC, et al. A review of mechanical drilling on fiber metal laminates. *J Compos Mater* 2021; 55(6), 843–869.
75. Park SY, Choi WJ, Choi CH, Choi HS. Effect of drilling parameters on hole quality and delamination of hybrid GLARE laminate. *Compos Struct* 2018; 185: 684–698.
76. Sinke J. Manufacturing of GLARE parts and structures. *Appl Compos Mater* 2003; 10(4): 293–305.
77. Pawar OA, Gaikhe YS, Tewari A, et al. Analysis of hole quality in drilling GLARE fiber metal laminates. *Compos Struct* 2015; 123: 350–365.
78. Feito N, Diaz-Álvarez J, López-Puente J, Miguelez MH. Numerical analysis of the influence of tool wear and special cutting geometry when drilling woven CFRPs. *Compos Struct* 2016; 138: 285–294.
79. Thirukumaran M, Jappes JW, Siva I, et al. Investigation of margin effect to minimize delamination during drilling of differently stacked GFRP-aluminum fiber metal laminates (3/2 GLARE). *J Manuf Technol Res* 2018;10(1/2): 17–27.
80. Giasin K, Ayvar-Soberanis S, French T, Phadnis V. 3D finite element modelling of cutting forces in drilling fibre metal laminates and experimental hole quality analysis. *Appl Compos Mater* 2017; 24(1): 113–137.
81. Giasin K, Ayvar-Soberanis S, Hodzic A. An experimental study on drilling of unidirectional GLARE fibre metal laminates. *Compos Struct* 2015; 133: 794–808.
82. Tyczyński P, Lemańczyk J, Ostrowski R. Drilling of CFRP, GFRP, glare type composites. *Aircr Eng Aerosp Technol: An Int J* 2014; 86(4): 312–322.
83. Kuo CL, Soo SL, Aspinwall DK, et al. Development of single step drilling technology for multilayer metallic-composite stacks using uncoated and PVD coated carbide tools. *J Manuf Process* 2018; 31: 286–300.
84. Ekici E, Motorcu AR, Yıldırım E. An experimental study on hole quality and different delamination approaches in the drilling of CARALL, a new FML composite. *FME Trans* 2021; 49(4): 950–961.
85. Ekici E, Motorcu AR, Uzun G. Multi-objective optimization of process parameters for drilling fiber-metal laminate using a hybrid GRA-PCA approach. *FME Trans* 2021; 49(2): 356–366.
86. Kumar J, Verma RK, Debnath K. A new approach to control the delamination and thrust force during drilling of polymer nanocomposites reinforced by graphene oxide/carbon fiber. *Compos Struct* 2020; 253: 112786.
87. Kumar J, Verma RK. Experimental investigation for machinability aspects of graphene oxide/carbon fiber reinforced polymer nanocomposites and predictive modeling using hybrid approach. *Def Technol* 2021; 17(5): 1671–1686.
88. Dubey AD, Kumar J, Kesarwani S. et al. Investigation on thrust and torque generation during drilling of hybrid laminates composite with different stacking sequences using multiobjective optimization module. *J Multiscale Model* 2021; 12(03): 2150009.

89. Kumar J, Verma RK, Mondal AK. Taguchi-Grey theory based harmony search algorithm (GR-HSA) for predictive modeling and multi-objective optimization in drilling of polymer composites. *Exp Tech* 2021; 45(4): 531–548.

90. Kumar J, Verma RK. A new criterion for drilling machinability evaluation of nanocomposites modified by graphene/carbon fiber epoxy matrix and optimization using combined compromise solution. *Surf Rev Lett* 2021; 28(09): 2150082.

91. Brinksmeier E, Janssen R. Drilling of multi-layer composite materials consisting of carbon fiber reinforced plastics (CFRP), titanium and aluminum alloys. *CIRP Annals* 2002; 51(1): 87–90.

92. Qi Z, Zhang K, Li Y, et al. Critical thrust force predicting modeling for delamination-free drilling of metal-FRP stacks. *Compos Struct* 2014; 107: 604–609.

93. Giasin K. The effect of drilling parameters, cooling technology, and fiber orientation on hole perpendicularity error in fiber metal laminates. *Int J Adv Manuf Technol* 2018; 97(9): 4081–4099.

94. Giasin K, Hawxwell J, Sinke J, et al. The effect of cutting tool coating on the form and dimensional errors of machined holes in GLARE® fibre metal laminates. *Int J Adv Manuf Technol* 2020; 107(5): 2817–2832.

95. Kumar J, Verma RK. A novel methodology of Combined Compromise Solution and Principal Component Analysis (CoCoSo-PCA) for machinability investigation of graphene nanocomposites. *CIRP J Manuf Sci Technol* 2021; 33: 143–157.

96. Jin SY, Pramanik A, Basak AK, et al. Burr formation and its treatments—a review. *Int J Adv Manuf Technol* 2020; 107(5): 2189–2210.

97. Abdelhafeez AM, Soo SL, Aspinwall DK, et al. Burr formation and hole quality when drilling titanium and aluminium alloys. *Procedia CIRP* 2015; 37: 230–235.

98. Kim J, Dornfeld DA. Development of an analytical model for drilling burr formation in ductile materials. *J Eng Mater Technol* 2002; 124(2): 192–198.

99. Dornfeld D, Min S. A review of burr formation in machining. In Aurich JC and Dornfeld D (eds.), *Burrs-Analysis, Control and Removal*. Springer, Berlin Heidelberg, pp. 3–11, 2010.

100. Giasin K, Gorey G, Byrne C, et al. Effect of machining parameters and cutting tool coating on hole quality in dry drilling of fibre metal laminates. *Compos Struct* 2019; 212: 159–174.

101. Giasin K, Ayvar-Soberanis S, Hodzic A. The effects of minimum quantity lubrication and cryogenic liquid nitrogen cooling on drilled hole quality in GLARE fibre metal laminates. *Mater Des* 2016; 89: 996–1006.

16 Micromachining of Polymer Composites and Nanocomposites

Kritika Singh Somvanshi, Saurabh Tiwari, and Prakash Chandra Gope

CONTENTS

16.1 INTRODUCTION

The revelation of carbon nanofiber, graphene, cellulose nanofiber, and other nanomaterials has fueled the growth of the nanoscience field in recent decades. In polymer nanocomposite, the expedient properties of nanofillers associated to polymers are discovered. When nanomaterials are reinforced in a polymer matrix, their mechanical, electrical, and thermal characteristics are improved over the primeval polymer [1,2]. Nanocomposites are in high demand, especially for advanced structures in numerous automotive, marine, and aerospace. The key benefit and purpose for using polymer

DOI: 10.1201/9781003343912-16

composites is their light weight. A smaller amount of weight saves additional fuel in transportation. In wind mill, the lightest possible weight of the blade generates supplementary power. Aside from load, nanocomposites are non-conductive, versatile, maintenance-free, and corrosion-resistant and have flexibility in design, and with extraordinary durability. Polymer composites reinforced with fillers have increased technology functionality in bioengineering, aviation, and structural applications that involve lightweight parts, sufficient heat absorption, and minimum coefficient of thermal expansion. However, apart from nanocomposites, microfabrication component requirements have increased in fields such as biomedical, telecommunications equipment, aerospace, mechatronics, photonic, avionics, and connectivity [3,4]. Microchannels, microcontrollers, electromechanical system goods, bioengineering, surgical implantations, angioplasty, microfluid channels, sensors, heat exchangers, micropropellers, and micro-actuators are some of the applications. Mechanical micromachining is a distinctive manufacturing method of making this micro-equipment, with attribute dimensions ranging from a few micrometers to millimeters. By creating specific functionality, mechanical cutting processes serve as a bridge between the macro- and microdomains [5].

Mechanical machining methods on a microscale, such as microdrilling, milling, and turning, are frequently used for tooling nanocomposite, ceramics, resins, alloy steels, and noble metals. Micromachining developments appear to be insufficient to create a finished result with adequate surface integrity and dimensional accuracy. As a result, mechanical micromachining methods such as drilling and milling have indeed been identified as possible post-processing approaches to meet these engineering needs. Mechanical micromachining has good surface finish and dimension consistency, and it can be used on a variety of materials. On the other hand, the introduction of reinforcements results in an intricate arrangement of and their enhanced mechanical and resistant properties like hardness, ultimate tensile strength, wear resistance, elongation, etc., may lessen their machinability. Furthermore, micromachining exacerbates the material removal process because it distinguishes it from conformist machining in terms of cutting-edge radius, MUCT (Minimum-Uncut Chip Thickness), and microstructure which are frequently discussed as size effects. Most of these conditions make the micromachining of nanocomposites complicated.

A considerable amount of research has been printed on the machining and manufacturing of composite materials. Figure 16.1 depicts the research conducted over the last 10 years, from 2011 to 2021. Different machining parameters and machinability investigations were recognized in the

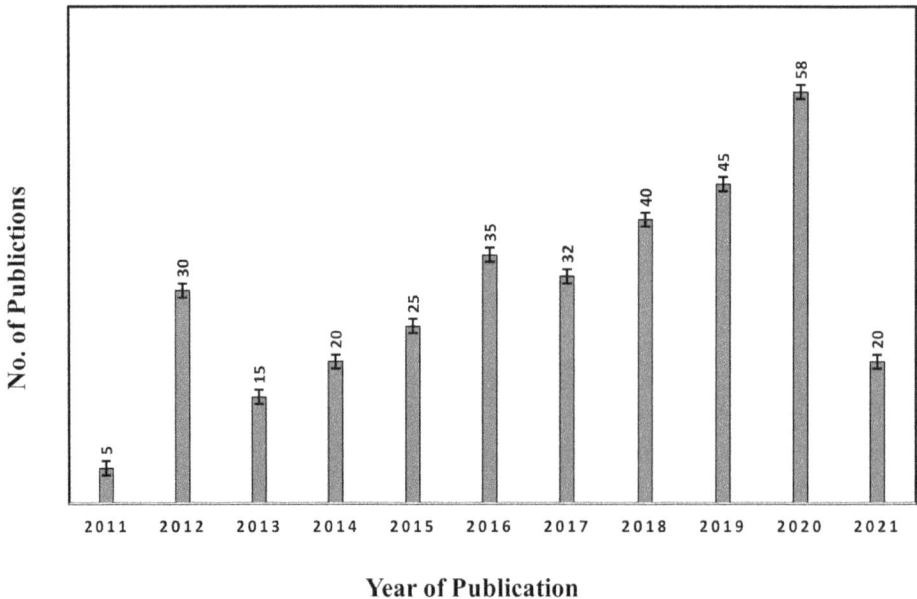

FIGURE 16.1 Number of publication on machining of polymer composites in the last ten years.

publications. It discovered that experiment into polymer composites has accelerated exponentially over the years. With over 75% of the research papers, plentiful machining processes of polymer-based composites are associated to micromilling, micro-electro discharge machining, and micro-drilling. This chapter discusses problems and obstacles in micromechanical machining methods like microturning, micromilling, microdrilling, and advanced micromachining.

16.2 POLYMER-BASED NANOCOMPOSITES

Nanotechnology and nanocomposites have seen significant growth in a variety of areas in the last few years. Polymer nanocomposite is therefore an effective combination of nanomaterials and polymers with at least one phase remaining in the nanoscale in the final material. The inclusion of nanoparticles in the matrix material not only greatly enhances almost all of the desirable characteristics (like thermal, physical, gas barrier, biodegradability, and flammability) of pure polymers, but also develops a different range of characteristics reliant on the nanomaterials incorporated [6,7]. The introduction of nanomaterials into polymer matrices has led to the discovery of a new class of materials, called nanocomposites, which can be used to fabricate new materials, changes the surface interaction as well as the physicochemical convolutions, which are actually associated with the efficiency of resulting synthetic polymers. The aspect ratio, geometrical profile, size of nanomaterials, and surface chemistry are vital factors in modifying such interactions and thus the characteristics of various processes. As a result, nanocomposites have carved an unparalleled niche in the arena of unconventional materials [8–10].

Polymer-based composites are categorized into following two types: intercalated and exfoliated. The categorization is based on the nanoparticles' state of distribution in polymeric matrix. The nanomaterials retain their original crystalline structure structures in the first type of polymer nanocomposites, though range among the lamellae or planes, are greater than in a natural position because nanoparticles are diffused via polymerization and therefore forming a sandwich-like framework. Exfoliated polymer nanocomposites are formed when nanofibers lose their existing structures due to impairment or slapdash structure in polymeric composites. Again, they are classified as 0-dimensional, 1-dimensional, and 2-dimensional nanopolymer composites depending upon the aspect of the nanocrystals incorporated into the matrix. Grounded on the basis of thermal properties, nanocomposites are further categorized as thermosetting and thermoplastic polymer composites. Polymer nanocomposites, on the other hand, can be distinguished on the basis of matrix used, which is determined by the primary interconnections present in the structures, such as pristine polymers. Depending upon the arrangement of the polymeric matrix, nanocomposites are categorized as dendritic and linear polymer composites as shown in Figure 16.2 [11,12].

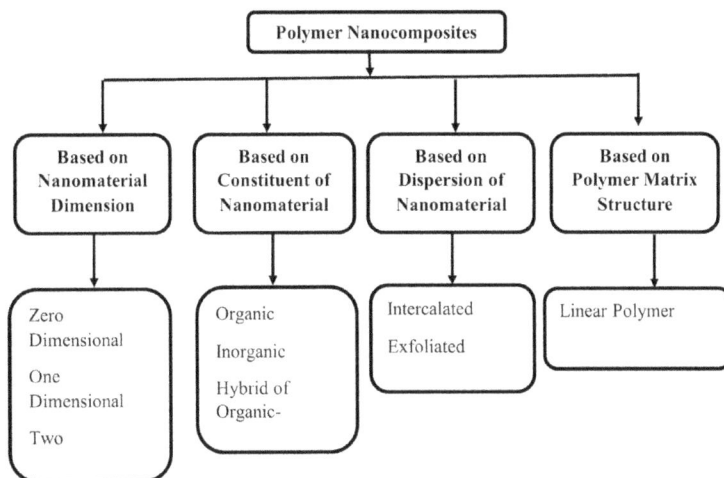

FIGURE 16.2 Classification of polymer nanocomposites.

16.3 MICROMACHINING OF NANOCOMPOSITES

The objective of extremely precise product micromachining is to create multipurpose products that are lighter in weight and mobile, consume less power, and are more efficient. Although with the revelation of new materials with exceptional strength-to-weight ratios, the advent of sophisticated tooling methods having ultra-high accuracy enables component miniaturization. Taniguchi [13] has altered machining advancement by increasing machining reliability from traditional (up to 100 m) to ultra-precision (1 nanometer) as shown in Figure 16.3 [14]. In another viewpoint, micromachining includes product dimensions (at least 2D scale between 1 and 500 m) which are too small to be manufactured by traditional machining. Highly precise machining focuses on machining accuracy with a dimension ratio more than 10,000:1. Actual goal of micromachining is to create microparts which are generally 1–100 m in range. Miniaturization is also described via the quantification of uncut thickness of chip whenever it is less than the average particle sizes or in the span of 0.1–200 m. Moreover, as machining technology advances, this gap will be shortened. Numerous micromachining methods are available, each with its own range of abilities in contexts of machining accuracy, the material of workpiece, part size, and geometrical ambiguity [15,16].

Categorization is an effort to determine the unique concepts that underpin some fundamental micromachining processes, typically established on whether the method depends on MEMS (Micro-Electromechanical Systems) and non-MEMS. Machining techniques that utilize lithography-based methods to create net shape in semiconductor devices like detectors, transceivers, actuators, and some more Si (silicon) substrate electronic equipment that are used to fabricate MEMS. MEMS-based approaches are mainly used to manufacture goods size ranging between 1 and 100 m. Phenomenal benefits of this type of microprocessing include good precision of machined goods with no burr generation on the machined surface, cheap with bulk implementations that are viable due to the shorter processing time. MEMS micromachining suffers a disadvantage due to the geometrical complexity

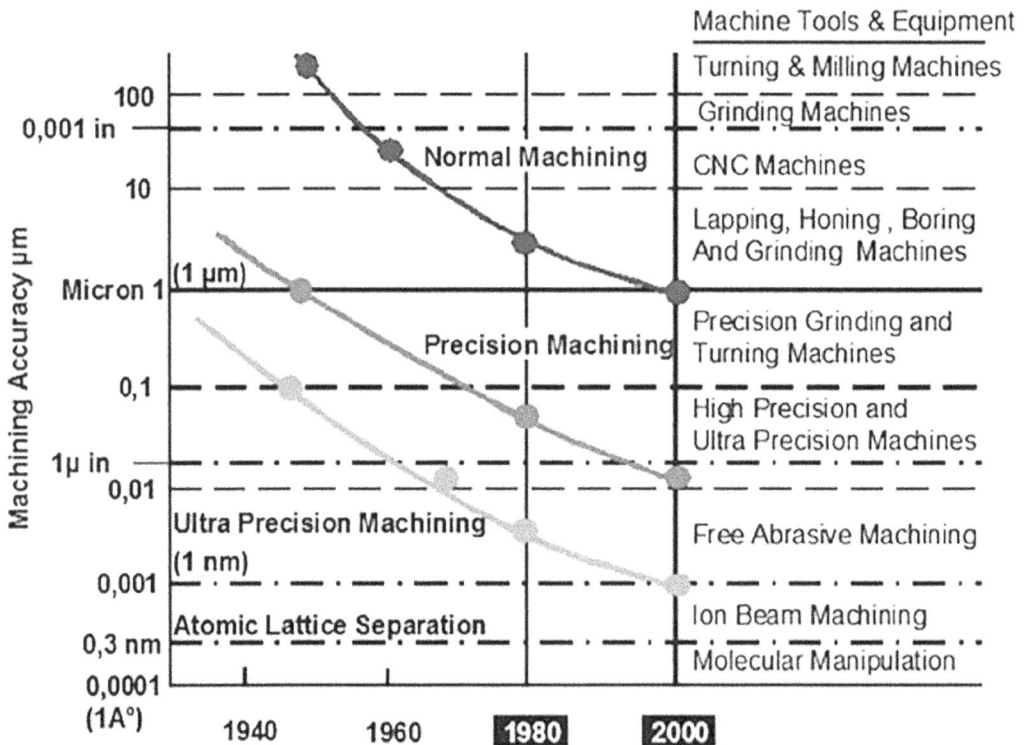

FIGURE 16.3 Model developed by Taniguchi showing advancement of machining from traditional to ultra-precision. (Reprinted from Ref. [14] copyright 2003, with permission from CIRP.)

of microcomponents, which will be managed efficiently having 2-dimentional or 2.5-dimensional workpiece material [17–19]. Despite the fact that MEMS can accomplish finer grain sizes, their adaptation in microfabrication has a strong potential for the development of cutting tools' accuracy, surface finish, a diverse choice of workpiece constituents, as well as massively complicated device geometry, that is, 3-dimensional. Furthermore, by discussing various size-effect problems, the disparity among micro- and macromachining is reconciled while investigating this strategy.

16.3.1 INFLUENCE OF GLASS TRANSITION TEMPERATURE OF POLYMER NANOCOMPOSITE ON MACHINABILITY

Properties of materials such as transition temperature, viscosity, melting temperature, and molar mass all influence polymer machinability. Machinability is affected by machining method constraints like cutting speed, feed rate, depth of cut, and cutting-edge radius, in addition to the above-mentioned properties. As shown in Figure 16.4, the stiffness of polymer is determined primarily by temperature [20]. When polymers are cooled below their glass transition temperature, their stiffness rises exponentially, generally by several degrees. The cold flow zone, that possesses excellent elastic behavior compared to a high range of temperature, where material demonstrates rubber-like behavior with reduced stiffness, achieves the best metal cutting and surface finish. Machining is characterized in a rubbery zone by significant shredding and surface roughness of composites. Bioplastics show superior brittle and crystal behavior at low temperatures, and tooling is categorized by microcracking and rupturing [21,22]. The relationship between glass transition temperature (T_g) and ambient temperature governs natural polymer machining. Acrylic-based polymers have T_g values close to or just below room temperature, making them hard to machine because of the absence of rigidity. PMMA (polymethyl methacrylate), on the other hand, has a T_g of about 110°C, and it is not so difficult to machine because of its extraordinary rigidity at ambient temperature [23].

16.3.2 BURR GENERATION IN MICRODRILLING

Excessive material generated due to plastic deformation of the workpiece is referred to as burr. Burrs develop on the workpiece's entrance and exit edges during the drilling operation. Entrance burrs are developed by plastic flow, whereas exit burrs are developed by the material extending

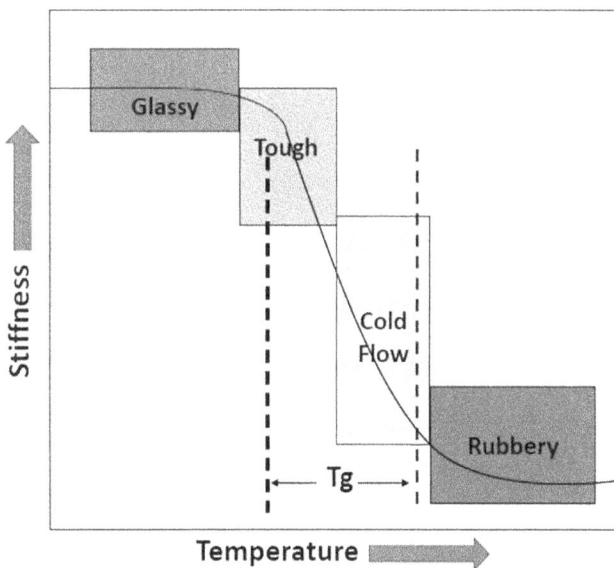

FIGURE 16.4 Stiffness curve and different machinability zone in polymers. (Reproduced from Ref. [20].)

from the workpiece's exit side. Due to improper arrangement of the components, exit burrs are often more intense than entrance burrs [24]. Drill bit materials, drilling conditions, and drill design all have a big impact on burr generation in drilling. The thrust force controls the burr generation brought on by plastic deformation at the exit edge. The thrust force rises as the feed rate does as well. The orthogonal cutting theory, which is depicted in Figure 16.5, can be used to understand the relationship amid rate of feed and thrust force with modifying diameter of drill bit. The thrust force is calculated by Equation 16.1:

$$\Delta F_t = \frac{kf \, \sin k \sin (\lambda - \alpha)}{2 \sin \varphi \cos (\varphi + \alpha - \lambda)} \, \Delta W \tag{16.1}$$

where F_t=Thrust Force, α=Tool Rake Angle, λ=Friction Angle, φ = Shear Plane Angle, and k=Shear Strength.

16.3.3 BURR GENERATION IN MICRO-END MILLING

As shown in Figure 16.6, burrs are categorized as Poisson burr, tear burr, and rollover burr based on the engagement of end mill cutter and workpiece material. Poisson burr is generated by the pressing material near the cutting verge [25]. Burr contains protuberance that is offset from or in the opposite direction from the movement of the cutting tool. Contrary to shearing, tear burr develops from the material of the workpiece being torn. The material that was left on the side of the workpiece after bending is known as a rollover burr.

The positions of the burrs in milling and slot milling on the face peripheral are depicted in Figure 16.7. The following requirements are used to identify burrs: the orientation of burr generation (En, Ex, and S), the comparative motion between the cutter and the workpiece (UM, DM), and the damaged edge (specified as (P, F)). The edges of the Poisson burr are highlighted in blue as En-F-UM and En-P-UM at the entry and Ex-F-DM and Ex-P-DM at the exit. Burrs highlighted in yellow that are located at En-P-DM and En-F-DM or at Ex-P-UM and Ex-F-UM should be rolled over. Poisson burrs are identified by the red-marked S-P-UM and S-P-DM burrs in both slot and circumferential milling. Burrs with green markings designated S-F-DM and S-F-UM as combinations of the poison/tear burr and rollover burr, accordingly [26].

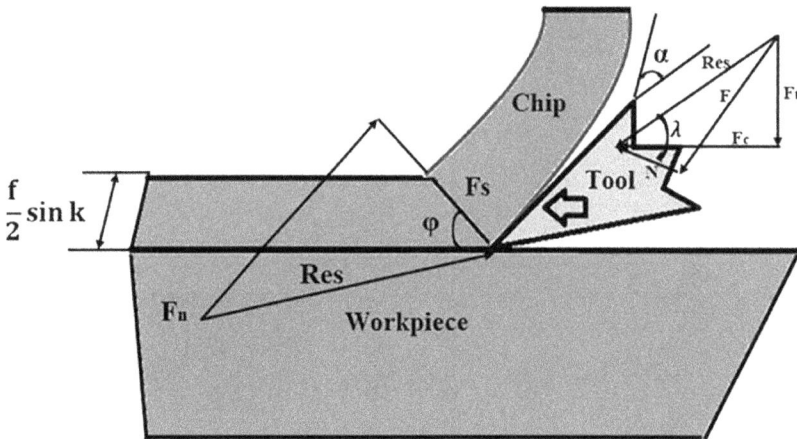

FIGURE 16.5 Orthogonal cutting theory. (Reproduced from Ref. [24].)

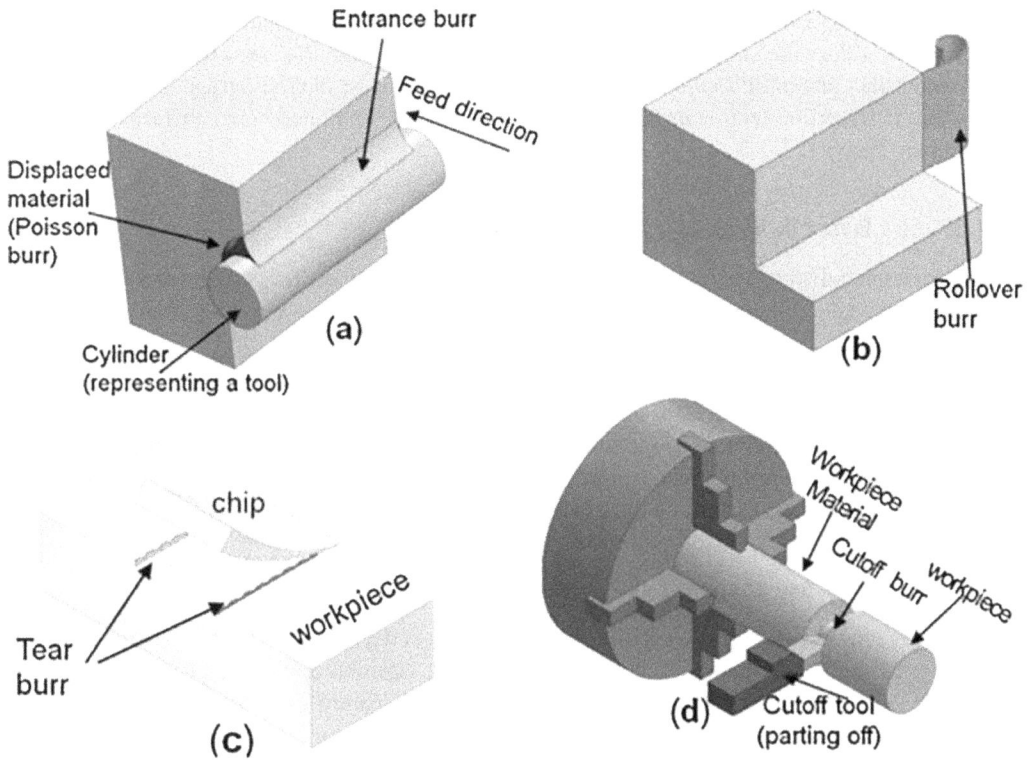

FIGURE 16.6 (a–d) Types of burr. (Reprinted from Ref. [25] copyright 2015, with permission from CIRP.)

FIGURE 16.7 Location of burr in slot milling and face-peripheral milling. (Reproduced from Ref. [26].)

16.3.4 SURFACE FINISH IN POLYMER NANOCOMPOSITES

Machine tool factors, including feed rate, depth of cut, and cutting speed, have big impact on surface abrasion. Surface finish assessment in GFRP (Glass Fiber-Reinforced Polymer) is challenging because intruding fibers might produce inaccurate results. Surface roughness changes with different fiber having 0°, 30°, 45°, and 90° orientation angles, and its intensification through accumulative fiber orientation angle is because of an upsurge in compressive strain [27]. Due to carbon nanotube protrusion on the workpiece surface, PS/MWCNT (polystyrene/ multi-walled carbon nanotube) displayed higher surface roughness than plain PS (polystyrene). It was found that when the speed of the spindle is amplified from 50,000 to 100,000 round per minute, the surface roughness of PMMA reduces from 360 to 320 nm [28]. Owning to the Feed Per Tooth (FTP) condition matching, the least chip thickness value, the surface quality of epoxy/GPL (Graphene Platelets) nanocomposite rises with an intensification in FPT value. Due to the reinforcing process of GNP (graphene nanoparticle) and the rise in transition temperature due to GNP, epoxy/GNP exhibited better surface finish compared to pure epoxy [29–31].

16.3.5 SIZE EFFECT IN MICROMACHINING

The approximated distortion from macro- to micromachining causes size consequences. Whenever the size of the uncut chip diminishes, such impacts are demonstrated by a sharp, nonlinear rise in specific energy [32–34]. Particularly at achieving grinding levels, that is, $\dfrac{f}{\text{Re}} = 0.1$, the specific cutting force might reach to 70 GPa. In micromilling, the specific cutting forces (k_c) are hyper-proportionally intensified when feed per tooth (f) rate is much lower than tool edge radius (Re) [35, 36]. The specific cutting energy is greater in micromachining when equated to macromachining is the clearest illustration of the exposition of the size effect as shown in Figure 16.8 [37–39].

16.3.6 EFFECT OF MICROSTRUCTURE

The homogeneous and isotropic criteria are no longer correct during micromachining with microtools because the tool edge radius reaches the particle size. Assembling discrete grains with random distribution inside the composite structure or an anisotropic feature is currently thought of as the solid material of a workpiece [40–42]. The cutting mechanism in this instance, therefore, entails breaking each particular grain separately, which calls for higher precise cutting forces and energies as well as average flow stress because of the bonding of atoms, as shown in Figure 16.9 [43–45].

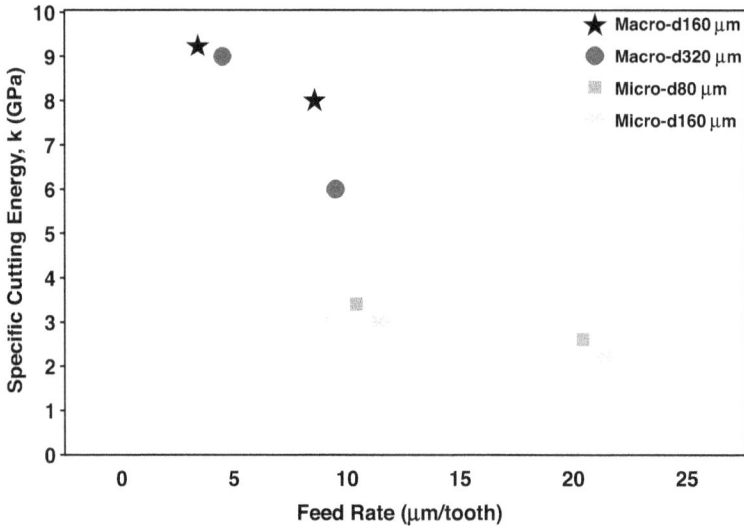

FIGURE 16.8 Effect of size on specific cutting energy in macromachining and micromilling. (Reproduced from Refs. [38,39].)

Vogler et al. [47] investigated a kinetic force modeling for micro-end milling having multi-phase ductile iron utilizing microstructure mapping. When micro-end milling is done in multi-phase ductile iron, it was observed that the cutting energy increases by more than 35% as a result of microstructural impacts and superior cutting force frequency in accordance with macromachining [48,49]. When a cutting tool passes over two neighboring grains with dissimilar mechanical characteristics, hardness, surface finish, cutting forces, and other cutting circumstances change as shown in Figure 16.10 [50–52].

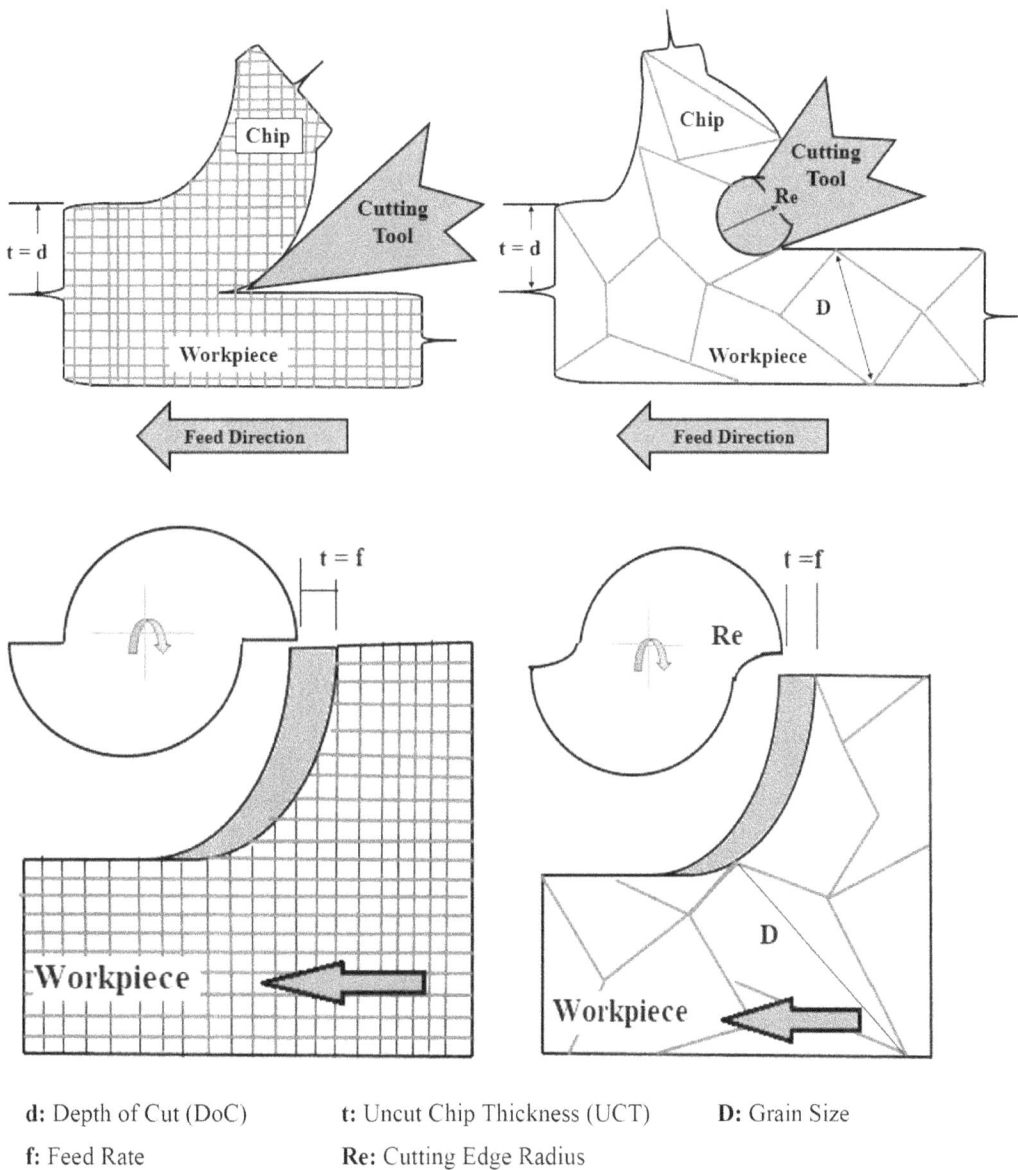

d: Depth of Cut (DoC)   **t:** Uncut Chip Thickness (UCT)   **D:** Grain Size

f: Feed Rate   **Re:** Cutting Edge Radius

FIGURE 16.9 A pictorial description representing differences between micro- and macromachining in terms of microstructure. (Reproduced from Ref. [46] copyright 1994, with permission from CIRP.)

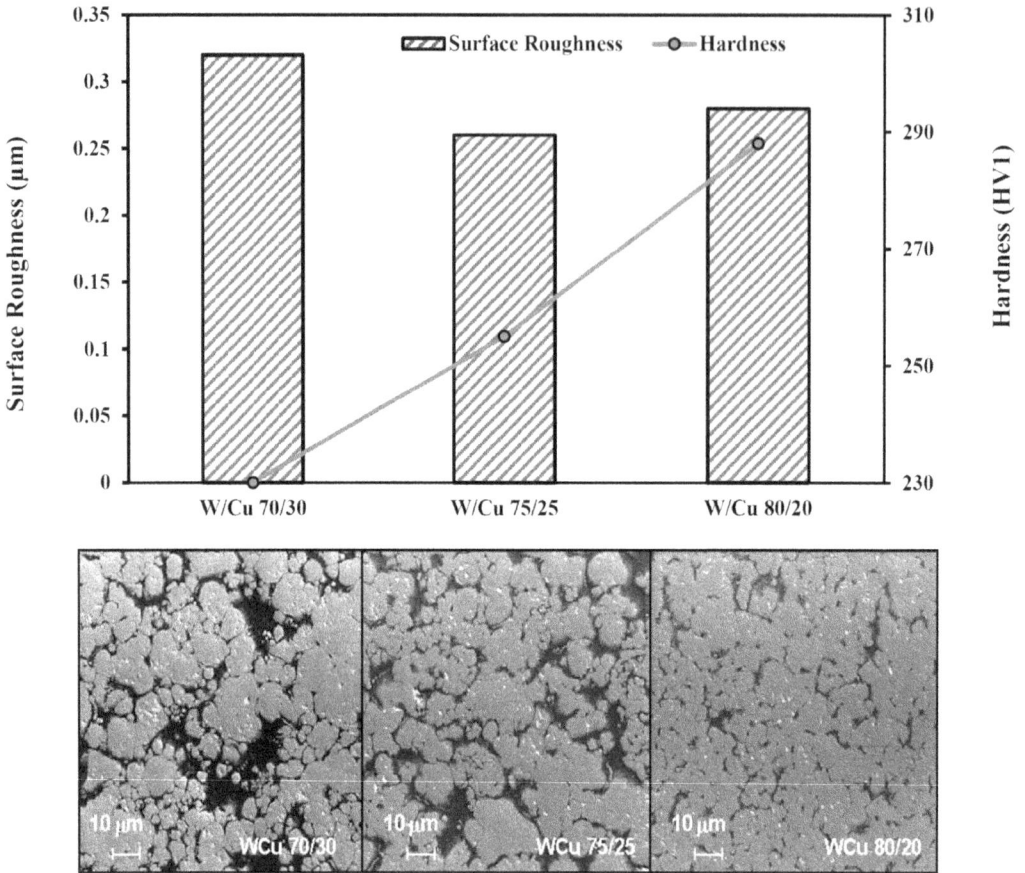

FIGURE 16.10 Effect of grain size on hardness and surface finish during micromachining of composite. (Reproduced from Ref. [38] and Reprinted from Ref. [53] copyright 2005, with permission from *Journal of Materials Processing Technology*.)

16.3.7 EFFECT ON RADIUS OF CUTTING-EDGE AND MINIMUM UNCUT CHIP THICKNESS

Radius of cutting edge and size distribution are becoming similar in miniaturization since the minimum uncut chip thickness can vary between submicron and just few microns, whereas the feed rate and depth of cut can vary between a few microns and conceivably 100m [54–56]. In that circumstance, the feed rate and tool edge radius now become main variables affecting the surface quality [57–59]. Additionally, the minimum chip thickness begins to reduce with a decrement in feed rate, which significantly increases surface roughness, as depicted in Figure 16.11. For a different range of cutting speeds, that is, 10 and 150 m/min, a reduction in UCT (Uncut Chip Thickness) under the range of the edge radius of about 10–60 nm results in a considerable rise in surface imperfection [60]. Because once cutting depth and feed rate drop under a predetermined level known as "minimum uncut chip thickness," shear stress in micromachining rises from around cutting edge in place of along shear plane and workpiece is reinvested or elastic deformation occurs somewhat than cut/sheared. Whenever the fraction of UCT and edge radius is decreased, it is seen that the surface finish suffers for two purposes: one being the plastic buildup of workpiece substance on the machine face, and the other is the adhesive growth of workpiece material on the processed surface brought on by material reinforcing [61–63].

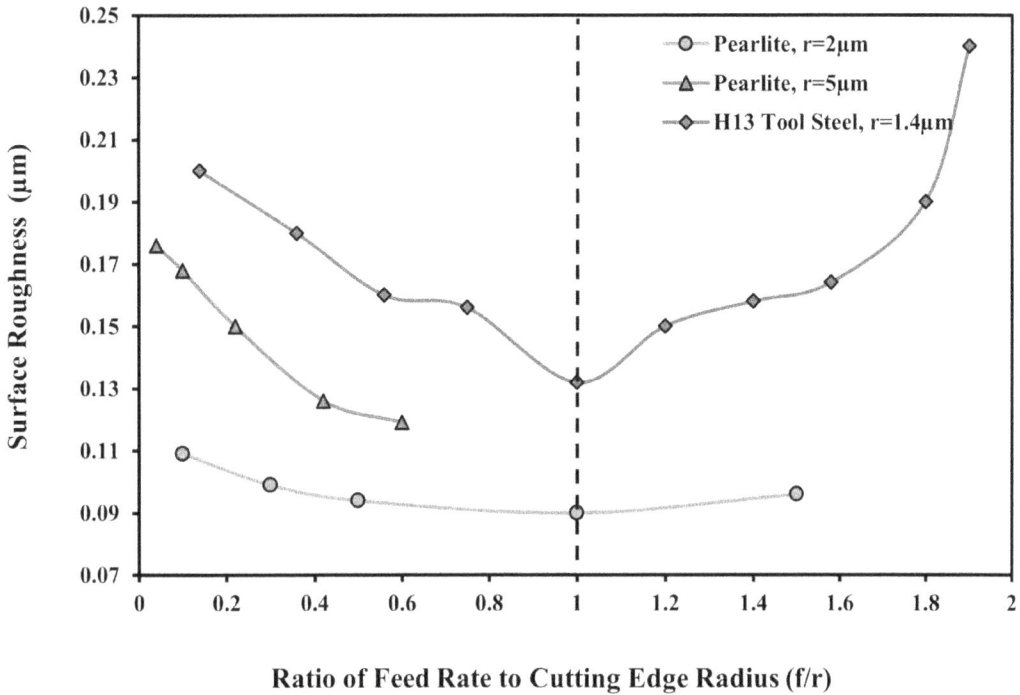

FIGURE 16.11 Effect of radius of tool edge and feed rate on roughness of surface in micromilling. (Reproduced from Refs. [38,47].)

16.4 MICROMACHINING PROCESS

Micro-electromechanical system (MEMS) methods are primarily employed for the manufacturing of bioengineering. However, micromachining technologies devise some advantages over micro-electromechanical system-based procedures, including the ability to manufacture 2-dimentional, 2.5-dimensional, and 3-dimensional structures even deprived of dangerous organic masks and suitability in lieu of a variety of compositions. The two main categories of microfabrication are lithographic and non-lithographic approaches. Some limitations of lithography-based methods include lesser aspect ratio arrangements, material restrictions, and extreme startup costs. As illustrated in Figure 16.12, non-lithographic-based methods like micromachining can be divided into traditional, enhanced, and hybrid methods [64]. There seem to be numerous production procedures for creating microfeatures on polymeric material, including both traditional and cutting-edge techniques. Several of these procedures call for additional steps to render the final result usable. Total cost of manufacturing goes higher as a result of certain procedures requiring expensive early design costs and specialized environments like vacuum chambers and neat rooms. Owing to substance ablation, laser scanning results in uneven polymer composition and non-uniform microchannel substrate. Toxic substances are used in the chemical process. As a result, traditional micromachining techniques like microdrilling and micromilling may efficiently and quickly generate complicated forms. Nevertheless, a comprehensive analysis of the depth of cut, substrate composition, tool edge curvature, specific cutting force, surface quality, and dynamic instabilities is also necessary. The fishbone figure illustrates different processing variables that affect the manufacturing of polymeric materials [65].

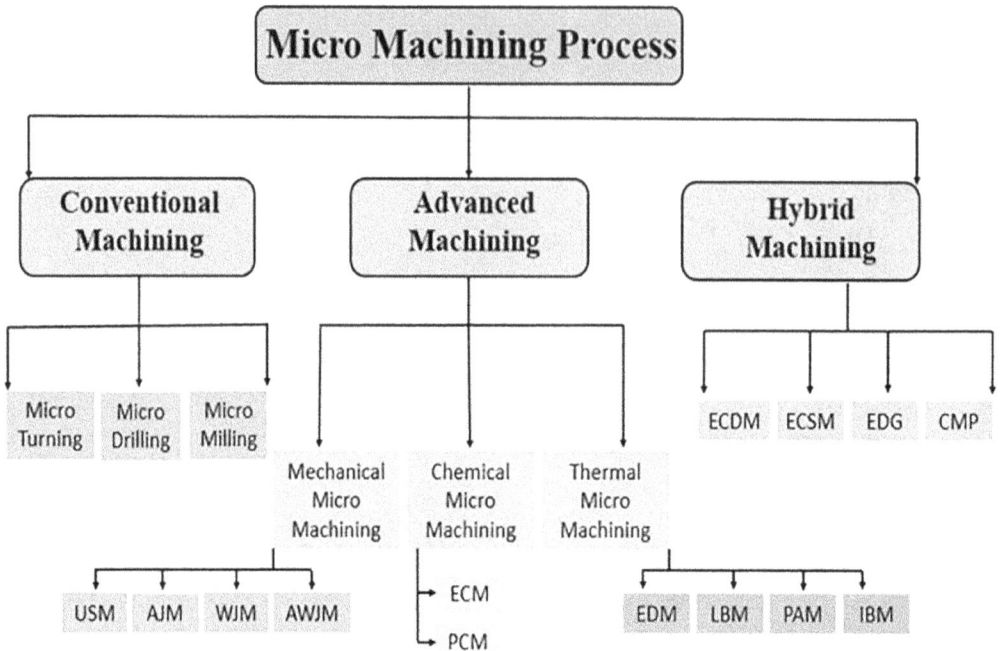

FIGURE 16.12 Systematic classification of micromachining process.

16.4.1 MECHANICAL MICROMACHINING

Polymer matrix composites (PMCs) reinforced with fiber or nanofillers provide improved mechanical qualities than traditional materials, such as workability, high stiffness, and resistance to corrosion. Although during the manufacturing process, most polymer nanocomposites go through some basic processing. Such polymers are difficult to machine due to their anisotropy, nonhomogeneous, and fibrous structure, which can result in significant wear rate and machined surface damage [66]. Mostly with right process parameters and defined characteristics, composite material can be machined using traditional machining techniques such milling, drilling, grinding, sawing, and turning. Ductile materials yield coiled and unbroken chips, whereas fiber-reinforced PMCs yield shattered and fragmented pieces. When PMCs are being machined, fibers break as well as the fiber and matrix debonded, resulting in the creation of chips. Deformation, fiber pullout, and fracturing are common PMC faults that occur during the processing of fiber-reinforced materials [67–69].

16.5 MICRODRILLING

Metals, plastics, alloys, ceramic materials, and hybrids can all be successfully drilled with small gaps. Depending on the prevailing conditions in geometric factors anywhere along leading edge, drilling is indeed a delicate procedure. The integrity of such holes is influenced by input factors like cutting force, feed rate, cutting depth, tool geometry, tool type, workpiece material, and kind of drilled activity. In drilling, factors such as geometrical parameters, feed rate, cutting speed, and tool wear affect torque and thrust forces. Numerous failures modes, including delamination, polymer distortion, interface debonding, breaking, fiber pulling out, fragmenting, and hole shrinking, were used during the hybrid drilling [70].

16.5.1 MECHANISM OF MICRODRILLING

Usually, polymer slicing involves orthogonal slicing because polymer trimming is complicated. When a polymeric material is in an amorphous condition, cracks spread, and when it is in an elastic range, material distortion takes place. Based mostly on minimum chip thickness that is not distorted, there are three separate areas in microdrilling: plowing, transitional, and shear. The plowing zone is taken into consideration up to the minimal chip thickness point, in which the actual rake angle is very adverse, and the chip thickness is undistorted and much lesser than the radius of blade edge [71]. The uncut depth of chip is nearly equivalent to the radius of cutting edge in the transitional zone, when there is change in rake angle from minus to plus. While the thickness of uncut chip exceeds the radius of tool edge and also the rake angle controls the cutting force, shear failure occurs. Drill removes material in different places: the cutting lip (main), the chisel edge (minor), and the indent. Cutting motion removes material from the slicing lip and chisel side areas. Shearing does not occur along a plane at shallow cut depths. Again, for majority of microdrilling applications, mechanical model can be utilized to forecast effective cutting force. As seen in Figure 16.13, the normal force goes up to a fixed value that corresponds to consistent drilling throughout sheet thickness. Even as tool exits on the contrary direction, it then displays a severe drop [72,73].

16.5.2 MICRODRILLING OF NANOCOMPOSITES REINFORCED WITH GFRP/CFRP NANOMATERIALS

Anand et al. [74] examined the microdrilling process of carbon fiber-reinforced polymer laminate and also studied its variation on cutting speed. The proportion of the size of uncut chip and radius of cutting edge was used in the suggested theoretical model to define the particular cutting force. Findings indicate that tool geometry radius had an impact on the thrust, radical, and definite cutting forces. When the ratio of the thickness of undeformed chip and the tool geometry is smaller than one, size effect is discovered to be an essential factor. Findings indicate that the layered structure of the composite material had an impact on the cutting force. By using three different techniques for microdrilling, Basso et al. [75] evaluated the diameter of hole of a CFRP/PPS (carbon fiber-reinforced polymer/polyphenylene sulfide) nanocomposite with direct, supported, and pilot drilling method. Just at hole entrance and escape, defects including uncut fiber/matrix or intralaminar fractures were found. It was found that the size effect and damaged generation were most significantly impacted by tiny drill feed rates. The analysis indicates that to lessen delamination effects, smaller feeds per revolution of drill and smaller thrust values were needed.

In CFRP- and GFRP-reinforced polymers, microdrilling torque and thrust force were examined by Rahamathullah et al. [76]. Due to the obvious growing chip concentration just at drill cutting edge, rate of feed enhances the thrust force. Owing to the structure's thermal softening, spindle speed reduces both thrust force and delamination loss. Because both thrust force and torque cause the dissociation of the laminate layer and circumferential shearing, they have an impact on delamination. Decreased feed rates result in larger holes because the hybrid composites in CFRP laminates engage with each other more slowly. When peck drilling GFRP plastic laminate as opposed to direct drilling, torque and thrust force are reduced. Unidirectional carbon fiber-reinforced plastic (UD-CFRP) nanocomposite microdrilling was investigated by Kim et al. [77]. Holes' performance can be improved, and tool wear can be decreased using the nano-solid air spray approach. Analysis was done on the issue of delamination, uncut nanofiber, and inward surface finish of the drilled surface. Outcomes demonstrated that, as illustrated in Figure 16.14, nano-solid lubricant dramatically lowered delamination, surface roughness, and uncut fiber. By decreasing the friction between the drill and the CFRP, tool wear was also decreased. Leading to improved sliding action of 2D thin sheets, tool life of the Multi-Wall Carbon Nanotube was recorded more than that of graphene nano-platelets (GNP). Especially, GNP-based lubrication was discovered to be more efficient than MWCNT in every instance [78,79].

(a)

(b)

(c)

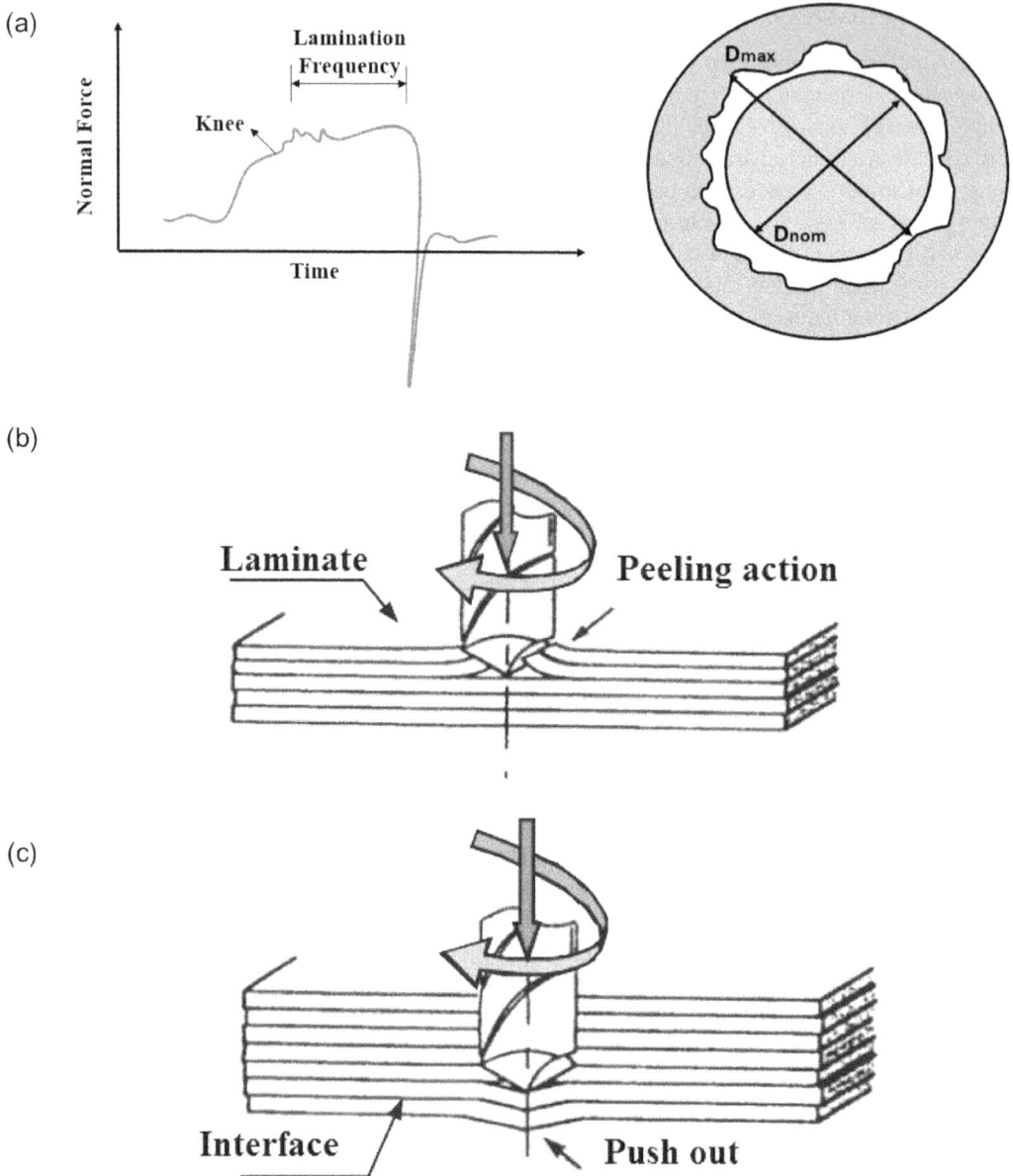

FIGURE 16.13 (a) Normal force variation which is responsible for delamination (Reproduced from Ref. [72]), (b) Schematic diagram of delamination geometry (Reproduced from Ref. [3]), and (c) Schematic diagram of peel-up and push-out. (Reprinted from Ref. [73] copyright 2014, with permission from Procedia Technology.)

Anand and Patra [80] used statistical techniques like ANOVA (analysis of variance) and RSM (response surface methodology) to examine the relationship between feed rate, cutting speed, and hole size and forces in cutting process in precision drilling of carbon fiber-reinforced polymer nanocomposites. Because of tool distortion, microhole sphericity inaccuracy grows with feed rate. With increased feed rate, the thrust force and delamination factor decline. At such a rate of feed near to the tool edge radius, excellent surface finish of hole and minimal forces in cutting process were recorded [81]. Shunmugesh and Panneerselvam [82] investigated the improvement of process variables in carbon fiber-reinforced polymer (CFRP) microdrilling utilizing the analysis of Taguchi and relation developed by Grey. Phadnis et al. [83] experimentally and theoretically examined the drilling of

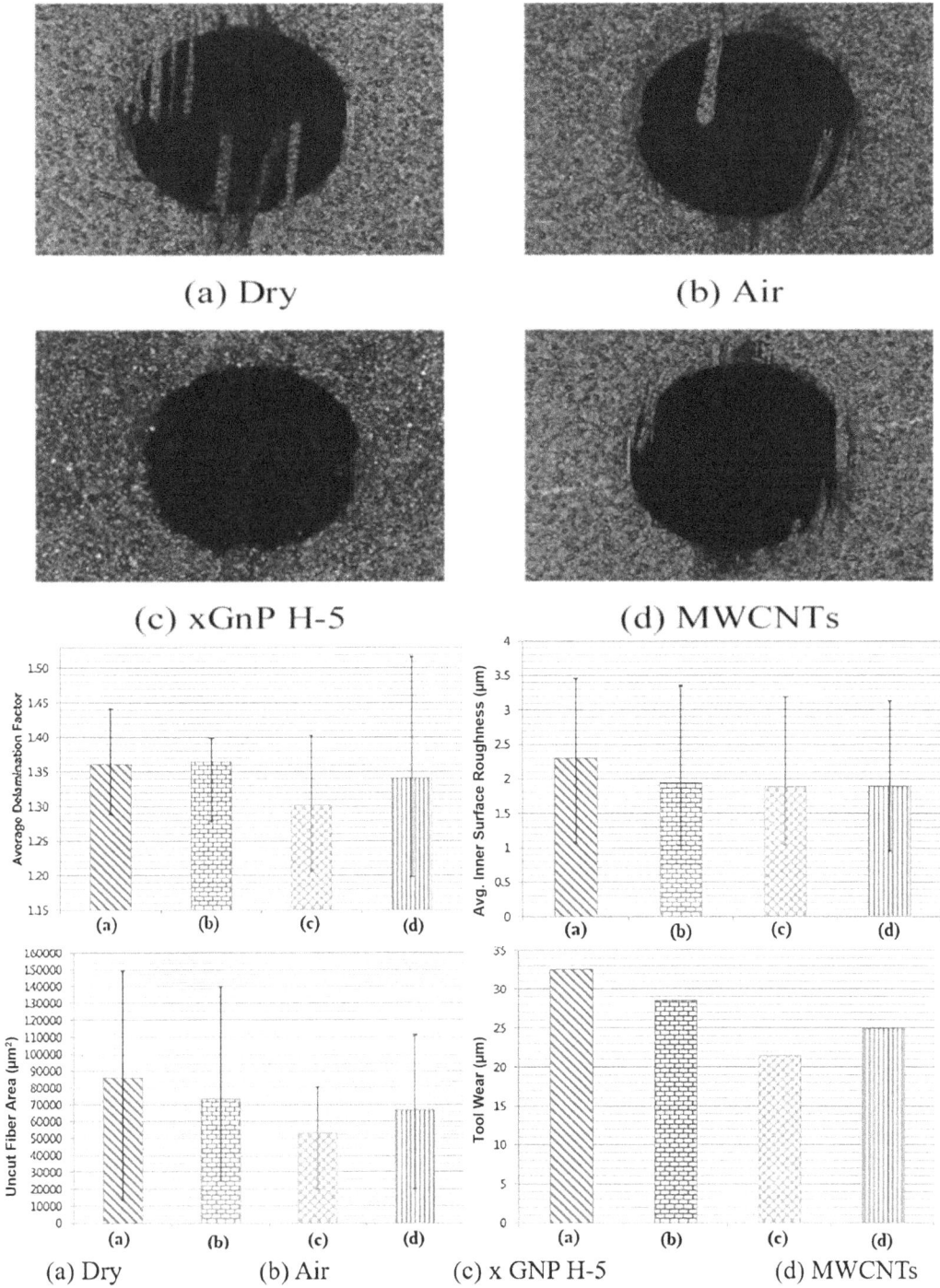

FIGURE 16.14 Drilled hole and delamination factor, norma roughness of surface, fiber uncut area, and wear of tool for different cases of graphene nanofiller. (Reprinted from Ref. [77] copyright 2019, with permission from *Journal of Manufacturing Processes*.)

CFRP/epoxy composite. For the purpose of examining the complex mechanics at the drill/work-piece, a three-dimensional finite element method was established. The size and shape of delamination at drill entry were predicted using the cohesive zone component. Damage in delamination, thrust force, and torque rise and fall with the rate of feed rate and cutting speed, respectively. The

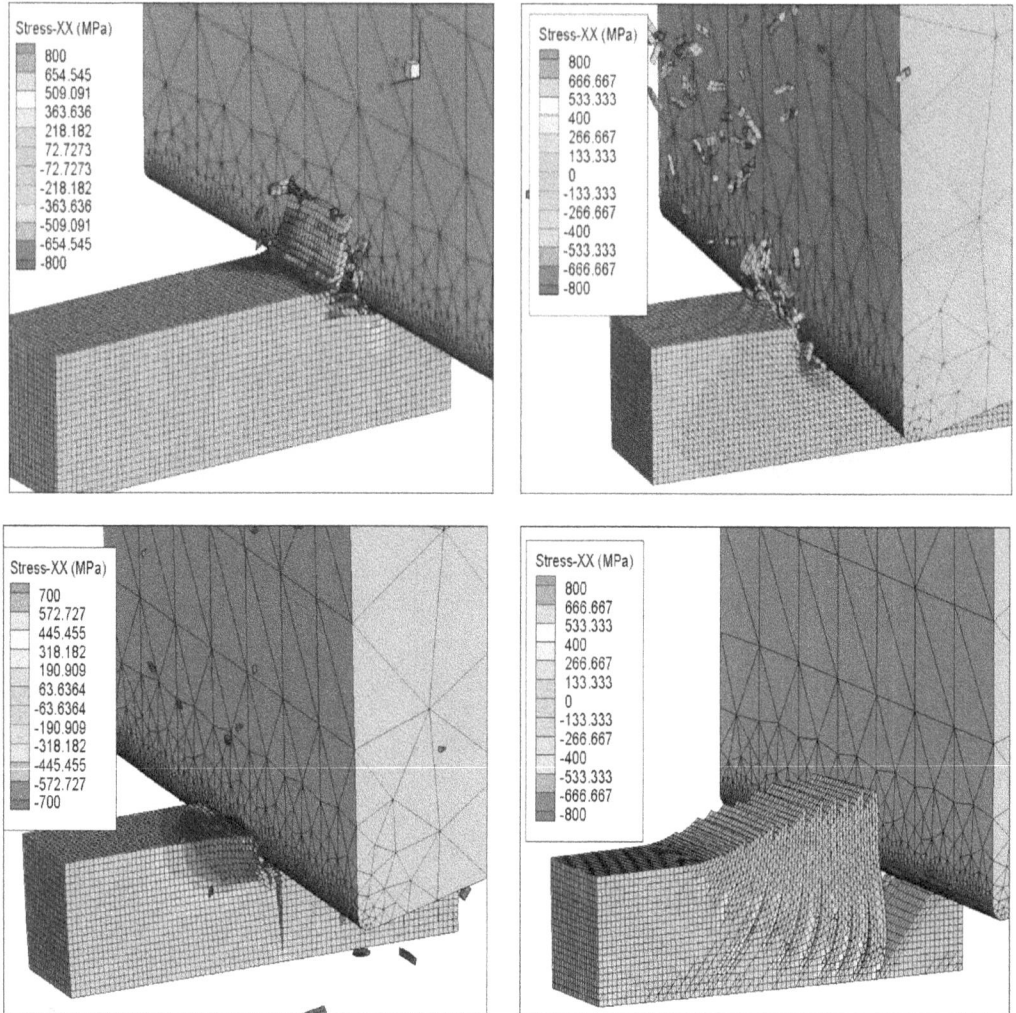

FIGURE 16.15 FEM simulation of orthogonal cutting of UD-CFRP at 0°, 45°, 90°, and −45°. (Reprinted from Ref. [84] copyright 2014, with permission from Procedia CIRP.)

Lagrangian nonlinear finite element method for machining was created by Usui et al. [84] to use an explicit temporal integrating, a structured mesh, and the inclusion of cohesive elements. Orthogonal processing was done at different fiber orientations, such as the 0°, 45°, −45°, and 90° depicted in Figure 16.15. At 0° of alignment, peel failure is seen; at 45°, mode II fracture; and at 90°, crack can be seen. Additionally, it was found that the 90° position damaged the object more than 0° position.

16.6 MICROMILLING

Micromilling enables the creation of three-dimensional parts from a range of objects, including ceramics, polymers, and metals and non-metallic materials. Innovations in biology and medicine have used AFM (atomic force microscopy) and tip-based micromilling [85–87]. The ability to cover all the three perpendicular directions makes micromilling superior to other microprocedures because it enables the construction of suitable design with intricate geometry. The radius of cutting edge varies between 5 and 20 m, and the diameters utilized in micromilling vary from 25 m to 1 mm. The radius of cutting edge in micromilling is similar to the material particle sizes and uncut chip size. Exact rake angle that participates in crack propagation is strongly negative and cutting force rises during crack growth because the edge of cutting is soft, and the thickness of minimum uncut chip is always lesser than the radii of cutting edge [88–90]. To reduce the effect of plow and achieve appropriate material removal, it is needed to ascertain the minimum thickness of chip and ratio of tool cutting radius. According to minimum chip thickness theory, afore the formation of chip, the feed rate and cutting depth must be superior than a critical thickness of chip. The complete cutting depth can be made from chip depicted in Figure 16.16 when the thickness of uncut chip is larger than the minimum undeformed thickness of chip ($h > h_m$) [91,92].

16.6.1 Mechanism of Micromilling Material Removal

When using an end mill, thickness firstly rises to the highest point before falling to zero. Depending on tool dimensions and machining conditions, the above process happens in the majority of single and multiple cutting-edge tools. Figure 16.17 depicts the milling procedure for two-edge micromilling. The chip starts to develop when the undeformed thickness of chip (h) surpasses the minimal undeformed chip thickness (h_m). The creation of the chips just would not take place if instant (h) was less than (h_m). In this situation, a component of the work material will recover elastic deformation while another part of such material will deform plastically. The minimum thickness of undeformed

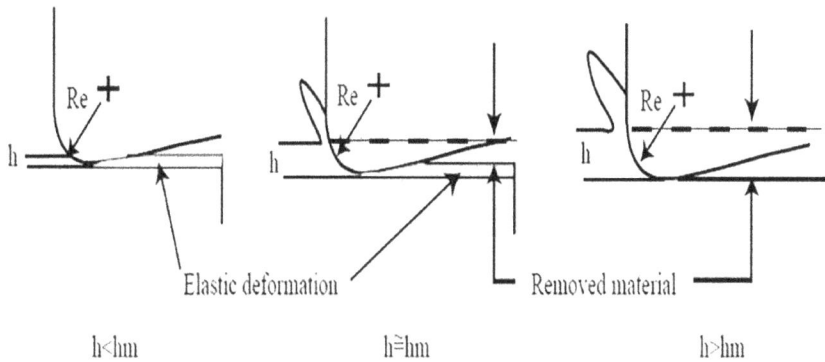

FIGURE 16.16 Schematic diagram of minimum chip thickness effect. (Reprinted from Ref. [18] copyright 2006, with permission from *International Journal of Machine Tools and Manufactures*.)

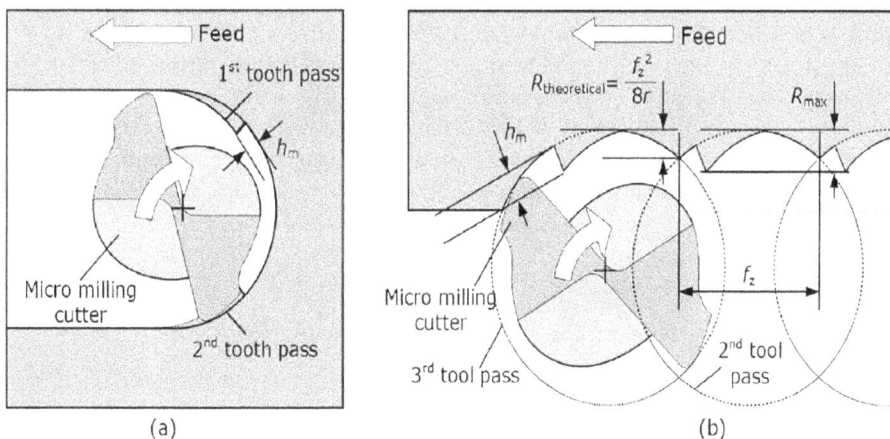

FIGURE 16.17 (a and b) Tool pass in the process of full immersion and side micromilling operation. (Reprinted from Ref. [93] copyright 2021, with permission from *International Journal of Machine Tools and Manufacture*.)

chip (h_m) is grounded on the basis of the bond among the geometry at the point of stagnation, angle of stagnant (θ_m), and radius of cutting edge (Re).

Theories of h_m or the stationary angle could be classified into three groups depending on several hypotheses. Depending upon that cutting force that resulted in the normal direction to the surface of the workpiece, the cutting force approach specified the h_m. The moment where the force shifts from being upward to being downward is referred to as (h_m). The stationary point during which material deforms or flows with the least amount of deformation or kinematic energy is described by material behavior pattern models. Since the two processes employ lateral comparative movement and standard pressure on the interaction, the scratched concept made the assumption that the micromachining operation is identical to the scratching operation.[93].

16.6.2 MICROMILLING OF NANOCOMPOSITES REINFORCED WITH GRAPHENE/MWCNT NANOFILLERS

Graphene/epoxy composites' surface morphology was investigated by Thakur et al. [94] utilizing analytical models during end milling. Outcomes demonstrated that surface texture is enhanced when rate of feed increases and diminishes with content of graphene and speed of spindle. Fast spindle speed enhances elevated surface polish by lowering contact friction and built-up edge development. So at contact, graphene presence acted as a lubricating oil and decreased built-up accumulation. They also explored the properties of milling of CFRP/GFRP/graphene nanocomposite laminates. Graphene/CFRP laminates are made using a vacuum bagging process after solution casting and analysis of surface finish and delamination was done in relation to graphene concentration, rate of feed, and speed of the spindle. The roughness of surface increases by increasing feed rate and decreases by increasing graphene content and spindle velocity. Because of plastic deformation and interlaminar stress, delamination parameters rise as spindle speed and depth of cut. Because of how lubricants behave and because of the enhanced interlaminar shear strength, graphene's presence decreases surface roughness and delamination [95,96].

Owing to a lubricating characteristic of graphene, a lower delamination factor and roughness of surface were demonstrated by graphene/CFRP composites than pure CFRP. Arora et al. [95] experimentally tested the processability of GPL-reinforced epoxy resin nanocomposites. They also studied the wear of tool, forces during cutting, chip shape, and quality of surface. As a consequence of clumping at higher loading, the analysis indicates that the value of minimum thickness of chip declined with graphene platelets' concentration. The value of the minimum thickness of chip was found between scale of 2–4 μm for 0.1 wt.%, 1–2 μm for 0.2 wt.%, and 0.3–1 μm for 0.3 wt.%. Kumar et al. [96] fabricated a nanocomposite of Polycarbonate/Multi-Wall Carbon Nanotubes/Graphene Nanoplatelet (PC/MWCNT/GNP) by the process of injection molding and done micromilling. There is a necessity of higher cutting force for PG/GNP nanocomposites than the pure PC because of nanocomposite heterogeneity. Extreme thermal conductivity is shown in PC/MWCNT/GNP because of presence of linkage between GNP and CNT, due to which a discontinuous chip is formed. Moreover, there is an enhancement in cutting force with FPT as the material is removed in zone of shear as depicted in Figure 16.18. Also the change from discontinuous to continuous chips was visible as GPL and FTP concentration increased [97].

16.6.3 MICROMILLING FOR MANUFACTURING OF MICROFLUIDIC CHANNEL

Another technique for creating microfluidic equipment for a variety of purposes is micromilling. Micromilling is superior to other methods in that it requires less process time, costs less to operate, and does not require corrosive acid or sterile conditions. Christ et al. [98] published upon that production of microchannels in polystyrene (PS) via micro-end milling. Three streams were created and PDMS (polydimethylsiloxane)-bonded, each with a different depth along its length. With

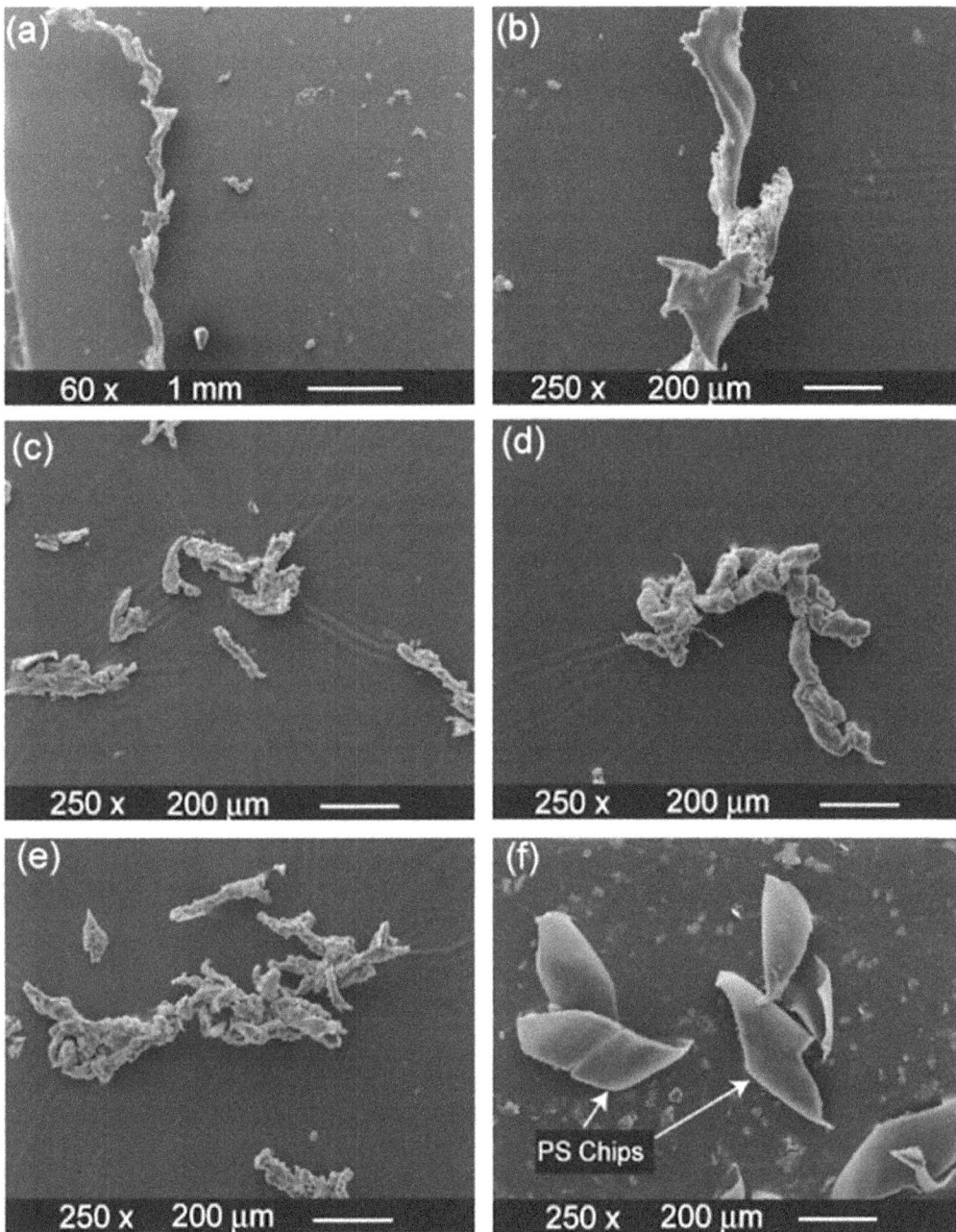

FIGURE 16.18 Morphology of Chips (a) and (b) Plain PC, (c) xGNP-M-5 / PC, (d) xGNPM-25 / PC, (e) xGNP- M-5 / MWCNT / PC, and (f) Plain PS at FPT of 3 μm/flute. (Reprinted from Ref. [96] copyright 2017, with permission from *Journal of Materials Processing Technology*.)

respect to process variables, microchannel width, depth, surface roughness, and burr development were examined [99–101]. The result revealed that great quality microchannels are produced by a high feed rate, shallow cut, and slow cutting speed. Rotation velocity of spindle has the biggest impact on dimensional inaccuracy, next by axial depth of cut. Microchannels with a superior surface quality can be created using small rate of feed and axial DoC (Depth of Cut) [102,103].

16.6.4 MICROMILLING OF GRAPHENE-BASED NANOCOMPOSITES

Since graphene offers a great possibility for strengthening different matrix composites like polymers, alloys, or ceramics, it has already been observed that just a few research has recently looked into such graphene-based composites [104–106]. Graphene platelets were used as the auxiliary filler from one of the earliest research mostly on micromachinability of nanocomposite materials based on graphene [107–109]. Main goal of this work was to compare overall micromachining reactions of two-phase and three-phase composite materials of epoxy/GF (glass fiber) and epoxy/GF/GPL, respectively. In comparison with the foundational composite, the inclusion of GPL has astonishingly shown that the hierarchical material has better micromachinability in relation with smaller cutting forces, surface finish, and wear life of tool [95]. They defined that the introduction of GPL improved the thermal properties of epoxy and lowered friction at tool-chip contact, decreased polymer buildup edge, and wear of tool [110–112].

16.7 MICROTURNING

Typical tool-based manufacturing procedure used to create axially symmetric products is called microturning. The above method is used to create microneedles for biological devices, pins for products used in electronic devices, and PCBs (Printed Circuit Boards) microjacks. Machinability of turning procedure is influenced by speed of spindle, cutting depth, rate of feed, material of tool, and workpiece. Researchers investigated the microturning process ability of glass fiber-reinforced polyamide (PA66/GF30). The findings demonstrated that the PCD and diamond-coated carbide machine tools generated the maximum and minimum cutting forces, respectively [113–115]. Davim *et al.* [116] examined the micromachining of polyether ether ketone and glass fiber (PEEK/GF30) using microturning parts made of cemented carbide (K20) and polycrystalline diamond (PCD). Like a result of the analysis, power increased for both cutting velocity and feed rate. Surface roughness rises as feed rate rises while falling as cutting speed. Pallapothu et al. [117] looked into the microdrilling on GFRP using a lathe. The findings of the statistical methodologies demonstrated that GFRP microdrilling is possible at lower spindle velocity and feed rates. The biggest contributions to MRR (Material Removal Rate) and circularity fault come from the spindle speed and feed rate, correspondingly.

16.8 ADVANCED MICROMACHINING

Modern micromachining techniques are being used to create tiny details that are challenging to create using traditional machining techniques. Wan et al. [118] investigate micro-EDM (Electrical Discharge Machining) of such a solution-cast PMMA/MWCNT nanocomposite. With respect to machining performance, the quality of the holes and the rate of metal removal were examined. The outcomes demonstrated that power supply and pulse on-time had a substantial impact on the MRR and hole diameter, respectively.

The impact of one- or two-directional CFRP manufactured by EDM was examined by Habib et al. [119]. MRR was shown to be greater in two-directional CFRP than that in one-directional CFRP. MRR was discovered to be rising and fixed for perpendicular and parallel directional machining, respectively. Dutta et al. [120] utilized copper foil electrode as a supplement to fabricate the CFRP of lower conductivity and also examined the microchannels through micro-EDM. Channel width was most affected by energy input followed by speed of tool and feed rate. Development of microcracks during machining operation is due to generation of thermal stress which in return leads to breaking of matrix and fiber as represented in Figure 16.19.

Chern et al. [121] examined the burr development in micromachining utilizing microtools. Combined micromilling and grinding operations might be carried out using the tool made from WEDG (Wire Electro Discharge Grinding). In up-milling, needle, primary, and feather burrs were

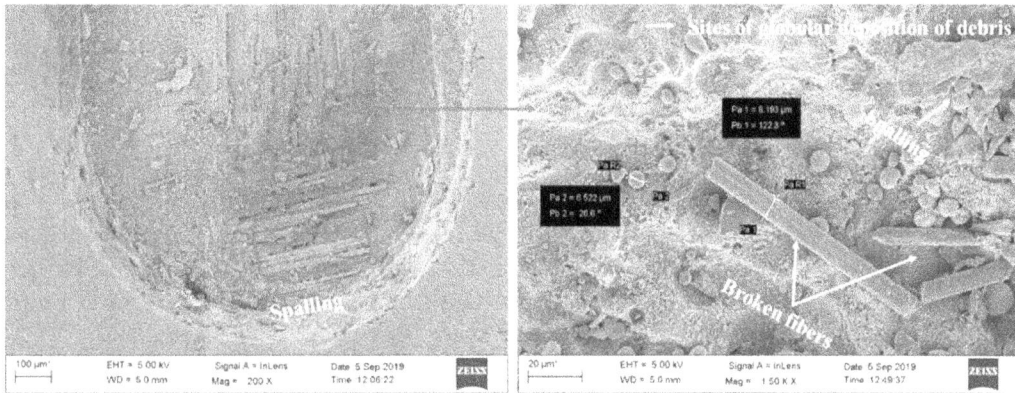

FIGURE 16.19 FESEM images presenting spalling and deposition of debris in machined surface. (Reprinted from Ref. [120] copyright 2020, with permission from *Materials Today: Proceedings.*)

created here on sidewall of the small slot. Owing to the reduced feed rate and axial alignment, minor burrs were created. Jain et al. [122] created the microchannels, micro-holes, and tools using electrochemical micromachining (ECMM). With consideration of different electrolyte concentrations, voltages, and pulse on time, microhole and microchannel were explored.

16.9   CONCLUSIONS

Recent research done in the previous 10–15 years on different polymer reinforcement materials, composite manufacturing techniques and its characterization, tooling, and diverse profile creation is covered in this chapter. Nowadays, polymer-based materials are used in place of metallic ones due to their superior strength-to-weight ratio, corrosion resistant, affordability, and low weight. The current study showed that great dimensional tolerance and surface quality can only be achieved by micromilling and microdrilling those composite materials. Filler materials cause differences in rate of material removal, tool wear, and surface roughness. Numerous academic studies have demonstrated that filler materials, which are lubricant to some extent, promote machinability. Filler material has a considerable impact on burr creation, chip shape, and surface morphology in microdrilling and milling. Use of different fillers, such as MWCNT, graphene, graphite, SiC (silicon carbide), and BN (boron nitride) filler materials, has a major impact on machining parameters including cutting force, thrust force, and torque, surface quality, delamination factor, fiber pullout, minimum chip thickness, cutting force, and surface roughness.

Any use of different fillers, such as MWCNT, graphene, BN, and SiC nanofiller, has a major impact on machining parameters including various forces like torque, cutting, and thrust, quality of surface, delamination, fiber pullout, minimum thickness of chip, and cutting force. Compared with bulk metals and processing processes, micromachining nanocomposite materials is much more difficult. The production process is affected by the interactions of numerous factors, especially the impact of microstructure (anisotropy, homogeneity, grain size), enhancements of thermal and mechanical characteristics especially fiber loading and distribution of nanofiller, and visible changes of micromachining, which also greatly reduce the process ability of these substances. Polymer–filler interaction is crucial for the formation of polymer-based composites. The surface chemistry of a contact region needs to be better understood. Several of the topics on which researchers are concentrating are the impact of substitute material configurations, the alignment of the fiber, and their relationship with polymer composite characteristics. The majority of studies on microdrilling has already been concentrated on size effect, chip shape, delamination study, and machining parameters study.

Future research can concentrate on different composite mixtures, increased tool life, process parameter improvement, hole features, energy consumption evaluation, and heat influence zone modeling. In the majority of the research, modern machining techniques (EDM, laser, etc.) were used to process metals and alloys in order to create tiny holes. It is imperative to research conventional device microdrilling. Although a noise model has been created for conventional drilling, there is really currently no research reported on chattering in microdrilling. It is essential to discuss the thermo-mechanical behavior of polymers during micromachining of diverse polymer composites with varied filling components. It is necessary to investigate cryogenic-based processing for polymer composites. There are not many publications on micromachining polymer composites. Additionally, although micromachining of nanocomposites has enormous potential for industrial use, the top of the line in this field has only demonstrated a very small number of applications, the most of which were in the prototyping phase. Additional research may therefore concentrate on industrial uses of nanocomposites' machining operations.

REFERENCES

1. Das, T. K., & Prusty, S. (2013). Graphene-based polymer composites and their applications. *Polymer-Plastics Technology and Engineering*, 52(4), 319–331.
2. Rawal, S., Sidpara, A. M., & Paul, J. (2022). A review on micro machining of polymer composites. *Journal of Manufacturing Processes*, 77, 87–113.
3. Lekkala, R., Bajpai, V., Singh, R. K., & Joshi, S. S. (2011). Characterization and modeling of burr formation in micro-end milling. *Precision Engineering*, 35(4), 625–637.
4. Saleem, W., Ijaz, H., Alzahrani, A., Asad, M., & Zhang, J. (2019). Numerical modeling and simulation of macro-to microscale chip considering size effect for optimum milling characteristics of AA2024T351. *Journal of the Brazilian Society of Mechanical Sciences and Engineering*, 41(8), 1–17.
5. Gao, S., & Huang, H. (2017). Recent advances in micro-and nano-machining technologies. *Frontiers of Mechanical Engineering*, 12(1), 18–32.
6. Ray, S. S., & Bousmina, M. (2006). *Polymer Nanocomposites and Their Applications*. American Scientific Publishers, New York.
7. Somvanshi, K. S., & Gope, P. C. (2021). Effect of ultrasonication and fiber treatment on mechanical and thermal properties of polyvinyl alcohol/cellulose fiber nano-biocomposite film. *Polymer Composites*, 42(10), 5310–5322.
8. Mergen, Ö. B., Umut, E., Arda, E., & Kara, S. (2020). A comparative study on the AC/DC conductivity, dielectric and optical properties of polystyrene/graphene nanoplatelets (PS/GNP) and multi-walled carbon nanotube (PS/MWCNT) nanocomposites. *Polymer Testing*, 90, 106682.
9. Choi, Y. J., Hwang, S. H., Hong, Y. S., Kim, J. Y., Ok, C. Y., Huh, W., & Lee, S. W. (2005). Preparation and characterization of PS/multi-walled carbon nanotube nanocomposites. *Polymer Bulletin*, 53(5), 393–400.
10. Somvanshi, K. S., Gope, P. C., & Dhami, S. S. (2017). Tensile properties characterization of rice husk fiber reinforced bio-composite. *International Journal of Engineering Research and Applications*, 7, 1–4.
11. Gusev, A. I., & Rempel, A. A. (2004). *Nanocrystalline Materials*. Cambridge International Science Publishing, Cambridge.
12. Karak, N., & Maiti, S. (2008). *Dendrimers and Hyperbranched Polymers–Synthesis to Applications*. MD publication Pvt. Ltd., New Delhi.
13. Taniguchi, N. (1983). Current status in, and future trends of, ultraprecision machining and ultrafine materials processing. *CIRP Annals-Manufacturing Technology*, 32, 573–582.
14. Byrne, G., Dornfeld, D., & Denkena, B. (1994). Advancing cutting technology. *CIRP Annals-Manufacturing Technology*, 52, 483–507.
15. Groover, M. P., Belson, D., Kusiak, A., Sánchez, J. M., Priest, J. W., & Burnell, L. J. (1994). *Handbook of Design, Manufacturing and Automation*, John Wiley Publication, University of Michigan, Ann Arbor.
16. Somvanshi, K. S., & Gope, P. C. (2021). A review on properties of nano biocomposite film for packaging applications from cellulose nano fiber. *International Journal of Engineering Research and Applications*, 11(1), 29–39.
17. Masuzawa, T., & Tönshoff, H. 1997. Three-dimensional micromachining by machine tools. *CIRP Annals-Manufacturing Technology*, 46, 621–628.

18. Chae, J., Park, S., & Freiheit, T. (2006). Investigation of micro-cutting operations. *International Journal of Machine Tools and Manufacture*, 46, 313–332.

19. Huo, D. 2013. *Micro-cutting: Fundamentals and Applications*. John Wiley & Sons, London.

20. Ghosh, R., Knopf, J. A., Gibson, D. J., Mebrahtu, T., & Currie, G. (2008). Cryogenic machining of polymeric biomaterials: An intraocular lens case study. In *Medical Device Materials IV: Proceedings of the Materials &Processes for Medical Devices Conference, Palm Desert, California, USA*, 54.

21. Jadhav, N. R., Gaikwad, V. L., Nair, K. J., & Kadam, H. M. (2009). Glass transition temperature: Basics and application in pharmaceutical sector. *Asian Journal of Pharmaceutics (AJP)*, 3(2), 82–89.

22. Wang, S., Tambraparni, M., Qiu, J., Tipton, J., & Dean, D. (2009). Thermal expansion of graphene composites. *Macromolecules*, 42(14), 5251–5255.

23. Friedrich, K. (2018). Polymer composites for tribological applications. *Advanced Industrial and Engineering Polymer Research*, 1, 3–39.

24. Bhandari, B., Hong, Y. S., Yoon, H. S., Moon, J. S., Pham, M. Q., Lee, G. B., & Ahn, S. H. (2014). Development of a micro-drilling burr-control chart for PCB drilling. *Precision Engineering*, 38(1), 221–229.

25. da Silva, L. C., da Mota, P. R., da Silva, M. B., Ezugwu, E. O., & Machado, Á. R. (2015). Study of burr behavior in face milling of PH 13–8 Mo stainless steel. *CIRP Journal of Manufacturing Science and Technology*, 8, 34–42.

26. Reichenbach, I. G., Bohley, M., Sousa, F. J., & Aurich, J. C. (2018). Micromachining of PMMA—manufacturing of burr-free structures with single-edge ultra-small micro end mills. *The International Journal of Advanced Manufacturing Technology*, 96(9), 3665–3677.

27. Filiz, S., Xie, L., Weiss, L. E., & Ozdoganlar, O. B. (2008). Micromilling of microbarbs for medical implants. *International Journal of Machine Tools and Manufacture*, 48(3–4), 459–472.

28. Thakur, R. K., Sharma, D., & Singh, K. K. (2019). Optimization of surface roughness and delamination factor in end milling of graphene modified GFRP using response surface methodology. *Materials Today: Proceedings*, 19, 133–139.

29. Ogawa, K., Nakagawa, H., Hirogaki, T., & Aoyama, E. (2012). Micro-drilling of CFRP plates using a high-speed spindle. In *Key Engineering Materials*. Trans Tech Publications Ltd, 523, pp. 1035–1040.

30. Thakur, R. K., Singh, K. K., & Kumar, K. (2020). Investigation of milling characteristics in graphene-embedded epoxy/carbon fibre reinforced composite. *Materials Today: Proceedings*, 33, 5643–5648.

31. Mian, A., Driver, N., & Mativenga, P. (2011). Identification of factors that dominate size effect in micromachining. *International Journal of Machine Tools and Manufacture*, 51, 383–394.

32. Sun, X., & Cheng, K. (2010). Micro-/nano-machining through mechanical cutting. In Qin, Y (Ed.), *Micromanufacturing Engineering and Technology*, Elsevier. pp. 24–38.

33. Dornfeld, D., Min, S., & Takeuchi, Y. (2006). Recent advances in mechanical micromachining. *CIRP Annals-Manufacturing Technology*, 55, 745–768.

34. Shaw, M. C. 2003. The size effect in metal cutting. *Sadhana*, 28, 875–896.

35. Câmara, M., Rubio, J. C., Abrão, A. & Davim, J. (2012). State of the art on micromilling of materials, a review. *Journal of Materials Science & Technology*, 28, 673–685.

36. Vollertsen, F. (2008). Categories of size effects. *Production Engineering*, 2, 377.

37. Kang, I., Kim, J., & Seo, Y. (2011). Investigation of cutting force behaviour considering the effect of cutting edge radius in the micro-scale milling of AISI 1045 steel. *Proceedings of the Institution of Mechanical Engineers, Part B: Journal of Engineering Manufacture*, 225, 163–171.

38. Le, B., Khaliq, J., Huo, D., Teng, X., & Shyha, I. (2020). A review on nanocomposites. Part 2: Micromachining. *Journal of Manufacturing Science and Engineering*, 142(10), 100802.

39. De Oliveira, F. B., Rodrigues, A. R., Coelho, R. T., & De Souza, A. F. (2015). Size effect and minimum chip thickness in micromilling. *International Journal of Machine Tools and Manufacture*, 89, 39–54.

40. Lu, X., Jia, Z., Liu, S., Yang, K., Feng, Y., & Liang, S.Y. (2019). Chatter stability of micro-milling by considering the centrifugal force and gyroscopic effect of the spindle. *Journal of Manufacturing Science and Engineering*, 141, 111003 (1–10).

41. Weule, H., Hüntrup, V., & Tritschler, H. 2001. Micro-cutting of steel to meet new requirements in miniaturization. *CIRP Annals-Manufacturing Technology*, 50, 61–64.

42. Moriwaki, T. 1989. Machinability of copper in ultra-precision micro diamond cutting. *CIRP Annals-Manufacturing Technology*, 38, 115–118.

43. Vogler, M. P., DeVor, R. E., & Kapoor, S. G. (2003). Microstructure-level force prediction model for micro-milling of multi-phase materials. *Journal of Manufacturing Science and Engineering*, 125, 202–209.

44. Venkatachalam, S., Fergani, O., Li, X., Yang, J. G., Chiang, K. N., & Liang, S. Y. (2015). Microstructure effects on cutting forces and flow stress in ultra-precision machining of polycrystalline brittle materials. *Journal of Manufacturing Science and Engineering*, 137, 021020.

45. Shimada, S., Ikawa, N., Tanaka, H. & Uchikoshi, J. (1994). Structure of micromachined surface simulated by molecular dynamics analysis. *CIRP Annals-Manufacturing Technology*, 43, 51–54.

46. Yuan, Z., Lee, W., Yao, Y., & Zhou, M. (1994). Effect of crystallographic orientation on cutting forces and surface quality in diamond cutting of single crystal. *CIRP Annals-Manufacturing Technology*, 43, 39–42.

47. Vogler, M. P., DeVor, R. E., & Kapoor, S. G. (2004). On the modeling and analysis of machining performance in micro-endmilling, part I: Surface generation. *Journal of Manufacturing Science and Engineering*, 126, 685–694.

48. Mian, A. J., Driver, N., & Mativenga, P. T. (2010). A comparative study of material phase effects on micro-machinability of multiphase materials. *The International Journal of Advanced Manufacturing Technology*, 50, 163–174.

49. Simoneau, A., Ng, E., & Elbestawi, M. (2006). Surface defects during microcutting. *International Journal of Machine Tools and Manufacture*, 46, 1378–1387.

50. Furukawa, Y., & Moronuki, N. (1988). Effect of material properties on ultra-precise cutting processes. *CIRP Annals*, 37(1), 113–116.

51. Popov, K. B., Dimov, S. S., Pham, D. T., Minev, R., Rosochowski, A., & Olejnik, L. (2006). Micromilling: Material microstructure effects. *Proceedings of the Institution of Mechanical Engineers, Part B: Journal of Engineering Manufacture*, 220, 1807–1813.

52. Lauro, C. H., Ribeiro Filho, S. L. M., Christoforo, A. L., & Brandão, L. C. (2014). Influence of the austenite grain size variation on the surface finishing in the micromilling process of the hardened AISI H13steel. *Matéria*, 19, 235–246.

53. Uhlmann, E., Piltz, S., & Schauer, K. (2005). Micro milling of sintered tungsten–copper composite materials. *Journal of Materials Processing Technology*, 167, 402–407.

54. Liu, X., DeVor, R. E., Kapoor, S., & Ehmann, K. (2004). The mechanics of machining at the microscale: Assessment of the current state of the science. *Journal of Manufacturing Science and Engineering*, 126, 666–678.

55. Ng, C. K., Melkote, S. N., Rahman, M., & Kumar, A. S. (2006). Experimental study of micro-and nano-scale cutting of aluminum 7075-T6. *International Journal of Machine Tools and Manufacture*, 46, 929–936.

56. Ducobu, F., Rivière-Lorphèvre, E., & Filippi, E. (2013). Chip formation in micro-cutting. *Journal of Mechanical Engineering and Automation*, 3, 441–448.

57. Woon, K., & Rahman, M. (2010). Extrusion-like chip formation mechanism and its role in suppressing void nucleation. *CIRP Annals-Manufacturing Technology*, 59, 129–132.

58. Ikawa, N., Shimada, S., & Tanaka, H. (1992). Minimum thickness of cut in micromachining. *Nanotechnology*, 3, 6.

59. Liu, K., & Melkote, S. N. (2007). Finite element analysis of the influence of tool edge radius on size effect in orthogonal micro-cutting process. *International Journal of Mechanical Sciences*, 49, 650–660.

60. Bissacco, G., Hansen, H. N., & De Chiffre, L. (2006). Size effects on surface generation in micro milling of hardened tool steel. *CIRP Annals-Manufacturing Technology*, 55, 593–596.

61. Lvov, N. (1969). Determining the minimum possible chip thickness. *Machines & Tooling*, 4, 45.

62. Kim, C. J., Mayor, J. R., & Ni, J. (2004). A static model of chip formation in microscale milling. *Transactions of the ASME-B-Journal of Manufacturing Science and Engineering*, 126, 710–718.

63. Liu, X., DeVor, R., & Kapoor, S. (2006). An analytical model for the prediction of minimum chip thickness in micromachining. *Journal of Manufacturing Science and Engineering*, 128, 474–481.

64. Leo Kumar, S. P., Jerald, J., Kumanan, S., & Prabakaran, R. (2014). A review on current research aspects in tool-based micromachining processes. *Materials and Manufacturing Processes*, 29(11–12), 1291–1337.

65. Anand, R. S., & Patra, K. (2014). Modeling and simulation of mechanical micro-machining—A review. *Machining Science and Technology*, 18(3), 323–347.

66. Yousefpour, A., Hojjati, M., & Immarigeon, J. P. (2004). Fusion bonding/welding of thermoplastic composites. *Journal of Thermoplastic Composite Materials*, 17(4), 303–341.

67. Vaidya, U. K., & Chawla, K. K. (2008). Processing of fibre reinforced thermoplastic composites. *International Materials Reviews*, 53(4), 185–218.

68. Kuram, E. (2016). Micro-machinability of injection molded polyamide 6 polymer and glass-fiber reinforced polyamide 6 composite. *Composites Part B: Engineering*, 88, 85–100.

69. Wan, M., Li, S. E., Yuan, H., & Zhang, W. H. (2019). Cutting force modelling in machining of fiber-reinforced polymer matrix composites (PMCs): A review. *Composites Part A: Applied Science and Manufacturing*, 117, 34–55.

70. Jain, N., Somvanshi, K. S., Gope, P. C., & Singh, V. K. (2019). Mechanical characterization and machining performance evaluation of rice husk/epoxy an agricultural waste based composite material. *Journal of the Mechanical Behavior of Materials*, 28(1), 29–38.

71. Ervine, P., O'Donnell, G. E., & Walsh, B. (2015). Fundamental investigations into burr formation and damage mechanisms in the micro-milling of a biomedical grade polymer. *Machining Science and Technology*, 19(1), 112–133.

72. Abrate, S., & Walton, D. A. (1992). Machining of composite materials. Part I: Traditional methods. *Composites Manufacturing*, 3(2), 75–83.

73. Kavad, B. V., Pandey, A. B., Tadavi, M. V., & Jakharia, H. C. (2014). A review paper on effects of drilling on glass fiber reinforced plastic. *Procedia Technology*, 14, 457–464.

74. Anand, R. S., Patra, K., & Steiner, M. (2014). Production engineering research and development. *Production Engineering*, 8(3), 301–307.

75. Basso, I., Batista, M. F., Jasinevicius, R. G., Rubio, J. C. C., & Rodrigues, A. R. (2019). Micro drilling of carbon fiber reinforced polymer. *Composite Structures*, 228, 111312.

76. Rahamathullah, I., & Shunmugam, M. S. (2014). Mechanistic approach for prediction of forces in micro-drilling of plain and glass-reinforced epoxy sheets. *The International Journal of Advanced Manufacturing Technology*, 75(5), 1177–1187.

77. Kim, J. W., Nam, J., & Lee, S. W. (2019). Experimental study on micro-drilling of unidirectional carbon fiber reinforced plastic (UD-CFRP) composite using nano-solid lubrication. *Journal of Manufacturing Processes*, 43, 46–53.

78. Rajamurugan, T. V., & Shanmugam, K. (2013). Analysis of delamination in drilling glass fiber reinforced polyester composites: Experimental investigations and finite element implementation. *Materials*, 45, 80–87.

79. Rahamathullah, I., & Shunmugam, M. S. (2013). Analyses of forces and hole quality in micro-drilling of carbon fabric laminate composites. *Journal of Composite Materials*, 47(9), 1129–1140.

80. Anand, R. S., & Patra, K. (2018). Cutting force and hole quality analysis in micro-drilling of CFRP. *Materials and Manufacturing Processes*, 33(12), 1369–1377.

81. Anand, R. S., & Patra, K. (2017). Mechanistic cutting force modelling for micro-drilling of CFRP composite laminates. *CIRP Journal of Manufacturing Science and Technology*, 16, 55–63.

82. Shunmugesh, K., & Panneerselvam, K. (2016). Optimization of process parameters in micro-drilling of carbon fiber reinforced polymer (CFRP) using Taguchi and grey relational analysis. *Polymers and Polymer Composites*, 24(7), 499–506.

83. Phadnis, V. A., Makhdum, F., Roy, A., & Silberschmidt, V. V. (2013). Drilling in carbon/epoxy composites: Experimental investigations and finite element implementation. *Composites Part A: Applied Science and Manufacturing*, 47, 41–51.

84. Usui, S., Wadell, J., & Marusich, T. (2014). Finite element modeling of carbon fiber composite orthogonal cutting and drilling. *Procedia Cirp*, 14, 211–216.

85. Wang, J., Yan, Y., Geng, Y., Gan, Y., & Fang, Z. (2019). Fabrication of polydimethylsiloxane nanofluidic chips under AFM tip-based nanomilling process. *Nanoscale Research Letters*, 14(1), 1–14.

86. Altintas, Y., & Jin, X. (2011). Mechanics of micro-milling with round edge tools. *CIRP Annals*, 60(1), 77–80.

87. Deng, B., Zhou, L., Peng, F., Yan, R., Yang, M., & Liu, M. (2018). Analytical model of cutting force in micromilling of particle-reinforced metal matrix composites considering interface failure. *Journal of Manufacturing Science and Engineering*, 140(8), 081009.

88. Dib, M. H. M., Duduch, J. G., & Jasinevicius, R. G. (2018). Minimum chip thickness determination by means of cutting force signal in micro endmilling. *Precision Engineering*, 51, 244–262.

89. Klocke, F., Gerschwiler, K., & Abouridouane, M. (2009). Size effects of micro drilling in steel. *Production Engineering*, 3(1), 69–72.

90. Özel, T., & Liu, X. (2009). Investigations on mechanics-based process planning of micro-end milling in machining mold cavities. *Materials and Manufacturing Processes*, 24(12), 1274–1281.

91. Lai, X., Li, H., Li, C., Lin, Z., & Ni, J. (2008). Modelling and analysis of micro scale milling considering size effect, micro cutter edge radius and minimum chip thickness. *International Journal of Machine Tools and Manufacture*, 48(1), 1–14.

92. Aramcharoen, A., & Mativenga, P. T. (2009). Size effect and tool geometry in micromilling of tool steel. *Precision Engineering*, 33(4), 402–407.

93. Chen, N., Li, H. N., Wu, J., Li, Z., Li, L., Liu, G., & He, N. (2021). Advances in micro milling: From tool fabrication to process outcomes. *International Journal of Machine Tools and Manufacture*, 160, 103670.

94. Thakur, R. K., Singh, K. K., & Sharma, D. (2019). Modeling and optimization of surface roughness in end milling of graphene/epoxy nanocomposite. *Materials Today: Proceedings*, 19, 302–306.

95. Arora, I., Samuel, J., & Koratkar, N. (2013). Experimental investigation of the machinability of epoxy reinforced with graphene platelets. *Journal of Manufacturing Science and Engineering*, 135(4), 041007 (1–7).

96. Kumar, M. N., Mahmoodi, M., TabkhPaz, M., Park, S. S., & Jin, X. (2017). Characterization and micro end milling of graphene nano platelet and carbon nanotube filled nanocomposites. *Journal of Materials Processing Technology*, 249, 96–107.

97. Azmi, A. I., Lin, R. J. T., & Bhattacharyya, D. (2012). Experimental study of machinability of GFRP composites by end milling. *Materials and Manufacturing Processes*, 27(10), 1045–1050.

98. Christ, K. T., Smith, B. B., Pfefferkorn, F. E., & Turner, K. T. (2010). Micro end milling polystyrene for microfluidic applications. *5th International Conference on Micro Manufacturing*.

99. Chen, P. C., Pan, C. W., Lee, W. C., & Li, K. M. (2014). An experimental study of micromilling parameters to manufacture microchannels on a PMMA substrate. *The International Journal of Advanced Manufacturing Technology*, 71(9), 1623–1630.

100. Aramcharoen, A., Sean, S. K. C., & Kui, L. (2012). An experimental study of micromilling of polymer materials for microfluidic applications. *International Journal of Abrasive Technology*, 5(4), 286–298.

101. Okuda, K., Tsuneyoshi, T., Li, W., & Shibahara, H. (2009). Study on cutting of micro channel by end mill with small diameter. In *Key Engineering Materials*. Trans Tech Publications Ltd, 407, pp. 351–354.

102. Korkmaz, E., Onler, R., & Ozdoganlar, O. B. (2017). Micromilling of poly (methyl methacrylate, PMMA) using single-crystal diamond tools. *Procedia Manufacturing*, 10, 683–693.

103. Wilson, M. E., Kota, N., Kim, Y., Wang, Y., Stolz, D. B., LeDuc, P. R., & Ozdoganlar, O. B. (2011). Fabrication of circular microfluidic channels by combining mechanical micromilling and soft lithography. *Lab on a Chip*, 11(8), 1550–1555.

104. Chu, B., Samuel, J., & Koratkar, N. (2015). Micromilling responses of hierarchical graphene composites. *Journal of Manufacturing Science and Engineering*, 137, 011002.

105. Gopalakrishna, H., Rao, J. S., Kumar, S. N., Shetty, V. V., & Rai, K. (2016). Effect of Friction on the Cutting Forces in High Speed Orthogonal Turning of Al 6061-T6. *IOSR Journal of Mechanical and Civil Engineering*, 11(2), 78–83.

106. Mahmoodi, M, Mostofa, M., Jun, M., & Park, S.S. (2013). Characterization and micromilling of flow induced aligned carbon nanotube nanocomposites. *Journal of Micro and Nano-Manufacturing*, 1(1), 011009 (1–8).

107. Gong, Y., Baik, Y. J., Li, C. P., Byon, C., Park, J. M., & Ko, T. J. (2017). Experimental and modeling investigation on machined surfaces of HDPE-MWCNT polymer nanocomposite. *The International Journal of Advanced Manufacturing Technology*, 88, 879–885.

108. Zinati, R. F., & Razfar, M. (2014). Experimental and modeling investigation of surface roughness in end-milling of polyamide 6/multi-walled carbon nano-tube composite. *The International Journal of Advanced Manufacturing Technology*, 75, 979–989.

109. Samuel, J., DeVor, R. E., Kapoor, S. G., & Hsia, K. J. (2006). Experimental investigation of the machinability of polycarbonate reinforced with multiwalled carbon nanotubes. *Journal of Manufacturing Science and Engineering*, 128(2), 465–473.

110. Marcon, A., Melkote, S., Kalaitzidou, K., & DeBra, D. (2010). An experimental evaluation of graphite nanoplatelet based lubricant in micro-milling. *CIRP Annals*, 59(1), 141–144.

111. Shyha, I., Fu, G. Y., Huo, D. H., Le, B., Inam, F., Saharudin, M. S., & Wei, J. C. (2018). Micro-machining of nano-polymer composites reinforced with graphene and nano-clay fillers. *Key Engineering Materials*, 786, 197–205.

112. Gao, C., & Jia, J. (2017). Factor analysis of key parameters on cutting force in micromachining of graphene-reinforced magnesium matrix nanocomposites based on FE simulation. *The International Journal of Advanced Manufacturing Technology*, 92(9), 3123–3136.

113. Gaitonde, V. N., Karnik, S. R., Silva, L. R., Abrão, A. M., & Davim, J. P. (2009). Machinability study in microturning of PA66 GF30 polyamide with a PCD tool. *Materials and Manufacturing Processes*, 24(12), 1290–1296.

114. Silva, L. R., Davim, J. P., Festas, A., & Abrão, A. M. (2009). Machinability aspects concerning micro-turning of PA66-GF30-reinforced polyamide. *The International Journal of Advanced Manufacturing Technology*, 41(9), 839–845.

115. Davim, J. P., Silva, L. R., Festas, A., & Abrão, A. M. (2009). Machinability study on precision turning of PA66 polyamide with and without glass fiber reinforcing. *Materials & Design*, 30(2), 228–234.
116. Davim, J.P., Reis, P. (2004). Machinability study on composite (polyetheretherketone reinforced with 30% glass fibre-PEEK GF 30) using polycrystalline diamond (PCD) and cemented carbide (K20) tools. *The International Journal of Advanced Manufacturing Technology*, 23, 412–418.
117. Pallapothu, H, Kumar, A., & Laxminarayana, P. (2020). Micro drilling of glass fibre reinforced polymer composites. *Mater Today Proceeding*, 46, 9252–9256.
118. Wan, Y., Kim, D., Park, Y. B., & Joo, S. K. (2008). Micro electro discharge machining of polymethylmethacrylate (PMMA)/multi-walled carbon nanotube (MWCNT) nanocomposites. *Advanced Composites Letters*, 17(4), 096369350801700401.
119. Habib, S., Okada, A., & Ichii, S. (2013). Effect of cutting direction on machining of carbon fibre reinforced plastic by electrical discharge machining process. *International Journal of Machining and Machinability of Materials*, 13(4), 414–427.
120. Dutta, H., Debnath, K., & Sarma, D. K. (2020). Improving the performance of µED-milling using assisting electrode for fabricating micro-channels in CFRP composites. *Materials Today: Proceedings*, 28, 755–760.
121. Chern, G. L., Wu, Y. J. E., Cheng, J. C., & Yao, J. C. (2007). Study on burr formation in micro-machining using micro-tools fabricated by micro-EDM. *Precision Engineering*, 31(2), 122–129.
122. Jain, V. K., Kalia, S., Sidpara, A., & Kulkarni, V. N. (2012). Fabrication of micro-features and micro-tools using electrochemical micromachining. *The International Journal of Advanced Manufacturing Technology*, 61(9), 1175–1183.

17 Grinding Nanocomposite Materials for Space Applications

Mark James-Jackson, Jason T. Brantley,
Rosemar Batista da Silva, Marcio Bacci da Silva,
and Alisson Rocha Machado

CONTENTS

17.1 NANOCOMPOSITE MATERIALS

Nanostructured polymer–matrix composites (PMCs) have excellent mechanical, chemical, physical, and electrical properties (surface-to-volume ratio > 3). The ratio is ~ 3 when particle size is ~ 1.25 nm, which changes mechanical properties drastically. When one or more aspect of the particle is nanoscale, then the composite is termed a nanocomposite. Figure 17.1 shows the general arrangement of fibers bonded in a matrix filled with carbon nanotubes. The reinforcement that is responsible for superior property enhancement include clays, carbon nanotubes (CNTs), or other nanoscale particles [1].

Nanoparticles are added to the matrix to reinforce it when strained. Nanoparticles have at least one dimension < 100 nm, which is classed as one-dimensional, two-dimensional for 2D nanoparticles (two aspects < 100 nm), and three-dimensional for 3D nanoparticles (all aspects < 100 nm). For space applications, composites are typically filled with 2D and/or 3D nanoparticles such as

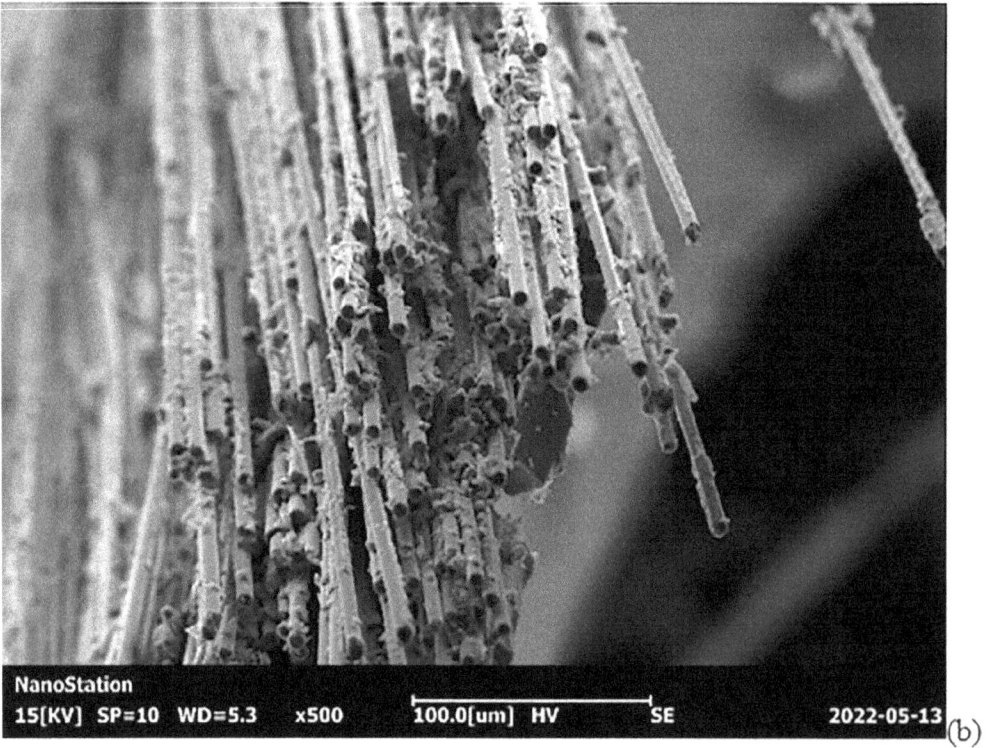

FIGURE 17.1 Nanostructured carbon fiber composite material showing: (a) carbon fibers embedded in a matrix composed of resins and CNTs; (b) magnified image showing the ends of sheared fibers. (Images by M. J. Jackson and J. T. Brantley.)

carbon nanotubes (CNTs), fibers, wires, rods, whiskers, and 2D graphene. CNTs are made using the arc-discharge method and have aspect ratios > 1000.

CNTs are single-wall (SWCNT) or multiwalled (MWCNT) and can be produced from graphene. SWCNT is a graphene layer rolled into a cylinder ~ 0.5–5 nm diameter (density, $\rho = 0.8$ g/cm^3, thermal conductivity, $k = 6000$ W/mK, and thermal stability in air > 600°C). MWCNTs are fullerenes (C_{60}) composed of two graphene sheets separated by a van der Waals force to form inner radii ~ 1.5–15 nm, and outer radii ~ 2.5–50 nm (density = 1.8 g/cm^3, thermal conductivity = 2000 W/mK, and thermal stability in air > 600°C). CNTs manufactured with specific aspect ratios and length(s) create varying mechanical properties including high surface area which is useful for PMCs. A high concentration of CNTs increase strain hardening and yield stresses in PMCs that makes grinding and finishing more difficult. The direct use of graphene sheets in the polymer–matrix increases shear moduli, Young's moduli, and hardness by 27%, 150%, and 35%, respectively [2]. A high concentration of carbon is prevalent within the resin binder surrounding the fibers (Figure 17.2).

The ability of carbon to hybridize allows atomic orbitals to form 2D CNTs, 3D nanoparticles, and nanostructured clays. The electronic configuration of carbon shows that electrons are strongly bound in 1s and weakly bound in the 2p and 2s orbitals (valence electrons). Therefore, defects can be found on the surfaces of CNTs in the form of heptagons and pentagons that form curved and spirals CNTs. The geometry and properties of CNTs are dominated by the diameter and chiral angle, θ. For SWCNTs, three conditions differentiate SWCNTs based on the chiral angle. SWCNTs are designated: armchair, zigzag, or achiral nanotubes, and chiral. The conditions are as follows:

1. When chiral indices = 0, or $((n, 0), (0, m))$, then chiral angle is 0°;
2. If the chiral angles are equal $(m = n)$, then the chiral angle is 30°; and
3. When the chiral angle is $0° < \theta < 30°$, then $(m \neq n)$.

3D nanomaterials are made up of particles with all three dimensions in the nanoscale range and are mostly cubic or spherical in shape. Fillers are usually used with 3D nanomaterials, and these are typically cellulose in particle, granule, or crystal forms. Mechanical properties are exceptional with CNT reinforcement ($E = 1.3$ TPa and σ_{uts} ~ 50–200 GPa). Clearly, the improvement of mechanical properties of nanostructured composites presents challenges to their processing and circularity, that is, reuse, recycling, and remanufacturing [3].

The processing of traditional composite materials is very well understood with research focused on grinding and finishing of PMCs [4], metal–matrix composites (MMCs) [5], and ceramic–matrix

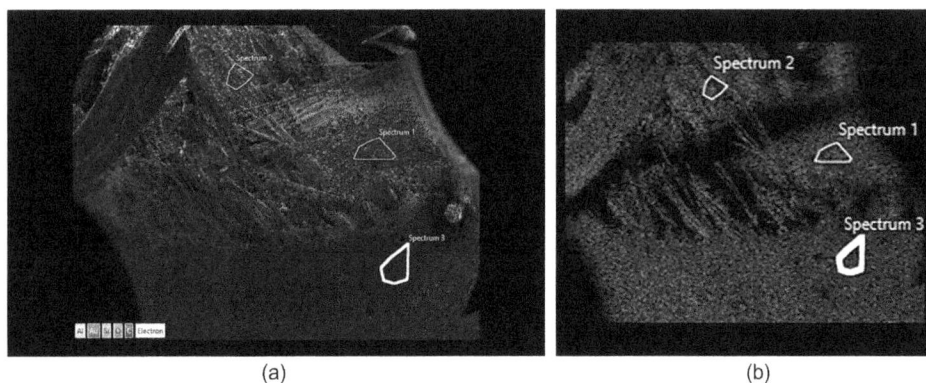

(a)   (b)

FIGURE 17.2 (a) Electron-dispersive spectrum of elements showing a high concentration of carbon (91.24 wt. %/93.28 at. %) and oxygen (8.76 wt. %/6.72 at. %) on the surface of the nanostructured composite material; (b) Concentration of carbon $k\alpha_{1,2}$ ~ 91.24 wt. % (93.28 at. %) for spectra 1, 2, and 3. (Images by M. J. Jackson and J. T. Brantley.)

composites (CMCs) [6]. However, for nanocomposite materials, processing is currently not well defined. The following section lays the foundation for understanding the grinding of nanocomposite materials.

17.2   GRINDING AND FINISHING PROCESSES FOR NANOCOMPOSITES

The grinding of nanocomposite materials is primarily focused on aerospace/space components made from a variety of nanocomposites in metals, polymers, and ceramics. Strain rates in grinding processes ($\dot{\gamma} > 10^7 \, s^{-1}$) generate thin plane zones based on microstructural alignment of grains, inclusions, and chemical species [7]. Abrasive grit geometry and direction reluctantly form combinations of plowing and sliding to take place between abrasive grit and nanocomposite material at variable depths accompanied by significant contact loads (normal forces) creating microcracks and chips. Rubbing and plowing dominate due to bond bridges interacting with chip formation processes. There are six dominant interactions between abrasive and nanocomposite material depending on the ratio of tangential and normal force ($\mu = F_t/F_n$): (1) abrasive/nanocomposite interactions: (1.1) cutting (chip production ($\mu > 1$) and surface cracking ($\mu < 1$)); (1.2) plowing ($\mu \sim 0.5$–1); (1.3) sliding ($\mu \sim 0.165$); (2) chip/bond sliding ($\mu \sim 0.3$–0.5); (3) chip/nanocomposite sliding ($\mu \sim 0.3$–0.5); and (4) bond/nanocomposite sliding ($\mu \sim 0.3$–0.5) (Figure 17.3).

Abrasive grits cut nanocomposites then become blunt as time ebbs away then plowing and sliding along the surface in a confused way [7]. Cutting and surface modifications were controlled through dressing, directing coolant flows, and altering cycle times to produce an envelope that creates precise nanocomposites structures (Figure 17.3).

Each grinding process has its own characteristics. However, models of the principal mode of grinding have been developed over many years, and the principal models for the process are described. The total grinding force encountered is:

$$F_{total} = F_{chip} + F_{sliding} + F_{plowing} \tag{17.1}$$

$F_{total} = F_{chip}$ (chip creation force) + $F_{sliding}$ (sliding force) + $F_{plowing}$ (plowing force). Subsequently, component forces in the tangential and normal directions:

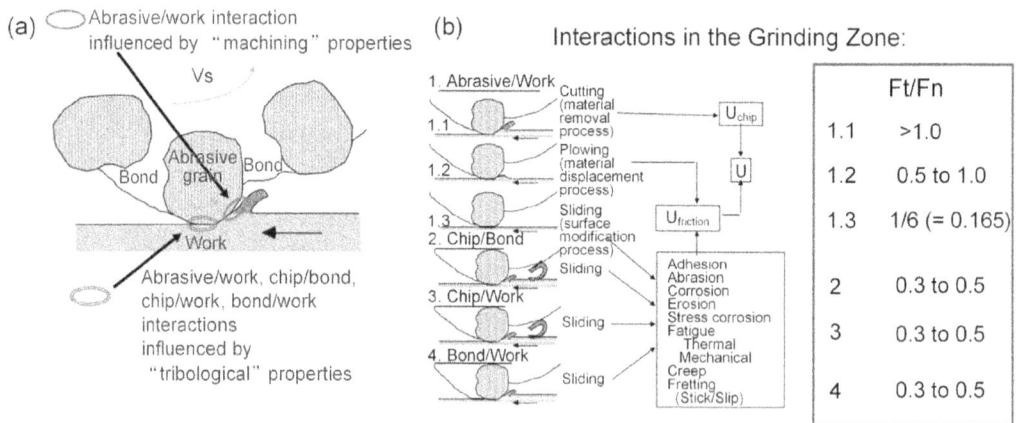

FIGURE 17.3   Abrasive and nanocomposite workpiece couple: (a) tribological interactions and (b) detailed interactions ($\mu = F_t/F_n$). (Figures drawn by S. Subramanian/M. P. Hitchiner (Saint-Gobain Abrasives), Reproduced with Permission.)

$$F_t = F_{t\ \text{chip}} + F_{t\ \text{sliding}} + F_{t\ \text{plowing}} \tag{17.2}$$

$$F_n = F_{n\ \text{chip}} + F_{n\ \text{sliding}} + F_{n\ \text{plowing}} \tag{17.3}$$

where Equations 17.2 and 17.3 are tangential and normal components of force. Forces magnify as surface area of active cutting edges increase. The magnitude of F_t and F_n depends on chip creation energy, u_{chip}; nanocomposite workpiece velocity, v_w; grinding wheel velocity, v_s; width of cut, b; depth of abrasive incision, a; stress in the contact zone, p_c; effective wear land area, $A_{\text{effective}}$; equivalent diameter, d_e; force of plowing, F_{plowing}; and coefficient of contact friction, μ:

$$F_t = \frac{u_{\text{chip}} v_w ba}{v_s} + \mu p_c A_{\text{effective}} b \sqrt{d_e a} + F_{t\ \text{plowing}} \tag{17.4}$$

$$F_n = \frac{k_{\text{chip}} u_{\text{chip}} v_w ba}{v_s} + p_c A_{\text{effective}} b \sqrt{d_e a} + k_{\text{plowing}} F_{t\ \text{plowing}} \tag{17.5}$$

where k_{chip} is a chip creation function, and k_{plowing} is a surface plowing function. The plowing force is caused by the abrasive grit moving at ~ 1 N/mm. The associated tangential and normal plowing forces:

$$F_{t\ \text{plowing}} = F_{\text{plowing}}(v_w a) \tag{17.6}$$

$$F_{n\ \text{plowing}} = k_{\text{plowing}} F_{\text{plowing}}(v_w a) \tag{17.7}$$

The tangential and normal sliding forces:

$$F_{t\ \text{sliding}} = \mu F_{n\ \text{sliding}} \tag{17.8}$$

$$F_{n\ \text{sliding}} = p_c\ A_{\text{effective}} b \sqrt{d_c a} \tag{17.9}$$

For chip creation, u_{chip} and k_{chip} ~ 2:

$$F_{t\ \text{chip}} = \frac{u_{\text{chip}} v_w ba}{v_s} \tag{17.10}$$

$$F_{n\ \text{chip}} = \frac{k_{\text{chip}} u_{\text{chip}} v_w ba}{v_s} \tag{17.11}$$

Contact stress, p_c, is based on curvature difference between abrasive grit and nanocomposite material:

$$p_c = c_c \Delta \tag{17.12}$$

Δ is difference in curvature (c_c = constant):

$$\Delta = \frac{2}{d_s} - \frac{1}{R} \approx \frac{4 v_w}{d_e v_s} \tag{17.13}$$

Here, diameter, d_s, cutting radius, R, and effective diameter, d_e, pertain to the grinding wheel. Grinding stresses are a function of area of wear for active abrasive grits:

$$A_{\text{effective}} = A_{0,\text{ effective}} + A_{\text{dull}} - A_{\text{sharp}} \tag{17.14}$$

The effective wear flat area ($A_{0,\text{ effective}}$), area of dulling (A_{dull}), and self-sharpening area (A_{sharp}) are generated by attrition:

$$A_{0,\text{ effective}} = -A_0 \ln\left(\frac{A_1\delta}{f(w)}\right) \tag{17.15}$$

The dressing severity, δ, is dependent on dressing parameters and lubricant selection:

$$\delta = a_0 s_d^x a_d^y \tag{17.16}$$

Severity of dressing depends on dressing depth (a_0), feed (s_d), and dressing depth of cut (a_d). Immediately after dressing, the wheel begins grinding and is worn by attrition of surface layers of the grit:

$$A_{\text{dull}} = k_l l_{\text{sliding}} \tag{17.17}$$

k_l is constant depending on abrasive wear (Figure 17.4), composite material, and sliding length (l_{sliding}):

$$l_{\text{sliding}} = \int_0^t \frac{v_s l_c}{\pi d_s} dt \tag{17.18}$$

Thus, $l_c = \sqrt{a \cdot d_e}$ and $d_e = \dfrac{d_w \cdot d_s}{d_w \pm d_s}$

Specific energy and power model for chip creation:

$$u = u_{\text{chip}} + u_{\text{sliding}} + u_{\text{plowing}} \tag{17.19}$$

$$P = P_{\text{chip}} + P_{\text{sliding}} + P_{\text{plowing}} = F_t v_s \tag{17.20}$$

chip creation energy and power functions:

$$u_{\text{chip}} = \frac{F_{t\text{ chip}} v_s}{b v_s a} \tag{17.21}$$

$$P_{\text{chip}} = u_{\text{chip}} v_w b a \tag{17.22}$$

Power of plowing and sliding interactions ($\mu \sim 3$):

$$P_{t\text{ plowing}} = F_{t\text{ plowing}} v_s \tag{17.23}$$

$$P_{\text{sliding}} = \mu p_c A_{\text{effective}} \sqrt{d_e a} v_s b \tag{17.24}$$

Total power consumed: $P = P_{\text{chip}} + P_{\text{sliding}} + P_{\text{plowing}}$:

$$P = u_{\text{chip}}(b v_w a) + \mu p_c A_{\text{effective}} b \sqrt{d_e a} v_s + F_{t\text{ plowing}} v_s \tag{17.25}$$

FIGURE 17.4 Coated abrasive grits bonded to woven fabric used as finishing agents for nanocomposite materials. (Image by M. P. Hitchiner (Saint-Gobain Abrasives). Reproduced with Permission.)

The anticipated rise in surface temperature is responsible for damaging the cutting process in grinding nanocomposite materials by softening/melting of the resin that bonds the fibers together. Maximum surface temperature:

$$\theta_{\text{maximum}} = \frac{1.13 q_w \sqrt{\alpha} a^\beta d_e^\alpha}{k \sqrt{v_w}} \tag{17.26}$$

Here, thermal conductivity (k), depth of cut (a), nanocomposite workpiece speed (v_w), thermal diffusivity (α), partitioned heat flux to nanocomposite workpiece (q_w), and heat flux to composite: $q_w = \dfrac{\varepsilon P}{b l_c}$. The specific energy that is critical at the softening temperature:

$$u^* = u_0 + B d_e^\alpha a^{-\beta} v_w^{-\gamma} \tag{17.27}$$

where B is a thermal parameter equal to $\left(\dfrac{k \theta^m}{\beta \sqrt{\alpha}} \right)$. The critical power at softening is:

$$P^* = u_0 b v_w a + B b d_e^\alpha a^{-\beta} v_w^{-\gamma} \tag{17.28}$$

When $P < P^*$, softening of the material does not happen. Heat is partitioned into the abrasive grit rather than the nanocomposite workpiece, and circular cutting tools carry away the heat via bond and body of the tool.

The predicted surface roughness:

$$\frac{\left(R_a - R_{a,\,\infty}\right)}{\left(R_{a,\,0} - R_{a,\,\infty}\right)} = \exp\left(\frac{-V_w'}{V_0'}\right) \tag{17.29}$$

where $R_{a,\,\infty} = R_g\left(Q_w'\right)^\varepsilon$, $R_{a,\,0} = R_0 s_d^x a_d^y\left(\dfrac{Q_w'}{V_s}\right)^\varepsilon$, V_0', R_0, x, y, and ε are constants. Eccentricity of nano-composites of variable thickness:

$$r = r_m + r_0 a \tag{17.30}$$

where eccentricity, $r = r_m$. Grinding machine systems must adapt to responses to operational parameters to minimize softening/melting of the nanocomposite:

$$u(t) - v(t) - w(t) = \frac{d\delta}{dt} \tag{17.31}$$

The normal grinding force, F_n, is dependent on the stiffness of the wheel, k_s, and associated deflections:

$$F_n = k_g \delta \tag{17.32}$$

$$k_g^{-1} = k_s^{-1} + k_w^{-1} + k_a^{-1} \tag{17.33}$$

The equations described are used to monitor grinding system performance when processing nano-composite materials. However, equations that describe the performance of grinding nanostructured composites need to be developed, especially in kinematics and grinding kinetics to minimize the subtraction of material and wear of the abrasive cutting tool. This will avoid softening/melting of the matrix and much better control of the grinding process.

17.2.1 KINEMATICS OF GRINDING NANOCOMPOSITES

Estimates of grinding forces are needed so that cutting and finishing of nanocomposite materials can be calculated in real time using grinding kinetics and kinematics [8], Jackson [7]. External, internal, and surface grinding produce general equations with factors applied to account for system peculiarities [7,9]. It is necessary to consider dynamic factors of grinding wheels especially stiffness when considering kinetics/kinematics of grinding applied to nanocomposites. For grinding wheels with low stiffness, grinding force variations cause a change of material removal mechanism [10] that affects grinding wheel eccentricity, chatter, and run-out [11,12]. The existence of cyclic variance of grinding wheel concentricity has important physical significance in grinding processes [13–15], especially when ideal kinematic construction imposes thermomechanical effects that will change the abrasive grit shape and dynamic chip thickness, the importance of which is discussed in detail [16].

17.2.1.1 Kinematics of Grinding Nanocomposites: Dynamic Chip Thickness

The dynamic undeformed chip thickness leads to vibration in grinding wheels. [13,17,18] derived the undeformed chip thickness during grinding as a static and stable entity:

$$h = \left[\frac{4}{C \cdot r}\left(\frac{v_w}{v_s}\right)\left(\frac{a_e}{d_e}\right)^{0.5}\right]^{0.5} \tag{17.34}$$

where C is active grit concentration, r is chip ratio (width-to-thickness), v_w is nanocomposite material speed, v_s is wheel speed, a_e is depth of cut, and d_e is equivalent wheel diameter. Equation 17.34 does not consider chip shape and should consider dynamic components of grinding. Orthogonal vibrations contribute to cross-coupling terms in grinding:

$$h = 2\pi \frac{v_t}{v_s N_r} \sin\theta + 2\pi \frac{v_n}{v_s N_r} \cos\theta \qquad (17.35)$$

where, N_r, active grits per revolution where contact deformations are dominant. Nanocomposite materials with high elastic moduli have a linear proportionality between chip thickness, h, and elastic modulus ratio:

$$h = \frac{E_{composite}}{E_{grinding\ wheel}} = \frac{E_c}{E_s} \qquad (17.36)$$

The influence of dynamic interference between grinding wheel and nanocomposite materials on the dynamic undeformed chip thickness must be known and involves calculating the number of active cutting grits, N_g [7].

17.2.1.2 Kinematics of Grinding Nanocomposites: Active Cutting Grits

The dynamic number and distribution density of cutting grits [7,8,19,20] directly influence specific cutting energy and total grinding force and are lower than the static density of grits:

$$N = C_{ds} b v_s \qquad (17.37)$$

where N is the dynamic number of grits, b is the grinding width, C_{ds} is the distribution density of grits, and v_s is the wheel speed. Static density of abrasive grits measured [21,22]:

$$C_s(z) = A z^k \qquad (17.38)$$

And when characterizing the removal of nanocomposite material [23]:

$$C_s(z) = C_s(z') \cdot \left(1 - \left(\frac{V_{sh}}{V_t}\right)\right) \qquad (17.39)$$

The dynamic abrasive grit quantity:

$$C_d(z) = \frac{C_s(z)}{1 + \dfrac{2}{3}\left(\dfrac{C_s(z)}{z} \cdot \dfrac{\tan\theta}{\tan\varepsilon} \cdot E h^3\right)} \qquad (17.40)$$

Here, z is the distance of abrasive grits from surface to cutting depth, parameters A and k are variables associated with wheel and dressing parameters, $C_s(z)$ = static abrasive grit density, v_t is total volume of wheel engaged with the nanocomposite during cutting, and v_{sh} is total kinematic volume generated by active grits. The dynamic number of abrasive grits for grinding feed angle, ε:

$$N_g = N \cdot (\tan \cdot \varepsilon)^m \qquad (17.41)$$

The number of effective abrasive grits (N_g) and their effect on grinding force(s) and chip thickness are not understood, and the number of grits operating at a specific moment in time is also unknown. Clearly, this type of information is important to know especially when grinding complex materials such as nanocomposites.

17.2.1.3 Kinematics of Grinding Nanocomposites: Instantaneous Undeformed Chip Thickness

Chip creation parameters are calculated in terms of kinematics [24]. To model undeformed chip thickness, engineers can estimate the parameter by average modeling, measuring probability distributions and performing a statistical analysis of abrasive grit height.

Undeformed chip thickness models based on average modeling are numerous (Table 17.1). Expressions of undeformed instantaneous chip thickness can be derived from random distribution models (uniform distributions and Gaussian distributions) or from Rayleigh distributions. Gaussian and Rayleigh distributions are probabilistic, and the Gaussian form has centralization, symmetry, and uniform variability, while Rayleigh distributions are defined by one notable parameter [25]. Rayleigh distributions can be used when grinding hard and brittle materials, metals, and alloys. The undeformed instantaneous chip thickness of brittle materials compared with metals is difficult to calculate but is being investigated [26].

The instantaneous models of undeformed chip thickness, h, are focused on statistical modeling of abrasive grit height differences, where the active grit spacing can be calculated for a stationary grinding wheel [27,28].

TABLE 17.1
Published Models to Calculate the Undeformed Chip Thickness

Investigator(s)	Equation	Year
Pahlitzsch and Helmerdig [29]	$h = 2\lambda \dfrac{v_w}{v_s}\sqrt{\dfrac{a_p}{d_s}}$	(1943)
Reichenbach et al. [18]	$h = \left[\dfrac{4v_w}{v_s N_d c}\sqrt{a_p/d_s}\right]^{1/2}$	(1956)
Werner et al. [30]	$h = \dfrac{1}{A}\left(\dfrac{2}{N_{st}}\right)^{1/\alpha+1}\left(\dfrac{v_w}{v_s}\right)^{1/\alpha+1}\left(\dfrac{a_p}{s_s}\right)^{1/2(\alpha+1)}$	(1971)
Malkin [8]	$h = 2s\left(\dfrac{a_{j-1}}{d_s}\right)^{1/2} - \delta_j$	(2008)
Hecker et al. [22]	$f(h) = \begin{cases}(h/\sigma^2)e^{-h^2/2\sigma^2} & h \geq 0 \\ 0 & h < 0\end{cases}$	(2007)
Agarwal and Rao [31]	$h = \sqrt{\dfrac{a_e v_w}{C r v_s}\dfrac{1}{l_c}}$	(2013)
Ding et al. [28]	$h = 2\lambda_j \dfrac{v_w}{v_s}\left(\dfrac{a_{j-1}}{d_s}\right)^{1/2} - 2\dfrac{v_w}{v_s}\left(1+\dfrac{v_w}{v_s}\right)\left[a_{j-1} - (a_j a_{j-1})^{1/2}\right]$	(2017)
Wu et al. [26]	$f(h) = \begin{cases}\dfrac{1}{\sqrt{2\pi}\sigma_g}e^{\left[-\dfrac{(h-\bar{h})^2}{2\sigma_g^2}\right]} & h \geq 0 \\ 0 & h < 0\end{cases}$	(2019)
Jamshidi and Budak [27]	$h = R_{n,j} - R_{m,j} + (n-m)f_t\sin(\theta_{i,j}) + n_{i,j}$	(2020)

The effects of tool run-out and deflections on the size and aspect of h are difficult to quantify. Instantaneous chip thickness, h, of single abrasive grits is composed of three properties; static undeformed chip thickness ($h_{cu,\,0}$), dynamic chip thickness ($\Delta h_{cu,\,i}$), and deviation excited by the spindle $\left(\partial h_{cu,\,i}\varnothing_i\right)$ [32]:

$$h = \pm h_{cu,\,0} + \Delta h_{cu,\,i} + \partial h_{cu,\,i}\varnothing_i = \pm f_t \sin\varnothing_i + \left[\left(H_{is}(t-T) - H_{is}(t)\right) - H_{iw}(t-T) - H_{iw}(t)\right]$$
$$+ R_i'(\varnothing_i') - R_{i-1}'(\varnothing_i') \tag{17.42}$$

To estimate the critical thickness (h_{crit}) during brittle–ductile transitions of nanocomposite materials, the following relationship is based on collecting indentation data from experiments:

$$h_{crit} = \beta \cdot \left(\frac{E}{H}\right) \cdot \left(\frac{K_{1c}}{H}\right)^2 \tag{17.43}$$

The standard deviation of chip thickness will affect dispersion of abrasive grit sliding forces [33]. To find prior behavior of h and wear of abrasive grits, kinetics of the grinding process must be defined.

17.2.2 Kinetics of Grinding Nanocomposites

Interactions between tool and nanocomposite materials result in understanding the grinding coefficient and the instantaneous chip thickness. When compared with ideal kinematic grinding conditions, coefficients v_w, v_s, and N_r remain static during dynamic grinding (Table 17.2). The relationships between grinding force models and the effect on grinding parameters processing nanocomposite materials require further study.

Kinetic equations become more sophisticated and complicated due to the exact shape and form of the abrasive grit prior to and during dressing, and this aspect has a large effect on the kinetics and kinematics of grinding. Advances in developing abrasive grits that can be dressed to a known geometry are clearly a very important aspect of grinding composite materials with known and predictable outputs.

17.3 MATERIALS FOR GRINDING NANOCOMPOSITES

For high-strength nanocomposite materials, cubic boron nitride (cBN) and diamond reduce fiber pullout owing to their continuous sharpness and elimination of softening the resin bond due to their high thermal conductivity.

Natural diamond is used in dressing tools, whereas crushed natural diamond is used in single-layered grinding wheels requiring durable sharpness and convexity (Figure 17.5). Large natural diamonds used in dressing form in the Earth's mantle, whereas microdiamonds (< 0.5 mm) form in lamproitic magma/kimberlite, which is low in CaO, Al_2O_3, and Na_2O and has a high K_2O/Al_2O_3 ratio, and a medium-to-high MgO percentage.

Artificial diamond is synthesized at incredible temperatures (> 2500 K) and pressures (> 100,000 bar pressure) using graphite precursors. Diamond is metastable under normal operating temperatures and pressure and is cubic (sp^3 covalent, space group = Fd3m, 8 atoms per unit cell, 1.54Å ionic distance, density = 3.515 g/cm^3, crystal habit = {111}, {100}, {113}, {100}, growth and deformation twins on {111} planes, with perfect cleavage on {111}). Diamond growth conditions can be aided using metal solvents (Ni or Co). Graphite is soluble in both Ni and Co, which is better than diamond in Ni and Co. Nucleation of diamond is improved at higher temperatures due to increased binding energy.

The growth rates on cubic {100}, dodecahedron {011}, and octahedron {111} planes are dominated by whether Ni or Co are used as solvents. At low temperatures, the planes are cubic {100}, and at high

TABLE 17.2
Grinding Force Models

Investigators	Model	Grinding Interaction
Drew et al. [34]	$\begin{cases} F_n = \dfrac{u_{ch}b\delta V_w}{V_g} \\[2mm] F_t = k_1 F_z \end{cases}$	Cutting
Inasaki et al. [35]	$F_n(t) = k_g a(t) + c_g \dot{x}(t)$	Cutting
Li and Shin [11]	$\begin{cases} F_n = K_r \dfrac{2\pi}{v_s N_r}(\upsilon_t \sin\theta + \upsilon_n \cos\theta) N_a A \\[3mm] F_t = K_t \dfrac{2\pi}{v_s N_r}(\upsilon_t \sin\theta + \upsilon_n \cos\theta) N_a A \end{cases}$	Cutting
Zhu et al. [36]	$F = \mp (K_c + K_0) h_w b [x(t) - \mu(t) - T]$	Cutting
Tahvilian et al. [37]	$F_t = k_e E \left(1 - \left\| \dfrac{\varphi}{\sqrt{2h_0/R_0}} \right\|^3 \right) + k_c \dfrac{v_f}{n_c N}\left(h_0 - \dfrac{R_0 \varphi^2}{2}\right)$	Cutting
Guo et al. [15]	$\begin{cases} F_n = K(V_w/V_s)a_p + K_1(V_w/V_s)d_e^{-0.5}a_p^{0.5} + K_4(V_w/V_s)^a d_g^{b} C_s d_e^{0.5} a_p^{0.5+c} \\[2mm] F_t = K'(V_w/V_s)a_p + (K_2 + K_3 V_w/V_s d_e)d_e^{0.5}a_p^{0.5} + K_5(V_w/V_s)^a d_g^{b} C_s d_e^{0.5} a_p^{0.5+c} \end{cases}$	Cutting, Plowing, Rubbing
Gasagara et al. [38]	$\begin{cases} F_n = \left(\Phi_1 K_1 + \Phi_2 K_2 \ln\dfrac{V_c^{1.5}}{a^{0.25}V_w^{0.5}}\right)\dfrac{V_w a}{V_c}b + \left(\dfrac{4bAp_0 V_w}{V_c}\right)\left(\dfrac{a}{D_e}\right)^{1/2} \\[3mm] F_t = \left(K_1 + K_2 \ln\dfrac{V_c^{1.5}}{a^{0.25}V_w^{0.5}}\right)\dfrac{V_w a}{V_c}b + bA\left(\beta + \dfrac{4\alpha p_0 V_w}{D_e V_c}\right)(D_e a)^{1/2} \end{cases}$	Cutting, Rubbing

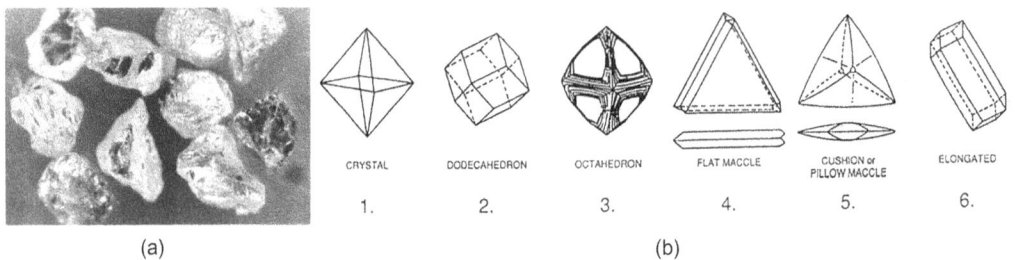

CRYSTAL   DODECAHEDRON   OCTAHEDRON   FLAT MACCLE   CUSHION or PILLOW MACCLE   ELONGATED

1.   2.   3.   4.   5.   6.

(a)   (b)

FIGURE 17.5 A collection of (a) natural diamonds with sharpness, convexity, and habit and (b) common shapes of diamonds: (1) Crystal; (2) Dodecahedron; (3) Octahedron; (4) Flat Maccle; (5) Pillow Maccle; and (6) Elongated. (Images by M. J. Jackson and M.P. Hitchiner (Saint-Gobain Abrasives). Reproduced with Permission.)

temperature, octahedra {111}. Intermediate cubo-octahedral grits used in metal bonds are used for grinding nanocomposite materials (Figure 17.6). Three types of press designs are used to make synthetic diamond: (1) belt press; (2) anvil press (tetrahedron/cubic); and (3) split-sphere (BARS) press. Belt presses (Figure 17.7a) make diamonds using an upper/lower anvil operating within a cylindrical cell.

FIGURE 17.6 Cubo-octahedral synthetic diamonds. (a) 150 # natural diamond no coating; (b) 150 # nickel coating; (c) 150 # titanium-nickel coating; (d) 150 # CVD copper coating. (Images by M.P. Hitchiner (Saint-Gobain Abrasives). Reproduced with Permission.)

Pressure is radially actuated using a steel belt [39]. Anvil presses have tetrahedral/cubic anvils applying pressure to cell faces (Figure 17.7b) and produces medium-to-high friability diamond. Split-sphere BARS presses produce very high-quality diamonds. Alternatively, cubic boron nitride (cBN or β-BN) can be used to cut nanocomposite materials. Boron nitride occurs normally in the hexagonal form (layered lattice atoms) known as h-BN (or α-BN). Cubic BN (cBN) is synthesized in the diamond range (space group = $F\overline{4}3m$, 4B and 4N atoms per unit cell, 1.57Å ionic distance, density = 3.45 g/cm^3, crystal habit = truncated tetrahedra {111}, or octahedra $\{111\}\{\overline{1}\,\overline{1}\,\overline{1}\}$, {111} and {100}, {111} and {110}, twinned on {111} planes, with perfect cleavage on {011}). cBN shows no affinity with transition metals. Therefore, Li$_3$N is used as the catalyst to produce cBN and its morphology is controlled by varying growth rates, changing temperature, and using appropriate dopants on $\{111\}\{\overline{1}\,\overline{1}\,\overline{1}\}$ and {100} planes. Growth on {111} dominates results in truncated tetrahedron morphology {111}, twinned plates {111}, and octahedra $\{111\}\{\overline{1}\,\overline{1}\,\overline{1}\}$. Grit shapes can be octahedral $\{111\}\{\overline{1}\,\overline{1}\,\overline{1}\}$, or cubo-octahedral. Grit shape variations are greater with cBN compared to diamond. Stoichiometrically stable cBN is colorless, whereas industrial cBN grits are amber, brown, and black depending on conditions of their synthesis.

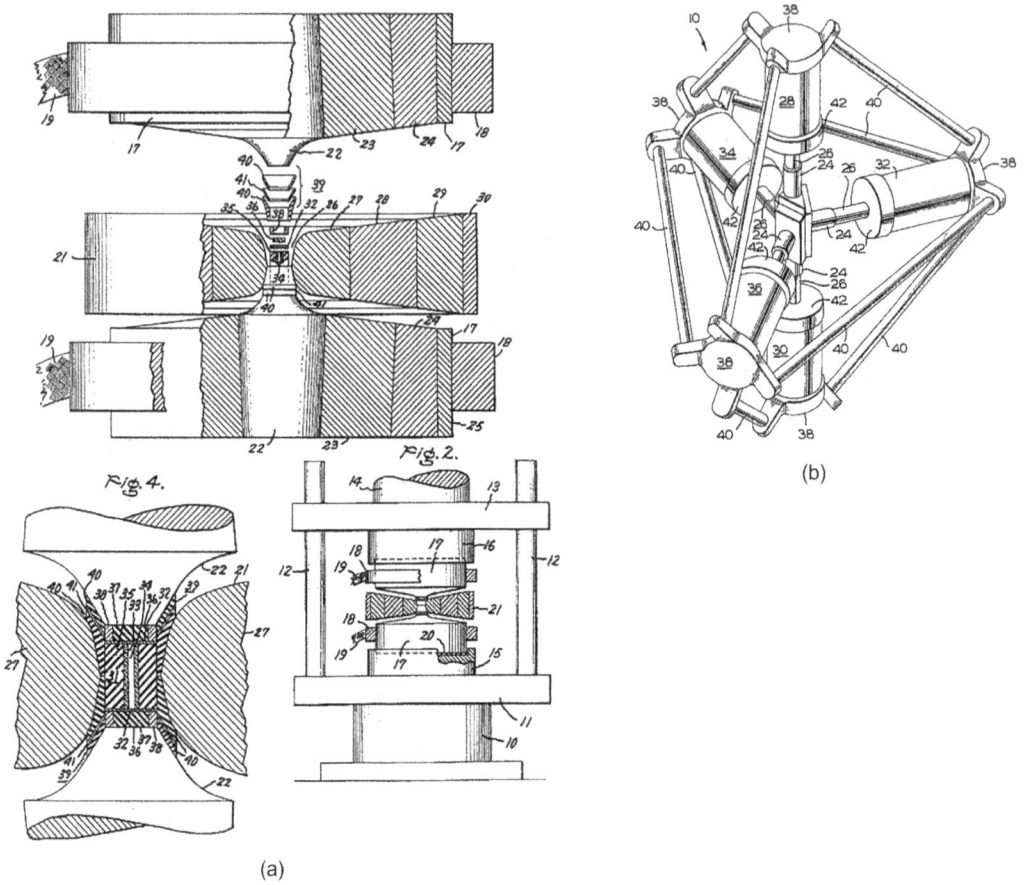

(a)

(b)

FIGURE 17.7 The belt press: (a) schematic diagram showing the cross-section of the apparatus (US Patent # 2,947,608) and tetrahedral anvil press: (b) schematic of the anvil press assembly (US Patent # 3,159,876). (US PTO—Public domain.)

17.3.1 EFFECTS OF ABRASIVE GRIT SHAPE

Grit shape changes grit strength, grinding performance, and grinding wheel structure. Shape affects grinding power, surface finish (roughness + waviness + error of form), and the force per grit. For synthetic diamond, blocky/rounded grits are stronger than angular/sharp grits. De Pellegrin et al. [40–42] describe shape/habit projections in relation to operating parameters (grit aspect ratio, projected area of grit, convexity of grit, and sharpness). Variations in diameter may be obtained but the shape of the grit governs its ability to cut, plow, and slide (tribological interactions):

1. **Grit Aspect ratio** (d_a/d_b): describes grit elongation and its effect on static and dynamic undeformed chip thickness;
2. **Projected area of grit**: used for the calculation of grit convexity;
3. **Convexity of grit**: function of grit strength and abrasiveness. Convexity implies reduced mechanical integrity and aggressiveness when angular:

$$C = \frac{A_f + A_p}{A_p}$$

(17.44)

where A_f is the area filled between grit projection and the idealized elastic membrane and A_p is the projected area of the grit; and

FIGURE 17.8 Extruded Al_2O_3 (~8:1 aspect ratio). (Image by M. P. Hitchiner (Saint-Gobain Abrasives). (Reproduced with Permission.)

4. **Grit Sharpness**: sharpness characterizes the nature of grit based on chip creation mode where abrasiveness is governed by the amount of penetration into the nanocomposite material.

Diamond grits were compared in shape (blocky-to-angular) with sharpness parameters quantifying differences in grit abrasiveness. Grit sharpness was correctly ordered for all diamond grits in terms of their wear rates, but convexity characterized the same performance in terms of the rates of wear [41,42].

However, integrity of grits changes wear rates due to differences in grinding parameters and type of nanocomposite material used for a particular application. Extruded alumina grits are very effective at grinding materials because their aspect ratios (~4:1 to 8:1) lend themselves to removing heat when grinding PMCs (Figure 17.8). More development is needed to create similar advances in aspect ratio for diamond and cBN that will significantly change the performance of diamond and cBN products when processing nanocomposite materials.

17.4 NANOCOMPOSITES FOR SPACE APPLICATIONS

The use of nanocomposites in aerospace and space applications is based on low weight, high strength and stiffness, durability, thermal stability, and performance that are critical in applications such as telescopes and antennas. In addition to the bodies of space vehicles, nanocomposites can be made as a monocoque shell, can be insulated with multiwall designs, can be wrapped with filaments, and can be designed as isotenoids, tension-shell, or tension-stringed structures. Expandable structures allow large payloads to be carried for transporting large or small structures, also made with nanocomposite materials. These can be classified as variable geometry and rigid structures for panels on satellites, elastic recovery bodies such as crew transfer tunnels, inflatable balloons such as paragliders, reentry vehicles, rigid membranes for solar collectors, or expanded honeycombs for large space bodies [43].

For aerospace applications, nanocomposites are used for radomes, landing gear doors, galley doors, tail fins, rudders, spoilers, wing tips, stabilizers, flaps, ailerons, cones, and dorsal fin assemblies [44]. The selection of the type of nanocomposite used for structural and propulsion applications is dependent on intended use, the nature of the material and its properties, its ability to be ground, and availability. The section details some of the studies recently conducted on the grindability of nanocomposite materials.

17.4.1 Polymer–Matrix Composites (PMCs)

The machining and grinding of PMCs are more complex than grinding metals [45–48]. The polymer bond is ~30% in volume to minimize fiber pullout by abrasive grits that have a tendency to soften the matrix by rubbing friction then plowing the fibers until they are pulled out by cutting grits. A soft, alumina grit wheel is recommended (WA46I8V, i.e., 46 grit size, I-hardness, 8-structure, and vitrified bond), and when compared to hard wheels, better surface finishes are recorded without melting the matrix [48]. This type of grinding wheel tends to sharpen effectively when grinding rather than dulling or blunting the sharp-edged cutting grits. Wang et al. [49] discovered that grinding forces are lower when using large abrasive grits at low concentrations. Surface finishes improved by employing smaller grit sizes that are highly concentrated [49] but tended to soften the matrix and create the conditions for fiber pullout. In general, PMCs are best ground with vitrified alumina wheels at low cutting speeds ($v_s < 30$ m/s). Typically, the specific material removal rate, Q'_w, is ~ 5–10 mm³/mm.s, and G-ratio typically 50–100.

17.4.2 Metal–Matrix Composites (MMCs)

It is reported that Al/SiC/Graphite MMCs ground with abrasive grinding grits subjected to electrical currents produce acceptable surface finish [50,51]. Kwak and Kim [52] noted that Q'_w in the range 20–50 mm³/mm.s and reasonable rotational oscillations (n_r ~1300 rpm) of the grinding wheel produce good grindability with MMCs. Resin-bonded diamond products at high speeds and depths of cut are also noted to produce excellent surface finish for MMCs [53]. Resin-bond wheels compared with single-layer electroplated wheels tended to wear excessively due to small grit clearances when grinding MMCs. Modeling techniques focusing on grinding parameters and their effect on MMC grindability showed that the sliding component of grinding energy was very small in magnitude, F_t and F_n were linear, and surface finish decreased as the MMC's hardness increased [53].

Wheel speed was affected by specific metal removal rate [54–57] by a small amount. Electrolyte concentrations affect Q'_w in the electrical discharge grinding of MMCs [50,55–58] with slotted grinding wheels [51]. Resin-bonded diamond wheels generated zero subsurface damage to nanocomposite materials and increases fatigue life compared with vitrified SiC [58]. Resin-bonded diamond grinding wheels are well matched to grind MMCs at high speeds ($v_s < 40$ m/s), with specific material removal rates in the range, Q'_w ~ 25–35 mm³/mm.s and G-ratios ~ 600–750.

17.4.3 Ceramic–Matrix Composites (CMCs)

CMCs are composites with properties dominated by hybrid construction (solid-state processing/ seeded gel formation, laser synthesis, Pechini processing, and melt synthesis, coprecipitation and hydrothermal synthesis, spray/plasma drying, ball milling and mixing, and sintering) and are well practiced in industry [59]. C/SiC CMCs are affected by fiber extrusion effects based of the direction of grinding, which dominates surface finish [60]. These problems tend to be solved when brittle grinding is promoted usually by cryogenic means [61].

Wang and Lin [62] discovered that components of grinding force (F_t and F_n) act differently in CMCs due to fiber orientation/direction. Wheel durability and stiffness studies are needed to fully understand the grinding of CMCs according to Tawakoli and Azarhoushang [63] who used fully segmented grinding wheels. Tribological interactions such as rubbing and plowing were changed due to intermittent cutting thereby improving surface finish/grinding ratio [63–65]. Polymer–concrete structures were studied for grindability with subsequent reductions in F_t and F_n [66].

The development of CMCs in the aerospace industry continues unabated [67]. Pratt and Whitney published a growth plan for geared turbofans with a focus on nanocomposite materials [68] and γ-titanium aluminides [69,70]. It should be noted that CMCs possess better grindability compared to monolithic ceramics (MC) using resin-bonded diamond grits (Figure 17.9). CMCs ground with

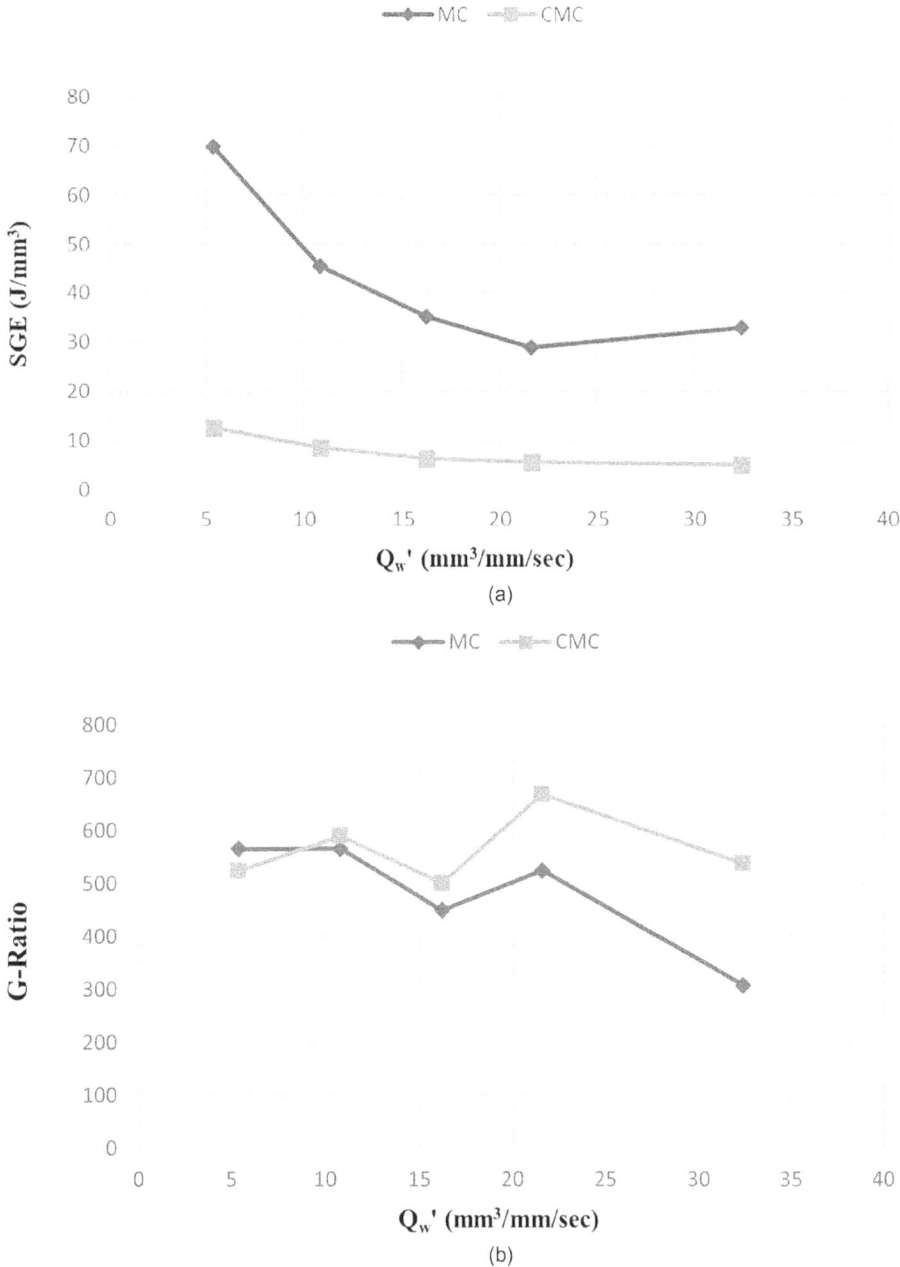

FIGURE 17.9 Grinding CMCs/MC with resin-bond wheels: (a) Grinding energy (SGE) versus specific MRR (Q'_w) and (b) grinding ratio (G-ratio) versus specific MRR (Q'_w). (Data graphs by M. J. Jackson based on data by M. P. Hitchiner (Saint-Gobain Abrasives). Reproduced with Permission.)

resin-bonded diamond wheels (v_s < 30 m/s) and specific MRR in the range, Q'_w ~ 20–23 mm³/mm.s produce G-ratios ~ 580–680.

17.5 FUTURE SCOPE OF WORK

After studying the various aspects and current knowledge in the field of grinding nanocomposites, the following future scope of work should be performed:

- Understand how the fundamental mechanisms of grinding apply to the grinding of nano-composite materials;
- Understand the effects of cryogenically cooling nanocomposite materials in terms of improving properties and grindability;
- Undertake experiments to fully understand how fibers are extruded or pulled from the bonding matrix;
- Re-examine the fundamental equations of grinding to account for fiber direction, matrix properties, and modifying effects of incorporating nanomaterials as fillers;
- Investigate the action of forces and their components on the mechanics of grinding;
- Develop machine tools for grinding nanocomposites that are dynamically stiff, powerful, and produce excellent surface finish;
- Study grinding wheel durability and create structures that can economically grinding CMCs and MMCs;
- Understand how waste can be minimized by studying the mechanisms of grinding nano-composite materials; and
- Study the grindability of nanofilled polymer–concrete composites used for machine tool structures.

REFERENCES

1. Shyha, I., Huo, D. (2021). *Advances in Machining of Composite Materials: Conventional and Non-conventional Processes*. 552 p. Springer Nature, Switzerland.
2. Gaikwad, A., Debnath, K., Gupta, M. K. (2021). Nano-structured polymer-based composites, in *Advances in Machining of Composite Materials - Conventional and Non-conventional Processes*, Edited by I. Shyha and D. Huo, Chapter 12, pp. 335–368, Springer Nature, Switzerland.
3. Shuaib, N. A. et al. (2021). Recycling of composite materials, in *Advances in Machining of Composite Materials - Conventional and Non-conventional Processes*, Edited by I. Shyha and D. Huo, Chapter 19, pp. 527–552, Springer Nature, Switzerland.
4. Jackson, M. J., Toward, M. J. (2021). Grinding and abrasive machining of composite materials, in *Advances in Machining of Composite Materials - Conventional and Non-conventional Processes*, Edited by I. Shyha and D. Huo, Chapter 16, pp. 459–484, Springer Nature, Switzerland.
5. Jackson, M. J. (2022). Fundamentals of metal-matrix diamond tools, in *The Fundamentals of Metal-Matrix Composites*, Edited by S. Ersoy (Series on Materials Science and Technologies), Chapter 2, pp. 37–84, NOVA Publishers, New York. ISBN: 978-1-68507-952-9. DOI: https://doi.org/10.52305/NQOM8549.
6. Jackson, M. J., Novakov, T. (2010). Machining ceramic matrix composites, in *Machining Composite Materials*, Edited by J. P. Davim, Chapter 6, pp. 213–256. ISTE-Wiley, London.
7. Jackson, M. J. (2020). Recent advances in ultraprecision abrasive machining processes. *SN Applied Sciences*. 2:1172.
8. Malkin, S. (2008). Grinding technology: Theory and applications of machining with abrasives. *Industrial Press/SME*. 2:1–372.
9. Warnecke, G., Zitt, U. (1998). Kinematic simulation for analyzing and predicting high-performance grinding processes. *CIRP Annals - Manufacturing Technology*. 47(1):265–270.
10. Akintseva, A. V., Pereverzev, P. P. (2017). Modeling the influence of the circle overtravel on the stability of internal grinding process. *Procedia Engineering*. 206:1179–1183.
11. Li, H., Shin, Y. C. (2006). A time-domain dynamic model for chatter prediction of cylindrical plunge grinding processes. *Journal of Manufacturing Science & Engineering*. 128:404–415.

12. Li, H., Shin, Y. (2007). A time domain dynamic simulation model for stability prediction of infeed centerless grinding processes. *Journal of Manufacturing Science and Engineering.* 129:539–550.
13. Guo, M. X., Li, B. Z., Ding, Z. S., et al. (2016). Empirical modeling of dynamic grinding force based on process analysis. *International Journal of Advanced Manufacturing Technology.* 86:3395–3405.
14. Guo, M. X., Li, B. Z. (2016). A frequency-domain grinding force model-based approach to evaluate the dynamic performance of high-speed grinding machine tools. *Machining Science & Technology.* 20:115–131.
15. Guo, M. X., Jiang, X. H., Ding, Z. S., et al. (2018). A frequency domain dynamic response approach to optimize the dynamic performance of grinding machine spindles. *International Journal of Advanced Manufacturing Technology.* 98(9–12):2737–2745.
16. Fang, C., Yang, C. B., Cai, L. G., et al. (2019). Predictive modeling of grinding force in the inner thread grinding considering the effect of grits overlapping. *International Journal of Advanced Manufacturing Technology.* 104:943–956.
17. Ma, Y. C., Yang, J. G., Li, B. Z., et al. (2017). An analytical model of grinding force based on time-varying dynamic behavior. *International Journal of Advanced Manufacturing Technology.* 89:2883–2891.
18. Reichenbach, G. S., Mayer, J. E., Shaw, M. C., (1956). The role of chip thickness in grinding. *Transactions ASME.* 18:847–850.
19. Zhang, N., Kirpitchenko, I., Liu, D. K. (2005). Dynamic model of the grinding process. *Journal of Sound & Vibration.* 280:425–432.
20. Wang, Y., Fu, Z. Q., Dong, Y. H., et al. (2018). Research on surface generating model in ultrasonic vibration-assisted grinding. *International Journal of Advanced Manufacturing Technology.* 96:3429–3436.
21. König, W., Lortz, W. (1975). Properties of cutting edges related to chip formation in grinding. *Annals of the CIRP.* 24:231–235.
22. Hecker, R. L., Liang, S. Y., Wu, X. J., et al. (2007). Grinding force and power modeling based on chip thickness analysis. *The International Journal of Advanced Manufacturing Technology.* 33:449–459.
23. Rogelio, L., Hecker, I, Ramoneda, M., Liang, S. Y. (2003). Analysis of wheel topography and grit force for grinding process modeling. *Journal of Manufacturing Processes.* 5:13–23.
24. Aurich, J. C., Kirsch, B. (2012). Kinematic simulation of high-performance grinding for analysis of chip parameters of single grits. *CIRP Journal of Manufacturing Science & Technology.* 5(3):164–174.
25. Chen, X., Öpöz, T. T. (2014). Characteristics of material removal processes in single and multiple cutting edge grit scratches. *International Journal of Abrasive Technology.* 6(3):226–242.
26. Wu, J., et al. (2019). A study on material removal mechanism of ultramicro grinding (UMG) considering tool parallel run-out and deflection. *International Journal of Advanced Manufacturing Technology.* 103:1–23.
27. Jamshidi, H., Budak, E. (2020). An analytical grinding force model based on individual grit interaction. *Journal of Materials Processing Technology.* 283. Article ID 116700. https://doi.org/10.1016/j.jmatprotec.2020.116700
28. Ding, W. F., et al. (2017). Grinding performance of textured monolayer CBN wheels: Undeformed chip thickness non-uniformity modeling and ground surface topography prediction. *International Journal of Machine Tools and Manufacture.* 122:66–80.
29. Pahlitzsch, G., Helmerdig, H. (1943). Determination and significance of chip thickness in grinding. *Workshop Technology.* 12:397–401.
30. Werner, G. (1971). Kinematic and mechanic during grinding processes. Doctoral Dissertation, TH Aachen, Germany.
31. Agarwal, S., Rao, P. V. (2013). Predictive modeling of force and power based on a new analytical undeformed chip thickness model in ceramic grinding. *International Journal of Machine Tools & Manufacture.* 65:68–78.
32. Chen, Y., et al. (2018). Quantitative impacts of regenerative vibration and abrasive wheel eccentricity on surface grinding dynamic performance. *International Journal of Advanced Manufacturing Technology.* 96:5–8.
33. Dai, C. W., et al. (2019). Grinding force and energy modeling of textured monolayer CBN wheels considering undeformed chip thickness non-uniformity. *International Journal of Mechanical Sciences.* 157:221–230.
34. Drew, S. J., Mannan, M. A., Ong, K. L., et al. (2001). The measurement of forces in grinding in the presence of vibration. *International Journal of Machine Tools and Manufacture.* 41:509–520.
35. Inasaki, I. et al. (2001). Grinding chatter – origin and suppression. *CIRP Annals - Manufacturing Technology.* 50:515–534.
36. Zhu, X. J., Wang, J. Q., Chen, Q., et al. (2011). Research on dynamic grinding force in ultrasonic honing chatter. *Key Engineering Materials.* 487:433–437.

37. Tahvilian, A. M., Hazel, B., Rafieian, F., et al. (2016). Force model for impact cutting grinding with a flexible robotic tool holder. *International Journal of Advanced Manufacturing Technology.* 85:133–147.
38. Gasagara, A., Jin, W. Y., Uwimbabazi, A., et al. (2020). Modeling of vibration condition in flat surface grinding process. *Shock and Vibration.* 2020: Article ID 3069895, 12 pages.
39. Ladd, R. (2005). Manufactured large single crystal diamond. *Finer Points - Wire Die Products and Applications.* 23–28.
40. De Pellegrin, D. V, Corbin, N. D., Baldoni, G., Torrance, A. A. (2002). The measurement and description of diamond particle shape in abrasion. *Wear.* 253:1016–1025.
41. De Pellegrin, D. V, Stachowiak, G. W. (2004) Evaluating the role of particle distribution and shape in two-body abrasion by statistical simulation. *Tribology International.* 37:255–270.
42. De Pellegrin, D. V., Corbin, N. D., Baldoni, G., Torrance, A. A. (2008). Diamond particle shape: Its measurement and influence in abrasive wear. *Tribology International.* 42:160–168.
43. Scipio, A. (1967). Structural Design Concepts – Some NASA Contributions. NASA (National Aeronautics and Space Administration), Report # NASA SP-5039. Washington D. C., USA.
44. Saha, P. K. (2017). Chapter 1 – Fundamentals of aerospace vehicles, in *Aerospace Manufacturing Processes*, pp. 1–24. CRC Press/Taylor and Francis, Boca Raton, FL.
45. Spur, G., Lachumund, U. (1998). Turning of fiber reinforced plastics, in *Machining of Ceramics and Composites*, Edited by Jahanmir et al, vol. 7, pp. 209–248. Marcel-Dekker, New York.
46. Klocke, F., et al. (1998). Milling of advanced composites, in *Machining of Ceramics and Composites*, Edited by Jahanmir et al, vol. 8, pp. 249–266. Marcel-Dekker, New York.
47. Puw, H., Hocheng, H., (1998). Milling of polymer composites, in *Machining of Ceramics and Composites*, Edited by Jahanmir et al, vol. 9, pp. 267–294. Marcel-Dekker, New York.
48. El Wakil, S. (2011). Grinding processes for polymer matrix materials, in H. Hocheng (Ed.), *Machining Technology for Composite Materials*. pp. 65–74, Woodhead Publishing, Cambridge, UK.
49. Wang, H., et al. (2016). Surface grinding of carbon-fiber reinforced plastic composites: Effect of tool variables. *Advances in Mechanical Engineering.* 8:1–14.
50. Yadav, R., Yadava, V. (2013). Influence of input machining parameters of slotted electrical discharge abrasive grinding of Al/SiC/Gr MMC. *Materials and Manufacturing Processes.* 28:1361–1369.
51. Yadav, R., Yadava, V. (2013). Multiobjective optimization of slotted electrical discharge abrasive grinding of MMCs using artificial neural network and non-dominated sorting generic algorithm. *Proceedings of the Institution of Mechanical Engineers*, Part B: Journal of Engineering Manufacture. 227:1442–1452.
52. Kwak, J., Kim, Y., (2008). Mechanical properties and grinding performance on aluminium-based MMCs. *Journal of Materials Processing Technology.* 201:596–600.
53. Anand Ronal, B., et al. (2009). Studies on the influence of grinding wheel bond material on the grindability of MMCs. *Materials and Design.* 30:679–686.
54. Shristastava, P., Dubey, A. (2013). Experimental modelling and optimization of electric discharge diamond face grinding of MMCs. *International Journal of Advanced Manufacturing Technology.* 69:2471–2480.
55. Yadav, R., Yadava, V. (2017). Experimental investigation of electrical discharge diamond peripheral surface grinding of hybrid MMCs. *Journal of Manufacturing Processes.* 27:241–251.
56. Yadav, R., Yadava, V. (2017). Performance study of electrical discharge diamond face surface grinding on hybrid MMCs. *Journal of Mechanical Science and Technology.* 31:317–325.
57. Liu, J., et al. (2013). Grinding aided electrochemical discharge machining of particulate reinforced MMCs. *International Journal of Advanced Manufacturing Technology.* 68:2349–2357.
58. Zong, Z. (2003). Grinding of aluminium based MMCs reinforced with alumina and SiC particles. *International Journal of Advanced Manufacturing Technology.* 21:79–83.
59. Singh, N., et al. (2017). Ceramic matrix composites: Processing techniques and recent advancements. *Journal of Materials and Environmental Sciences.* 8:1654–1660.
60. Du, J., et al. (2018). New observations of the fiber orientations effect on machinability in grinding C/SiC CMCs. *Ceramics International.* 44:13916–13928.
61. Singh, S., et al. (2010). Grindability improvement of composite ceramic with cryogenic coolant. *Proceedings of World Congress on Engineering*, Vol. II, 1–5, London, UK.
62. Wang, Y., Lin, B. (2012). Research on the grinding force and surface morphology of fiber-reinforced CMCs. *Advanced Materials Design and Mechanics.* 569:131–135.
63. Tawakoli, T., Azarhoushang, B. (2011). Intermittent grinding of CMCs utilizing a segmented wheel. *International Journal of Machine Tools and Manufacture.* 51:112–119.
64. Azarhoushang, B., Tawakoli, T. (2011). Development of a novel ultrasonic unit for grinding CMCs. *International Journal of Advanced Manufacturing Technology.* 57:945–955.

65. Cao, X., et al. (2013). A study on grinding surface waviness of woven CMCs. *Applied Surface Science*. 279:503–512.

66. Shamray, S., et al. (2016). High efficiency, high speed grinding of a composite wheel consisting of polymer concrete and steel structures. *Procedia CIRP*. 46:607–610.

67. Boyer, R. R., et al. (2015). Materials considerations for aerospace applications. *MRS Bulletin*, 40:1055–1066.

68. Anon., Pratt & Whitney's Geared Turbofan Growth Plan, Aviation Week & Space Technology, (2013). https://aviationweek.com/awin/pratt-whitney-s-geared-turbofan-growth-plan.

69. Hitchiner, M., Besse, J., Varghese, K. (2012). Grinding innovation in a growing defense market. *MFG4The Future Conference*, 1–9, Hartford, Connecticut, USA.

70. Jackson, M. J. (2021). Nanostructured grinding wheels for ultra-precision engineering applications. *Nanotechnology and Precision Engineering*. 4:035001.

18 Future Trends in Polymer Nanocomposites

Ankur Shukla and Karan Chandrakar

CONTENTS

18.1 INTRODUCTION

In order to create a new material with a variety of special features, one or multiple nanostructured materials (inorganic or organic) are combined with ceramic, metal, or polymer to create nanocomposites. Numerous scientific disciplines and industries use nanocomposites. Nanotechnology improvements and ongoing research could lead to more nanocomposites that can be used for many applications.

The market for polymer nanocomposites, which was estimated to be USD 8.66 billion in 2021 (Figure 18.1), would increase at a pace of over 19.1% compounded annual growth rate from 2022 to 2028 as a result of rising government initiatives for lightweight vehicles in developing nations like China, India, Japan, and Indonesia [1]. The increasing demand from the packaging and automotive industries worldwide will cause the market share for polymer nanocomposites to increase drastically. Polymer nanocomposites are used in the automobile sector to reduce component wear and corrosion, as well as weight and carbon emissions. Another significant aspect contributing to the growth of the nanomaterial sector is the rise of research and development efforts in the field

DOI: 10.1201/9781003343912-18

of nanotechnology. The U.S., Germany, and other developed nations are making investments to increase their capacity for producing and researching nanomaterials. The strict government rules relating to nanoparticles are the main obstacle expected to hinder the demand for the product.

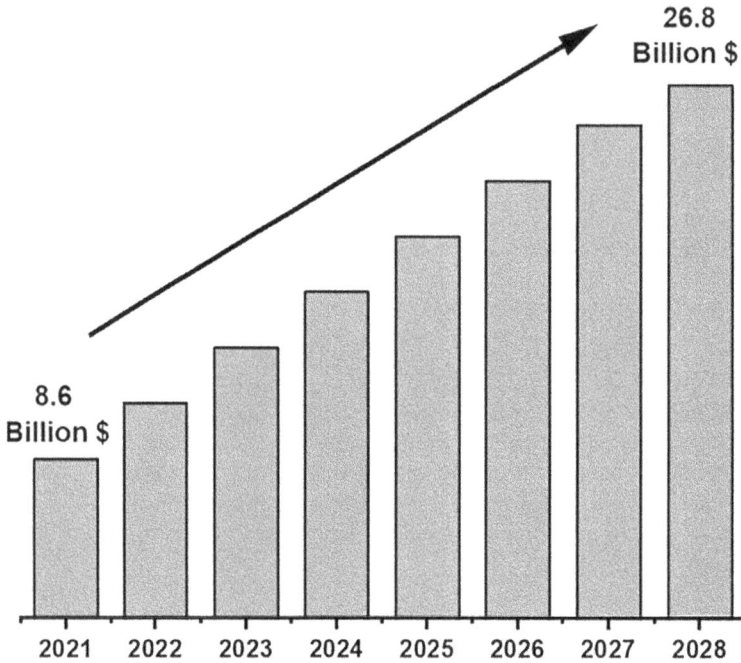

FIGURE 18.1   Estimated growth of nanocomposite market value.

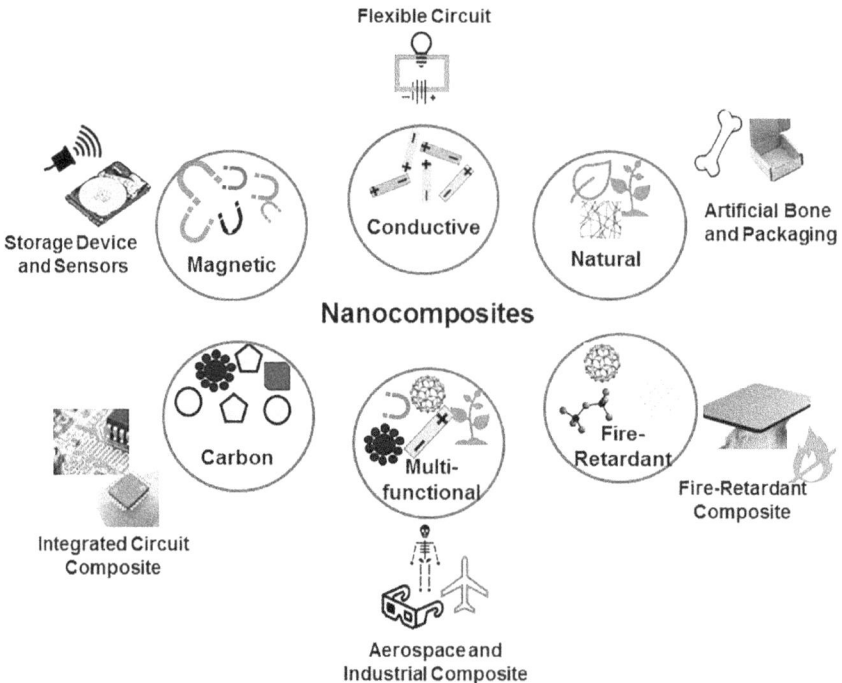

FIGURE 18.2   Future application areas for polymer nanocomposites.

Figure 18.2 shows various application areas for polymer nanocomposites such as multifunctional polymer nanocomposites, conductive polymer nanocomposites, magnetic devices and sensors, fire-retardant polymer nanocomposites, and natural polymer nanocomposites. Nanofibers, nanorods, nanoparticles, and carbon nanotubes (CNTs) are a few examples of nanocomposite materials' constituent parts. The synergistic properties of nanocomposite are one of the key characteristics of polymer nanocomposites. The multifunctional characteristics of nanocomposites include strong mechanical strength, excellent electrical conductivity, and improved optical qualities.

18.2 MAGNETIC NANOCOMPOSITES

Polymer nanocomposites have attracted significant scientific and technological attention in the last two decades. Polymers are frequently fortified with fillers of varying sizes to alleviate some limitations and expand their applications. Utilizing fillers in nanoscale to enhance polymers' physical and mechanical properties has resulted in a radical departure from conventional polymer composites [2,3]. The nanoscale fillers vary from isotropic to highly anisotropic sheet-like or needle-like morphologies and have at least one characteristic length scale of the order of nanometers. Nanoscience and nanotechnology allow for novel opportunities to produce polymer nanocomposites with interesting properties by combining nanoscale fillers and polymer materials in revolutionary ways. Recent developments in magnetic polymer nanocomposites have shown application potential in magnetic sensors, magnetic carriers, color imaging, magnetic storage, high-density magnetic recording, and electronic devices [4–9].

In situ nanoparticle synthesis in the presence of polymers, ex situ and in situ monomer polymerization in the presence of nanoparticles, in situ drop casting, dip coating, and spin coating have all been developed for the synthesis and processing of polymer nanocomposites. The synthesis and processing of magnetic polymer nanocomposites can also be carried out in various other ways. Electrical conductivity and magnetic properties are exhibited in the production of magnetic polymer nanocomposites. Polypyrrole nanocomposites filled with γ-Fe_2O_3 nanoparticles were fabricated by the oxidation method in an aqueous solution using ultraviolet radiation or using chemical oxidants and mechanical stirring [10]. A dispersion of Fe_2O_3 nanoparticles (5–25 nm) was produced by sonicating 20 mL of deionized water containing the necessary amount of Fe_2O_3. Under sonication, p-toluenesulfonic acid (6.0 mmol) and pyrrole (7.3 mmol) were added to the solution. Ammonium persulfate (3.6 mmol) was rapidly added to the previous solution at room temperature, and the mixture was sonicated for 1 hour. In another investigation, mechanical and ultrasonic stirring were utilized to examine the effect of nanocomposites' reaction time. The products were washed thoroughly with deionized water to remove unreacted components. The precipitated powder was dried thoroughly. Such methods involving the use of oxidants are known as solution-based oxidation methods. Typically, nanoparticles are disseminated in an oil phase in emulsions. Using ferric chloride as the oxidant, a microemulsion-based polymerization process was also reported to produce Fe_2O_3-filled magnetic PPy nanocomposites [11]. Surfactants and cosurfactants were mixed with the oxidant solution, and a pyrrole monomer was added. After the 24-hour polymerization process, the system was quenched with an organic solvent to remove the unreacted compounds. Electrochemical polymerization is another approach that can be utilized to create multilayered structures on conductive polymer films as substrates [12]. Yan et al. [12] used a highly doped PPy sheet as a substrate for multilayered Co and Cu formations. Usually, a two-step process is used to fabricate PPy–Fe_2O_3 nanocomposites [13]. In their study, superparamagnetic conductive polyester textile composites were fabricated using a two-step deposition technique, namely, magnetite and PPy deposition. While preparing magnetite–textile fiber composites, the polymer textile substrate (10- to 20-m fiber diameter) was immersed in the magnetite dispersion for 30 minutes and dried for several hours, followed by further drying. Magnetite-impregnated fibers were used for conductive polymer coating. A mixture of naphthalenedisulfonic acid, 5-sulfosalicylic acid, and ferric chloride was used to impregnate the fibers with the prepared solution, to which pyrrole was added later. But Jarjayes et al. [14] synthesized PPy–Fe_2O_3 nanocomposites in a single step electrochemical process. The electrolysis of a non-stirred aqueous solution of 0.5 mol/L pyrrole and ferrofluid formed ferrofluid-containing PPy sheets over an ITO anode. The films were dried at 60°C after leaving the solution aside for one night. The method was also used by Yan et al. [15] for the deposition of multiple layers of polypyrrole with magnetite (Fe_3O_4) over a stainless steel working electrode in the presence of p-toluenesulfonic acid. The galvanostatic method was used for polymerization, where a constant was applied across the electrodes in the solution. The passed electric charge controlled the thickness of the deposited film during the electrochemical polymerization. The polypyrrole-covered electrode was used for deposition of magnetite, and the process was repeated further to obtain a multilayer sandwich structure.

The polymerization is generally carried out using the UV irradiation technique [4]. The process involves the addition of a photoinitiator such as the cationic initiator Cyracure UV from Dow to the pyrrole solution. The monomer solution usually contains a salt such as silver nitrate, which acts as the electron acceptor for photopolymerization of pyrrole. The process is carried out overnight in UV light (365 nm) [16]. Another fast way to produce nanocomposites at a low cost is electrospinning. The process was first reported in 1934 [17]. Electrospinning has been investigated for producing pure polyacrylonitrile (PAN) fibers and magnetic PAN–Fe_3O_4 nanocomposite fibers. A consistent, bead-free fiber manufacturing method is obtained by adjusting a large number of electrospinning variables. Although the process is quite sensitive, slight modifications in operating settings might result in substantial changes in fiber shape [18].

Magnetic nanocomposite materials often respond to external magnetic fields, also known as the giant magnetoresistance effect. In the absence of an external field, weak antiferromagnetic coupling between adjacent ferromagnetic layers causes antiparallel magnetization orientations. Such materials have many applications, such as reading heads in modern hard drives for magnetic storage, magnetic random access memories, and biological detection [19–21].

18.3 MAGNETIC POLYMER NANOCOMPOSITES

Multicomponent materials called magnetic nanocomposites often comprise nanosized magnetic components to stimulate external stimuli (i.e., static externally or alternating magnetic field). Until now, the search for novel nanocomposites has resulted in the blending of a wide range of materials (such as liquid crystals, silica carbon, gels, metal–organic frameworks, renewable polymers) with different kinds of magnetic particles, opening up exciting possibilities for fundamental research as well as for use in a variety of fields, such as medical therapy and diagnosis, separations, actuation, or catalysis. To emphasize broad ideas and recent developments in the creation of magnetic nanocomposites and the synthesis of magnetic nanoparticles, we have chosen a few of the most recent instances for this overview (Figure 18.3).

Ferrimagnetic and ferromagnetic materials exhibit fascinating magnetic properties at the nanoscale. In order to create integrated functional systems with distinct magnetic properties, nanoparticles of diverse magnetic materials have been incorporated into extended matrix materials (i.e., inorganic and organic polymers). Magnetite (Fe_3O_4) and maghemite (Y-Fe_2O_3) nanoparticles have attracted a lot of curiosity in both fundamental research due to their distinctive features, and the diverse biological and technical applications (i.e., a magnetic moment coupled with minimal toxicity and strong chemical stability). Co-precipitating Fe^{2+} and Fe^{3+} salts in an aqueous, essential medium have historically been used to create iron oxide particles. Although this method is straightforward, affordable, and scalable, it is frequently challenging to manage Fe_3O_4 nanocrystal properties. Iron oleate was discovered to be an intriguing iron precursor that creates iron oxide nanoparticles that are crystalline, monodisperse, and spherical and come in a range of sizes when they are thermally broken down in highly flammable, organic solvents. These iron oxide nanoparticles' phase composition and crystal structure have recently been the subject of extensive research. While the shape anisotropy for nanorods increases with the aspect ratio and is zero for spherical nanoparticles, the energy of the shape magnetic anisotropy rises with the morphological anisotropy of the nanoparticles. Significant progress has been recently made in the shape-controlled synthesis

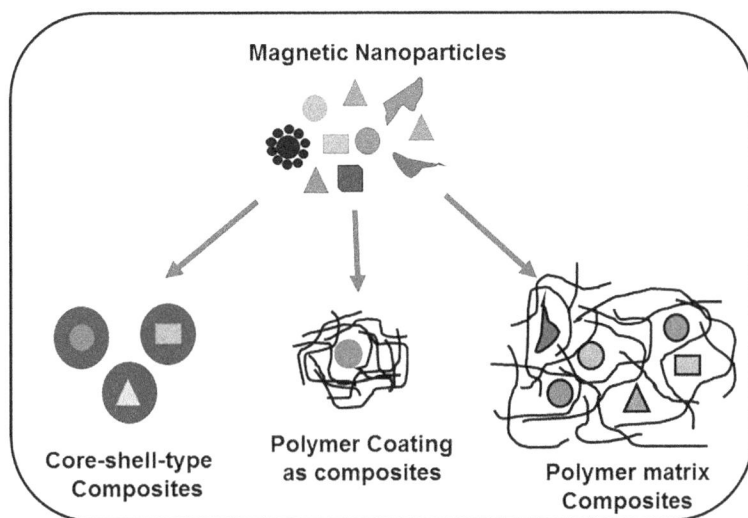

FIGURE 18.3 Magnetic nanoparticles and its nanocomposites.

of anisotropic Fe_2O_3 or Fe_3O_4 nanorods with controllable aspect ratios, also known as magnetic iron oxide nanocrystals [22]. Recent interest has also been generated by spindle-like hematite nanoparticles (i.e., Fe_2O_3) because of their mild ferromagnetic characteristics resulting from canted antiferromagnetism [23].

Organic ionomers and magnetic nanocomposites made of Fe_3O_4 and $CoFe_2O_4$ nanoparticles are intriguing possibilities for self-healing materials [24]. Ionic groups, either as side chains or as the backbone of the polymer, and unipolar polymer backbones are the main building blocks of organic ionomers. Segments of the polar ionic groups act as physical cross-linkers. The hyperthermal heating might expedite the self-healing process with the magnetic nanoparticles since these cross-links were thermally reversible. A nanocomposite's magnetic performance typically scales with the number of magnetic nanoparticles. However, particle aggregation may happen at high particle loadings (>10 vol%) because of the particles' poor compatibility with the polymer matrix.

18.4 CONDUCTIVE FILLER-BASED NANOCOMPOSITES

Conductive filler-based nanocomposites have gained a great deal of interest over the past few decades because they can give materials with superior processability and a variety of capabilities besides electrical conductivity. It is known that conductive networks in the polymer matrix dominate the electrical resistance of such nanocomposites. Synergistically combining the strengths of many constituents and optimizing the desired qualities are the theory behind producing high-performance nanocomposite materials. The properties of nanocomposites vary based on the type of fillers such as CNTs, multi-walled CNT, carbon black (CB), and graphene. Elastomers, thermoplastics, epoxy, block copolymers, and hydro/aerogels are some of the structural polymeric matrices for advanced nanocomposites utilized extensively due to their elastic physical and chemical properties [25–27]. The matrices may be elastomeric or sometimes rigid with high mechanical strength.

Nanocomposite materials containing nanoscale fillers benefit from significantly higher attainable loads, a larger specific interfacial area, controlled interfacial interactions, and greater overall compliance. CNTs have not yet been used to their full potential as reinforcements and conductors. The mechanical properties of nanocomposites, for instance, continue to be significantly inferior to their theoretical estimates. The most significant obstacles are obtaining a high-quality dispersion, proper alignment, and a strong interfacial bond between the CNT and the polymer. In addition, there are still several difficulties associated with preserving the original length of the CNT throughout the process steps. It has been demonstrated that nanotubes nucleate, producing an ordered CNT coating that maximizes CNT and polymer binding [28]. Changes in the polymer structure, such as the addition of reactive functionalities or the alteration of the pH of the solution, maybe promising research avenues. Peeterbroeck et al. [29] discovered that increasing the amount of polar vinyl acetate (VA) in ethylene vinyl acetate tends to promote the dispersion of CNTs marginally. Grunlan et al. [30] proposed a pH-sensitive polymer to separate and distribute CNTs throughout the polymer matrix, such that at low pH, the polymer is uncharged and has a helical structure. As pH increases, the polymer loses a proton and becomes negatively charged, causing extension of the polymer chains and repulsion, resulting in dispersion of the graft CNTs. Using conductive CNTs in conjunction with conductive polymer matrices such as polyaniline and polypyrrole could create materials that combine structural response with sensor functionality. Small amounts of CNTs added to polyamide 6 are known to improve the fire resistance of polymeric systems by increasing the viscosity of the melt, hence preventing dripping and flow under or near fire conditions. In addition, the CNT capacity to form a percolation network permits faster and more effective heat dissipation, hence boosting flame retardancy. Alternative energy sources that rely on high conductivity and charge capacitance open the way to developing new composite electrode materials, proton exchange membranes for fuel cells, and solar cells for photovoltaic devices. Integrating CNTs into conventional composites has great promise for enhancing functionality, compression and shear response, and even through-thickness properties. However, the fundamental obstacles are still those associated with processing, beginning with the processing of CNTs at scale, continuing through their dispersion in melts and solvents, and culminating in their alignment in actual products at the necessary levels and with the intended functionality. Resolving these concerns will unquestionably create new opportunities for creating innovative materials customized to specific uses. Another promising material is graphene. The Nobel Prize in Physics awarded for research in graphene has captivated the materials science community and attracted several new research organizations to the topic. Graphene is typically created through graphite or chemical vapor deposition (CVD), mechanical exfoliation, or chemical vapor deposition [31]. Graphene oxide is one of the most common graphene derivatives used to create polymer–graphene nanocomposites. In many instances, investigations indicate that graphene oxide compounds are biocompatible along with superior mechanical and regulated chemical characteristics [32]. Inducing edge and surface functions on graphene sheets, the bonding energy and shear strength have been greatly enhanced, according to molecular simulations, which is crucial for their incorporation into the polymer matrix [33]. The most often used nanomaterial in industry is CB, which is used for electrical modification of polymer matrices, to impart thermal performance of polymeric materials, and mechanical reinforcement and damping in synthetic rubbers [34–36].

18.5   MULTIFUNCTIONAL POLYMER NANOCOMPOSITES

Lan et al. have shown that the thermosetting styrene-based Shape Memory Polymer (SMP) exceeds the thermoplastic polyurethane-based SMP in terms of mechanical strength and moisture resistance. A new system of nanocarbon-filled thermosetting styrene-based SMP was presented [37,38]. High and consistent electrical conductivity is essential for the actual uses of Shape Memory Polymer Composites (SMPCs). Consequently, an SMP filled with nanocarbon particles at a 10 vol% concentration is usually employed for shape recovery properties. The electric current's heating effect helps quick recovery of the SMP nanocomposite. The conductivity of composites containing nanoparticles such as CB is greater than that of composites having microscale conductive filler. As the filler content increases, the volume resistivity lowers, and a synergy resulting from hybrid fiber-type systems results in lower resistivity as the high filler content occurs. In the case of short carbon fiber, the natural fibrillar structure has a greater propensity to form a three-dimensional network in composites, resulting in superior electrical responsiveness compared to particle fillers. The rate of shape recovery heavily depend on the size of the applied voltage and the electrical resistivity of the SMP nanocomposite. The short fibers are haphazardly dispersed. Numerous linkages between the fibers generate conductive networks that explain the composites loaded with short carbon fiber's superior electrical conductivity.

18.5.1 Polymer Nanocomposite-Based Fuel Cells

These nanocomposite membranes can utilize various inorganic materials, which can be classified as (a) hygroscopic metal oxide nanoparticles, which improve water retention at high temperatures, and (b) inorganic proton conductors with a good conductivity that is barely humidity sensitive. These include heteropolyacids, solid anhydrous acids, and layered metal phosphates or their phosphonates as nanoparticles or layered sheets. The use of metal oxides as inorganic fillers in polymer composites is widespread due to their high hygroscopicity and strength. Silicon dioxide (SiO_2), titanium dioxide (TiO_2), and zirconium dioxide (ZrO_2) are the most common and have various advantages. Metal oxides are incorporated into polymer membranes to (a) increase proton conductivity and maintain water retention, (b) raise fuel crossover resistance by putting up barriers in the flow channels at high temperatures and low relative humidity, and (c) enhance thermomechanical properties such as rigidity and thermal degradation at high temperatures. Due to increased water adsorption on the hygroscopic metal oxide surfaces, nanocomposites with a small number of metal oxides have enhanced water retention and improved membrane performance above 100°C. This improves the back-diffusion of water produced at the cathode and reduces the electro-osmotic drag from the anode to the cathode [39,40]. TiO_2 is also an excellent hydrophilic filler for polymer membranes, and its composite matrix has increased water-absorbent capabilities and enhanced mechanical and thermal properties. Too much TiO_2 inhibits component mixing, resulting in a fragile membrane [41–43]. Polymer electrolyte membranes must have a high proton conductivity at a low water content, long-term durability under fuel-cell working conditions, and a low fuel permeability. Nanocomposite membranes may be the most promising alternatives for resolving issues with currently utilized sulfonated membranes, such as Nafion and sulfonated hydrocarbons. Presently, most nanocomposites are composed of sulfonic acid moieties in polymer membranes and functionalized inorganic materials, necessitating the use of a solvent such as water to dissociate and conduct protons. To achieve a high functionalization level for organic and inorganic molecules without compromising membrane stability, additional research is required. Phosphoric acids or phosphonic acidic moieties may be excellent candidates for proton-conducting moieties due to self-ionization and dehydration in the absence of a solvent, as well as the ability to construct more complicated polymer topologies.

18.5.2 Polymer Nanocomposites for Aerospace

It is believed that blending a polymer with nanoscale fillers is the best way to create multifunctional characteristics. Fillers such as carbon nanofibers and CNTs offer novel opportunities for lightweight composites with exceptional mechanical, electrical, and thermal properties. Weakly interacting nanotube bundles and aggregates result in a dispersion state that drastically reduces the aspect ratio of the reinforcement. Gojny et al. [44] examined the interlaminar shear strength of nano-reinforced FRPs (Fiber Reinforced Polymers) and created a method (mini-calendering) for distributing epoxy resins include nanoparticles made of carbon. The fabrication of many nanocomposites was made possible by using a mini-calender to scatter CNTs (and CB). This technique was successful in obtaining a sufficient condition of dispersion. Even with a low nanotube percentage, the resulting nanotube/epoxy composites exhibit a considerable increase in fracture toughness and stiffness.

With as little as 0.3% of DWCNT-NH$_2$ (amino-functionalized double-wall CNTs) in the epoxy matrix, the interlaminar shear strength of a glass fiber-reinforced polymer composite increased by 19% [45]. Typical fiber-reinforced composite materials with excellent in-plane properties do not work well when out-of-plane through-thickness capabilities are required. Multifunctional materials might be created by combining a nanotube-modified matrix with fiber reinforcements (such carbon, glass, or aramid fibers). Spindler-Rantan and Bakis [46] demonstrated one application of CNT-reinforced polymers for filament-wound CFRP. One percent of single-wall nanotubes were incorporated into an epoxy matrix. Future benefits of using polymer nanocomposites in aerospace structures include faster, safer, and less expensive transportation [47–49]. Experiments on a CNT-based nanotube composite with complete integration have revealed a remarkable improvement in mechanical characteristics [50]. Examining the durability of these nanocomposites under varied environmental situations is vital to expand the utility of these hybrid materials.

18.6   NANOCOMPOSITES AND AIRCRAFTS

Nanocomposite made of poly(furfuryl alcohol), which was created utilizing renewable and bio-based resources, is acceptable for airplane interiors [51]. This novel nanocomposite has demonstrated strong mechanical and fire resistance. Additionally, this material has a higher peel strength, which considerably improves the product's integrity and durability. Previous research has shown that nanocomposites have aided in manufacturing lightweight, corrosion-resistant components without sacrificing the dependability and longevity of products related to airplanes. Some nanocomposites, such as multi-walled CNTs and acrylonitrile butadiene styrene/montmorillonite, are more durable, mechanically stronger, flame-resistant, chemically resistant, and thermally stable and utilized in airplanes. Nanocomposite coverings shield the spacecraft's structural components from the hostile conditions of the orbit. According to experts, the global nano market will grow as more industries, including packaging, aerospace, energy, automotive, and electronics, employ nanocomposites [52].

18.7   BIO-NANOCOMPOSITE AS A PLASTIC PACKAGING MATERIAL REPLACEMENT

Starch, carboxymethyl cellulose, and chitosan are examples of biopolymers that have been combined with nanomaterials to create bio-nanocomposite, which has the potential to address a significant environmental risk: the excessive use of non-biodegradable plastics. Nanoparticles give the nanocomposite its thermal, mechanical, and gas barrier qualities, while the biopolymer makes it biodegradable and non-toxic. Additionally, lightweight bio-nanocomposites have greater applicability. The challenging part is getting nanomaterials and polymer matrices together as part of creating innovative bio-nanocomposites. This nanomaterial is used in packaging materials, and due to its antibacterial properties, mechanical strength, thermal stability, and biodegradability, it has the potential to replace plastic packaging. The food industry might benefit significantly from this kind of intelligent packaging. Clay nanocomposites made of polylactic acid (PLA) are similar to ordinary polymers but have better tensile, mechanical, and thermal properties. Additionally, creating nanocomposites from more than two functional nanocomponents helps to successfully meet the design and strength criteria for a given packaging application [53].

18.8   MEMBRANES MADE OF NANOMATERIALS FOR SUSTAINABLE WATER PURIFICATION

Membrane technology offers a dependable and durable alternative for sustainable desalination to produce freshwater. Membrane biofouling is one issue; nevertheless, that hinders the efficacy of this device. The creation of a thin-film nanocomposite structure by incorporating nanoparticles into the polyamide layer has broadened the choices for the creation of antibiofouling membranes using nanocomposites [54].

18.9 ELECTRONICS AND POLYMER NANOCOMPOSITES

Modern electronic gadgets heavily rely on polymers. Polymer nanocomposites have recently received much attention from several scientific fields due to their excellent processing qualities and outstanding functionality. One of this nanocomposite's most crucial characteristics is electrical conductivity, which is used to create numerous new sensitive sensors that can detect critical physical parameters like pressure, temperature, solvent, or vapor. Hybrid nanocomposites have demonstrated exceptional performance with uncomplicated production, high mechanical flexibility, and low cost for creating non-volatile memory devices [55].

18.10 AGRICULTURAL APPLICATIONS

Synthetic fertilizers and insecticides are used by farmers to boost crop production and protect it from a pathogenic infestation to meet the rising demand for food brought on by the ongoing population rise. However, these contaminants negatively impact the soil and subsurface water source. Many of the substances have also reportedly been linked to cancer. By offering protection and boosting crop output, it has been demonstrated that nanocomposites have a considerable impact on encouraging sustainable agriculture [56].

18.11 NANOCOMPOSITES AND SENSORS

Researchers have created several enzyme sensors based on nanocomposite materials to detect diverse metabolites. In order to detect glucose with improved sensitivity and throughout a more extensive detection range of 0.08–12 mM, nanocomposite electrodes made of graphene and chitosan have been created [57]. An enzymatic biosensor for choline and acetylcholine detection was made using a different nanocomposite, one with incorporated CNTs and metal oxide nanocrystals [58]. These nanomaterials made of composite and metal oxide served as electrode modifiers, and this might be because of their excellent biocompatibility, high surface area, and high electrical conductivity, which accelerated the pace of the electron transit between choline and electrode surfaces. Considering the importance of choline for health and its drastic consequences when inadequate, it is essential to conduct additional research on choline sensors using various nanomaterials and real materials [59].

18.12 CARBON-BASED NANOCOMPOSITES

Composite materials with attachments for CNTs, nanodiamonds, and graphene have been seen as possible future developments. Carbon nanocomposites have attracted a lot of interest in a variety of fields, including biological applications, due to their amazing structural dimensions and exceptional mechanical, electrical, thermal, optical, and chemical properties. The significant developments in carbon nanocomposite technology over the past few years, along with the identification of new nanocomposite processing techniques, are done in various research work to enhance by providing suitable synthesis methods and facilitating the fabrication of diverse composites based on carbon nanomaterials. It is possible to increase the functional impact of nanotube and graphene composites. Recent studies have been focusing on fabrication of hybrid nanocomposite heterostructure electrodes for supercapacitor applications where graphene oxide is embedded in the metal–organic framework such as $MnCo_2O_4$/Ni-V-MOF [60]. In applications regarding lithium ion batteries, copper phosphide and rGO (reduced graphene oxide) composites as anode materials showed 716.8 mAh specific capacity for up to 200 cycles, which might offer new opportunities for the development of batteries [61].

Carbon nanocomposite is used in a number of industries, including aerospace, batteries, the chemical industry, fuel cells, optics, power generation, space, solar hydrogen, sensors, and thermoelectric devices. Active carbon, CB, graphene, nanodiamonds, and CNTs are some of the carbon nanocomposites currently being developed, produced, and used. It has been discovered that in contrast to conventional fiber composites, van der Waals force interfacial compounds significantly impact the mechanical performance of composites based on carbon nanomaterials.

Polymer nanocomposites have several expected advantages for aerospace constructions, chief among them the provision of future safer, quicker, and ultimately less expensive transportation. The most noticeable of these is a considerable reduction in the airframe weight.

Using actual items is still in its infancy, and much work needs to be done. The mechanical properties of a wholly integrated nanotube composite made of CNTs have been dramatically improved in experiments. O'Donnell et al. [62] investigated CNT reinforcement-based aircraft to illustrate potential airframe uses. In the low initial take-off mass category, every modeled airframe has seen an average weight reduction of 17.32%, whereas for the high initial take-off mass category, the lowest mass reduction was above 10%. All CNRP-structured airframes experienced an average fuel savings of 9.8%. Although it is doubtful that CNT-reinforced polymer structures would replace existing airframes, this analysis gives insight into a select number of advantages when a nanostructured material is used on a larger scale. Nanotubes can be utilized for sensitive electrochemical sensors by fusing them with conductive polymers.

Nanotubes can be utilized in electrochemical sensors by fusing them with conductive polymers, such as polyaniline (PANI). Even at high temperatures, increasing nanotube concentration increases PANI/nanotube layer conductivity and metal/semiconductor device current density. Coiled nanotubes as a mechanical resonance sensor by Volodin et al. [63] is a noteworthy accomplishment among numerous intriguing uses. Power devices like ultracapacitors benefit significantly from using CNTs. Utilizing CNTs to improve the performance of these devices beyond the state-of-the-art, or even just replacing conventional materials with them in some circumstances, is no longer considered an idea but is quickly becoming a reality.

Many aircraft and armored vehicle components have been given a performance boost by applying a variety of sophisticated composite coatings. In many prospective aircraft applications, coatings play a vital role in corrosion sensing of the outer layer of aircrafts. The creation of nanoparticles and nanocomposites, made up of nanoparticles and matrix materials, is needed to create protective coatings. Traditional paints frequently pose health risks and require much labor to create and apply. CNTs can be substituted for CB in powder paints to generate corrosion-resistant paints and surfaces. Other applications involve varied particles' impact on durability, thermal conductivity, and various stealth effects.

According to experimental findings, using CB enhances skid resistance. According to several researchers, adding nanoparticles to conventional FRPs improves properties at lower filler loadings compared to their macroscale and microscale counterparts. Since blending a polymer with nanoscale fillers gives it multifunctionality, nanofillers, especially high-aspect-ratio CNTs, may improve matrix properties by fusing between microscale fibers, especially in the z-direction. The literature on mechanical and functional qualities and manufacturing techniques, however, contains several contradictions and doubts. To provide a thorough knowledge and accomplish the required multifunctionality in these materials, examining these issues in details is vital.

Although CNTs have excellent strength, modulus, electrical, and thermal conductivities in addition to having a low density, their usefulness after being incorporated into polymers is yet to be fully realized. The majority of investigations have noted increases in stiffness and strength, while results for toughness are less solid and call for additional property optimization research.

In order to broaden the utility of these hybrid materials, examining the resilience of these nanocomposites in various environmental situations is imperative. Lack of understanding of polymer nanocomposites' degradation mechanisms has led to the development of more effective stabilizers to improve product performance and sensitizers to make degradable polymers to safeguard the environment. Both negative and positive features of polymer breakdown might occur. If left unchecked, it can harm polymer nanocomposites' performance, posing a fire safety risk or dangerous threat, but if correctly controlled, it can be employed to create new and improved materials. In order to broaden the utility of these hybrid materials, examining the resilience of these nanocomposites in various environmental situations is imperative.

CNTs, in particular, offer great potential for changing polymers. In terms of mechanical qualities, especially toughness, they are excellent fillers. They also make it possible to add functional qualities to polymeric matrices related to their electrical conductivity. Because conductive fiber can function as a sensor and a reinforcing fiber, electro-micromechanical evaluation has been studied as a cost-effective non-destructive method for damage sensing and assessing interfacial qualities.

18.13   NATURAL FIBER-BASED BIODEGRADABLE NANOCOMPOSITES

Growing environmental consciousness and worries about how to dispose of plastic trash have recently increased consumer demand for biodegradable polymers. These polymers are also made from renewable resources, such as bacterial fermentation of plant-based feedstocks. Thus, these materials address the need to identify substitutes for fossil fuel feedstocks, the rise in end-of-life disposal problems, and the requirement to develop landfill alternatives. The vast majority of biodegradable plastic uses are in short-life, disposable items, as expected [53]. Food packaging, including transparent films, trays, clam-shell packaging, and blister packs, is a significant market. Compost sacks and bags are additional packing materials.

Additionally, convenience items like disposable cutlery, plates, and cups are made with biodegradable polymers. Compostable materials have several uses in agriculture, such as plant pots, mulching film, and yarn. Biodegradable polymers are used in the medical industry for disposable gloves, sutures, and medication delivery systems. Biodegradable polymers perform poorly technically compared to many conventional polymers, especially in terms of mechanical and barrier properties. This is a significant hurdle to their adoption.

For this reason, biodegradable polymers have been treated with nanocomposite technology. This chapter discusses biodegradable nanocomposites, including their fabrication processes and how nanofillers affect mechanical, thermal, and barrier properties and impact biodegradability. Owing to its superior mechanical characteristics and nanoscale size, which produce incredibly high surface area-to-volume ratios, cellulose nanoparticles have received substantial interest over the past 15 years [64]. These characteristics, along with the exceptional capacity for surface functionalization, make nanocomposites the ideal candidates for increasing the mechanical properties of the active material. Other benefits of using cellulose nanoparticles are their low cost, low density, renewable nature, biodegradability, wide range of filler options, low energy consumption, good performance, and moderate abrasivity during processing [65]. By mechanically processing the biomass to produce nanofibrillated cellulose, or by acid hydrolyzing the biomass to produce cellulose nanocrystals, aqueous suspensions of cellulose nanoparticles can be created. Acid hydrolysis is used to dissolve low lateral order regions so that a subsequent vigorous mechanical shearing can transform the extremely crystalline, water-insoluble residue into a steady suspension. The resultant nanocrystals are rod-like particles, or whiskers, with nanometer-scale diameters that depend on the substrate's makeup and the hydrolysis conditions. These nanocrystals have an axial Young's modulus with a limit determined by theoretical chemistry of 167.5 GPa, which is near steel because they have a minimal number of defects (200 GPa).

Although there are still a lot of unanswered questions, nanocellulose has a wide range of uses. Despite the fact that the majority of studies have concentrated on its mechanical characteristics as a reinforcing phase and liquid crystal self-ordering properties, tens of scientific articles have proved its potential. It is challenging to distribute cellulose nanoparticles within a polymer matrix evenly.

Nanocellulose-reinforced polymeric films have the potential for generating films with excellent transparency and enhanced mechanical and barrier qualities, which makes packaging one use for these materials. Food and pharmaceutical packaging frequently should have a high oxygen barrier, so a change could be crucial for expanding into new markets. The use of cellulose nanoparticles is advantageous for the electrical device sector in addition to packaging. In various applications, including flexible displays, solar cells, electronic paper, and panel sensors, nanocellulosics are a viable reinforcing material due to their low thermal expansion, high strength, high modulus, and transparency. The Finnish Centre for Nanocellulosic Technologies, also known as Suomen Nanoselluloosakeskus, is a significant initiative for manufacturing nanocellulose that was launched in Finland in 2008.

The center aims to create new applications for cellulose as a material and an industrial-scale production technology for nanocellulose. The funding for the center comes from both public and private sources. The Finnish forest industry is undergoing a significant transformation, and new technologies are anticipated to help the industry become more competitive. The ArboraNano network is another significant initiative that is backed by the Canadian government and is represented by FPInnovations and NanoQuébec. The goal of this effort to value nanocellulose (as cellulose nanocrystals) is also to revitalize Canada's forestry industry, which has been negatively impacted by the rising competition from emerging nations in Asia and South America. In June 2011, Domtar Corporation and FPInnovations established the business CelluForce, which will be in charge of producing nanocrystalline cellulose in Windsor, Quebec's first plant of its sort. Production is anticipated to begin in the first quarter of 2012.

18.14 BONE REPAIR

Recent studies have shown that nanocomposites can be an effective tool for correcting abnormalities in bone and soft tissue. An essential component in achieving successful outcomes is the inherent ability of nanocomposites to enhance osteointegration and osteoinductivity. In addition, this beneficent influence further improves the reparative processes involved in soft tissue regeneration inside the osseous milieu. It has been demonstrated that nano-hydroxyapatite/poly(lactide) can enable appropriate cell adhesion to human cartilage [66].

18.15 APPLICATIONS OF BIODEGRADABLE POLYMER NANOCOMPOSITES

18.15.1 APPLICATIONS IN BIOMEDICINE

Due to the vast amount of research in this area, this section only provides a brief overview and refers the reader to the appropriate literature studies. Because other polymers were more popular at the time because of their durability PLA's biodegradability was generally overlooked when it was first discovered. Due to PLA's biocompatibility, interest in the material sprang up again in the 1960s. Most of the research done to date has focused on using these two polymers in biomedical applications because chitin shares this characteristic. The ability of PLA to be absorbed by live tissue also qualifies it for use as a temporary "scaffold" for bone and tissue repair. Similar to the food industry, legislation limits the types of nanofillers that can be used in nanocomposites for biomedical purposes. As a result, the primary filler employed in these materials is hydroxyapatite, which is biocompatible in and of itself. Thus, collagen/hydroxyapatite and PLA/hydroxyapatite nanocomposites are utilized in a variety of orthopedic applications. Utilizing PLA-based nanocomposites as agents for controlled medication release is another extensive research area. The quantity and rate of absorption by living tissue, as well as their mechanical properties, especially at body temperature, are the critical characteristics of biodegradable nanocomposites for biomedical applications [67].

18.15.2 PACKAGING

The issues with biodegradable and bioderived polymers for use in food, such as poor barrier and mechanical qualities, are frequently addressed via nanocomposite technology. If biodegradability is to be preserved, the alternative of polymer blending can only be used with other biodegradable polymers. In addition to having stronger solvent resistance and improved mechanical properties, biodegradable nanocomposites frequently exhibit better water and oxygen barrier qualities. While these factors pertain to the food product's shelf-life, nanocomposite technology can have an impact on the packaging's end of life, especially when it offers a workable solution to the issues that are frequently encountered with biodegradable and bioderived polymers for use in food, namely, weak mechanical properties and poor barrier properties. If biodegradability is to be preserved, the alternative of polymer blending can only be used with other biodegradable polymers. In addition to having stronger solvent resistance and improved mechanical properties, biodegradable nanocomposites frequently exhibit better water and oxygen barrier qualities. However, these concern the shelf-life of the food product and nanocomposite technology can contribute to the end of the packaging's useful life, mainly where composting or anaerobic digestion is concerned. By doing so, it outperforms polymer blends. Utilizing nanofillers can lead to controlled and increased biodegradability, instead of blending with non-biodegradable polymers, which can impair the composite's overall biodegradability. Food waste, leftovers from the food it protected, or food waste mixed with uneaten or rejected food is frequently present in post-consumer food packaging. The idea of disposing of food waste and packaging by composting or anaerobic digestion is increasingly considered a viable solution, especially if energy recovery in the form of biogas can be made simultaneously, even though recycling is a possibility. In conclusion, research on biodegradable nanocomposite materials is still very much in its infancy, and the "precautionary principle" still holds for the time being [68].

18.15.3 AUTOMOTIVE APPLICATIONS

The steadily rising cost of oil affects all industries, not the automotive sector, which has responded by looking for ways to make cars more fuel-efficient by replacing non-structural metal body parts with polymer nanocomposite materials. The main problem in this application is not the material's biodegradability but rather its sustainability in light of rising and erratic oil prices. Toyota was among the "early adopters" of the new technology, although a list of names would quickly become outdated as many other manufacturers followed suit. As the matrix polymer, research in automotive applications has tended to concentrate on PLA, PBS, and cellulose esters, while the nanofillers are either organoclays or plant fibers like kenaf, hemp, and cellulose fibers. The three industries—food packaging, automotive applications, and biomedical—all approach creating new products with a low-risk mindset. Therefore, it is anticipated that these applications will advance at a moderate but constant pace [69,70].

18.15.4 FLAME-RETARDANT NANOCOMPOSITES

The majority of research on nanocomposites and polymer flammability has focused on organo-modified layered silicates (OMLSs). To enhance the fire performance of nanocomposites, numerous academic organizations are now conducting research that primarily focuses on the following aspects: the creation of novel surface and interfacial alterations for nanoparticles, the impact of manufacturing processes on nanocomposites, the fire-retardant action mechanisms, and the synergistic effects in multicomponent flame-retardant systems incorporating nanoparticles.

When industry attempted to utilize nanoparticles as flame retardants, particularly with a view to replacing halogenated chemicals, the last of these became very prominent. Different combinations were looked into because the incorporation of such components is insufficient to meet fire standards like minimal limiting oxygen index values or V0 class in a vertical UL 94 test. The first topic covered in this chapter is the advantages and benefits of polymer nanocomposites for flame retardancy. Second, techniques for enhancing polymers' fire resistance are discussed, along with the fire qualities that nanoparticles make better and the mechanisms by which fire resistance works. The chapter will also pay particular attention to fresh experimental methods used to look at how well nanocomposites function in a fire. The inclusion of nanoparticles in flame-retardant polymer materials will be looked at in the section that follows, both for nanoparticles alone, taking into account surface and interfacial alterations, and for mixtures of nanoparticles with flame retardants [71]. Future developments will be covered in the conclusion, including the possibility of grafting particular polymers or oligomers onto the surface of nanoparticles as well as producing nanoparticles directly inside a polymer.

Future developments in the subject will be driven by the creation of novel types of nanoparticles, particularly those resembling flame-retardant additives like magnesium and aluminum nanohydroxides. Furthermore, it is anticipated that novel surface modifications of these nanoparticles may improve their reactivity with other elements of the flame-retardant system in which they are used. New techniques for creating these materials may also be taken into consideration, such as in situ synthesis of nano-oxides from precursors during polymer processing. Since OMLSs-based combinations have been the subject of the majority of investigations, it stands to reason that pairings of OMLSs with different flame-retardant intumescent systems could be particularly intriguing.

In this regard, mixtures of OMLSs, nano-oxides, nano-hydroxides, and phosphorus Flame Retardants (FRs) appear to be promising. Since they can react with phosphorus FRs and have an impact on heat stability, nano-oxides can be incorporated into a variety of FR systems, especially for polymers undergoing hydrolysis processes. They also have a high aspect ratio and a big specific surface area, and some nanometric particles (nano-hydroxides as opposed to nano-oxides) have the benefit of releasing water. The impact of barriers may be influenced by these particles. Therefore, creative FR systems might be created by maximizing the barrier effect by packing nano- and microparticles that are left at the surface following polymer ablation caused by combustion. The solution to this problem is systems. Tethering the FR agents to the nanoparticles might be a creative technique to achieve this goal. Some nano-oxides can have certain phosphorus compounds, like oligomers, bonded to them using grafting procedures. As a result, it is anticipated that more sophisticated FR systems will be created using novel nanoparticle surface modifications (perhaps blended), in conjunction with intumescent systems using a variety of polymer and interfacial agents. Control of the morphology during and after processing will be necessary to accommodate these developments. It is necessary to conduct more research on the relationship between the morphology and fire behavior of such complex systems [72–74].

18.16 CONCLUSIONS

The use of nanoscale fillers offers ample opportunities to produce nanocomposites with various properties having a wide range of applications from packaging to healthcare sectors. Generally, the properties of the filler define the application of nanocomposite such as Fe_2O_3-based composites, which respond to external magnetic fields and, thus, can be used for devices related to biological detection, magnetic storage, etc. Nanoparticles may be of different shapes and sizes, which alters their properties that can be tailored sometimes for a specific application. One of the major drawbacks while manufacturing of polymer nanocomposites is the aggregation of particles due to incompatibility with the matrix. The matrix or the particle surface need to be modified to increase the compatibility. Conductive fillers primarily carbon-based compounds such as CNTs, graphene, and CB are used for the preparation of polymer nanocomposites. Carbon-based compounds not just impart electrical conductivity to nanocomposites but also improve other properties such as increased mechanical properties and fire resistance of nanocomposites. Conductive fillers often need to be used above a specific concentration to impart electrical conductivity to nanocomposites; otherwise, they can be used as an additive to improve mechanical properties and to give a black carbon-like finish such as done using CB during manufacturing of vehicle rubber tires. From vehicles to aircrafts, CNTs and carbon nanofibers have found their applications in the production of composites due to their large surface area and outstanding thermal and electrical properties, thus providing multifunctional finishing to the products. The future technologies have been focusing on the development of polymer membranes containing inorganic metal oxide fillers having water retention properties that could help in the regulation of fuel flow in fuel cells. The use of chitosan or other sensitive material with a conductive electrode material such as graphene can be used for detection of analytes like glucose, choline, or acetylcholine. To counter the issues of environmental pollution, there have been research studies directed toward the development of biodegradable polymer nanocomposites. This involves the use of biodegradable matrixes and biodegradable fillers such as natural fibers or cellulose nanoparticles for the production of biodegradable gloves, sutures, etc. The application areas of polymer nanocomposites are vast, and there have been developments focused toward manufacturing low-cost, high-performance composites.

REFERENCES

1. O. Roy and A. Sharif, "Industrial implementation of polymer-nanocomposites," in *Advanced Polymer Nanocomposites*, M. E. Hoque, K. Ramar, and A. Sharif, Eds. Elsevier, 2022, pp. 537–546.
2. R. J. Young, I. A. Kinloch, L. Gong, and K. S. Novoselov, "The mechanics of graphene nanocomposites: A review," *Compos. Sci. Technol.*, vol. 72, no. 12, pp. 1459–1476, 2012.
3. E. Reynaud, C. Gauthier, and J. Perez, "Nanophases in polymers," *Rev. Métallurgie*, vol. 96, no. 2, 1999, doi: 10.1051/metal/199996020169.
4. S. Wei, J. Zhu, P. Mavinakuli, and Z. Guo, "Magnetic polymer nanocomposites: Fabrication, processing, property analysis, and applications," in *Multifunctional Polymer Nanocomposites*, J. Leng, A. Kin-tak Lau, Eds. CRC Press, Boca Raton, FL, 2010, pp. 135–159.
5. L. Díaz, M. Santos, C. Ballesteros, M. Maryško, and J. Pola, "IR laser-induced chemical vapor deposition of carbon-coated iron nanoparticles embedded in polymer," *J. Mater. Chem.*, vol. 15, no. 40, pp. 4311–4317, 2005.
6. H. Cao, G. Huang, S. Xuan, Q. Wu, F. Gu, and C. Li, "Synthesis and characterization of carbon-coated iron core/shell nanostructures," *J. Alloys Compd.*, vol. 448, no. 1, pp. 272–276, 2008.
7. N. Fan, X. Ma, Z. Ju, and J. Li, "Formation, characterization and magnetic properties of carbon-encapsulated iron carbide nanoparticles," *Mater. Res. Bull.*, vol. 43, no. 6, pp. 1549–1554, 2008.
8. Z. Guo, M. Moldovan, D. P. Young, L. L. Henry, and E. J. Podlaha, "Magnetoresistance and annealing behaviors of particulate Co–Au nanocomposites," *Electrochem. Solid-State Lett.*, vol. 10, no. 12, p. E31, 2007.
9. X.-W. Wei, G.-X. Zhu, C.-J. Xia, and Y. Ye, "A solution phase fabrication of magnetic nanoparticles encapsulated in carbon," *Nanotechnology*, vol. 17, no. 17, pp. 4307–4311, 2006.
10. Z. Guo, K. Shin, A. B. Karki, D. P. Young, R. B. Kaner, and H. T. Hahn, "Fabrication and characterization of iron oxide nanoparticles filled polypyrrole nanocomposites," *J. Nanoparticle Res.*, vol. 11, no. 6, pp. 1441–1452, 2009.
11. K. Sunderland, P. Brunetti, L. Spinu, J. Fang, Z. Wang, and W. Lu, "Synthesis of γ-Fe₂O₃/polypyrrole nanocomposite materials," *Mater. Lett.*, vol. 58, no. 25, pp. 3136–3140, 2004.
12. F. Yan, G. Xue, and F. Wan, "A flexible giant magnetoresistance sensor prepared completely by electrochemical synthesis," *J. Mater. Chem.*, vol. 12, no. 9, pp. 2606–2608, 2002.
13. C. Forder, S. P. Armes, A. W. Simpson, C. Maggiore, and M. Hawley, "Preparation and characterisation of superparamagnetic conductive polyester textile composites," *J. Mater. Chem.*, vol. 3, no. 6, pp. 563–569, 1993.
14. O. Jarjayes, P. H. Fries, and C. Bidan, "New nanocomposites of polypyrrole including γ-Fe₂O₃ particles: electrical and magnetic characterizations," *Synth. Met.*, vol. 69, no. 1, pp. 343–344, Mar. 1995.
15. F. Yan, G. Xue, J. Chen, and Y. Lu, "Preparation of a conducting polymer/ferromagnet composite film by anodic-oxidation method," *Synth. Met.*, vol. 123, no. 1, pp. 17–20, 2001.
16. P. Poddar, J. L. Wilson, H. Srikanth, S. A. Morrison, and E. E. Carpenter, "Magnetic properties of conducting polymer doped with manganese–zinc ferrite nanoparticles," *Nanotechnology*, vol. 15, no. 10, pp. S570–S574, 2004.
17. D. Li and Y. Xia, "Electrospinning of nanofibers: Reinventing the wheel?," *Adv. Mater.*, vol. 16, no. 14, pp. 1151–1170, 2004.
18. D. Zhang et al., "Electrospun polyacrylonitrile nanocomposite fibers reinforced with Fe₃O₄ nanoparticles: Fabrication and property analysis," *Polymer*, vol. 50, no. 17, pp. 4189–4198, 2009.
19. A. Moser et al., "Magnetic recording: advancing into the future," *J. Phys. Appl. Phys.*, vol. 35, no. 19, pp. R157–R167, 2002.
20. R. L. Edelstein et al., "The BARC biosensor applied to the detection of biological warfare agents," *Biosens. Bioelectron.*, vol. 14, no. 10, pp. 805–813, 2000.

21. M. M. Miller et al., "A DNA Array Sensor Utilizing Magnetic Microbeads and Magnetoelectronic Detection," Naval Research Lab Washington DC, Jan. 2001. Accessed: Jun. 28, 2022. [Online]. Available: https://apps.dtic.mil/sti/citations/ADA482516.

22. L. Bao, W.-L. Low, J. Jiang, and J. Y. Ying, "Colloidal synthesis of magnetic nanorods with tunable aspect ratios," *J. Mater. Chem.*, vol. 22, no. 15, pp. 7117–7120, 2012.

23. M. Ozaki, S. Kratohvil, and E. Matijević, "Formation of monodispersed spindle-type hematite particles," *J. Colloid Interface Sci.*, vol. 102, no. 1, pp. 146–151, 1984.

24. N. Hohlbein, A. Shaaban, and A. M. Schmidt, "Remote-controlled activation of self-healing behavior in magneto-responsive ionomeric composites," *Polymer*, vol. 69, pp. 301–309, 2015.

25. L. H. Sperling, *Introduction to Physical Polymer Science*. John Wiley & Sons, New Jersey, 2005.

26. F. Bovey, *Macromolecules: An Introduction to Polymer Science*. Academic Press, New York, 1979.

27. M. M. Coleman, *Fundamentals of Polymer Science: An Introductory Text*, Second Edition (2nd ed.). Routledge, New York, 1997..

28. J. N. Coleman et al., "High performance nanotube-reinforced plastics: Understanding the mechanism of strength increase," *Adv. Funct. Mater.*, vol. 14, no. 8, pp. 791–798, 2004.

29. S. Peeterbroeck et al., "The influence of the matrix polarity on the morphology and properties of ethylene vinyl acetate copolymers–carbon nanotube nanocomposites," *Compos. Sci. Technol.*, vol. 67, no. 7, pp. 1659–1665, 2007.

30. J. C. Grunlan, L. Liu, and Y. S. Kim, "Tunable single-walled carbon nanotube microstructure in the liquid and solid states using poly(acrylic acid)," *Nano Lett.*, vol. 6, no. 5, pp. 911–915, 2006.

31. M. J. Allen, V. C. Tung, and R. B. Kaner, "Honeycomb carbon: A review of graphene," *Chem. Rev.*, vol. 110, no. 1, pp. 132–145, 2010.

32. Y. Liu, D. Yu, C. Zeng, Z. Miao, and L. Dai, "Biocompatible graphene oxide-based glucose biosensors," *Langmuir*, vol. 26, no. 9, pp. 6158–6160, 2010.

33. C. Lv, Q. Xue, D. Xia, M. Ma, J. Xie, and H. Chen, "Effect of chemisorption on the interfacial bonding characteristics of graphene–polymer composites," *J. Phys. Chem. C*, vol. 114, no. 14, pp. 6588–6594, 2010.

34. C. Sirisinha and N. Prayoonchatphan, "Study of carbon black distribution in BR/NBR blends based on damping properties: Influences of carbon black particle size, filler, and rubber polarity," *J. Appl. Polym. Sci.*, vol. 81, no. 13, pp. 3198–3203, 2001.

35. D. Shamir, A. Siegmann, and M. Narkis, "Vibration damping and electrical conductivity of styrene–butyl acrylate random copolymers filled with carbon black," *J. Appl. Polym. Sci.*, vol. 115, no. 4, pp. 1922–1928, 2010.

36. K. Hu and D. D. L. Chung, "Flexible graphite modified by carbon black paste for use as a thermal interface material," *Carbon*, vol. 49, no. 4, pp. 1075–1086, 2011.

37. W. M. Huang, B. Yang, and Y. Q. Fu (2012). Polyurethane Shape Memory Polymers (1st ed.). CRC Press, Boca Raton, p. 383.

38. X. Lan, J. S. Leng, Y. J. Liu, and S. Y. Du, "Investigate of electrical conductivity of shape-memory polymer filled with carbon black," *Adv. Mater. Res.*, vol. 47–50, pp. 714–717, 2008.

39. P. L. Antonucci, A. S. Aricò, P. Cretì, E. Ramunni, and V. Antonucci, "Investigation of a direct methanol fuel cell based on a composite Nafion®-silica electrolyte for high temperature operation," *Solid State Ion.*, vol. 1–4, no. 125, pp. 431–437, 1999.

40. W. H. J. Hogarth, J. C. Diniz da Costa, and G. Q. (Max) Lu, "Solid acid membranes for high temperature (¿140° C) proton exchange membrane fuel cells," *J. Power Sources*, vol. 142, no. 1, pp. 223–237, 2005.

41. U. H. Jung, K. T. Park, E. H. Park, and S. H. Kim, "Improvement of low-humidity performance of PEMFC by addition of hydrophilic SiO_2 particles to catalyst layer," *J. Power Sources*, vol. 159, no. 1, pp. 529–532, 2006.

42. Y. Devrim, S. Erkan, N. Baç, and I. Eroğlu, "Preparation and characterization of sulfonated polysulfone/titanium dioxide composite membranes for proton exchange membrane fuel cells," *Int. J. Hydrog. Energy*, vol. 34, no. 8, 3467–3475, 2009.

43. Z. Wang and S. K. Saxena, "Raman spectroscopic study on pressure-induced amorphization in nanocrystalline anatase (TiO_2)," *Solid State Commun.*, vol. 118, no. 2, pp. 75–78, 2001.

44. F. H. Gojny, M. H. G. Wichmann, U. Köpke, B. Fiedler, and K. Schulte, "Carbon nanotube-reinforced epoxy-composites: enhanced stiffness and fracture toughness at low nanotube content," *Compos. Sci. Technol.*, vol. 64, no. 15, pp. 2363–2371, 2004.

45. F. H. Gojny, M. H. G. Wichmann, B. Fiedler, W. Bauhofer, and K. Schulte, "Influence of nano-modification on the mechanical and electrical properties of conventional fibre-reinforced composites," *Compos. Part Appl. Sci. Manuf.*, vol. 36, no. 11, pp. 1525–1535, 2005.

46. S. Spindler-Ranta and C. E. Bakis, "Carbon nanotube reinforcement of a filament winding resin: 47th International SAMPE Symposium and Exhibition," *Int. SAMPE Symp. Exhib.*, vol. 47 II, pp. 1775–1787, 2002.

47. J. Njuguna and K. Pielichowski, "Polymer Nanocomposites for Aerospace Applications: Fabrication," *Adv. Eng. Mater.*, vol. 6, no. 4, pp. 193–203, 2004.

48. J. Njuguna, K. Pielichowski, and J. Fan, "Polymer nanocomposites for aerospace applications," in *Advances in Polymer Nanocomposites*, F. Gao, Ed. Woodhead Publishing, Cambridge, 2012, pp. 472–539.

49. J. Njuguna and K. Pielichowski, "Polymer nanocomposites for aerospace applications: Properties," *Adv. Eng. Mater.*, vol. 5, no. 11, pp. 769–778, 2003.

50. S. I. Kundalwal, R. S. Kumar, and M. C. Ray, "Smart damping of laminated fuzzy fiber reinforced composite shells using 1–3 piezoelectric composites," *Smart Mater. Struct.*, vol. 22, no. 10, p. 105001, 2013.

51. H. Deka, M. Misra, and A. Mohanty, "Renewable resource based 'all green composites' from kenaf biofiber and poly(furfuryl alcohol) bioresin," *Ind. Crops Prod.*, vol. 41, pp. 94–101, 2013.

52. A. Bhat et al., "Review on nanocomposites based on aerospace applications," *Nanotechnol. Rev.*, vol. 10, no. 1, pp. 237–253, 2021.

53. N. Basavegowda and K.-H. Baek, "Advances in functional biopolymer-based nanocomposites for active food packaging applications," *Polymers*, vol. 13, no. 23, p. 4198, 2021.

54. L. Pang et al., "Antibiofouling thin-film nanocomposite membranes for sustainable water purification," *Adv. Sustain. Syst.*, vol. 5, no. 6, p. 2000279, 2021.

55. M. Tyagi and D. Tyagi, "Polymer nanocomposites and their applications in electronics industry," *Int J Electron Electr Eng*, vol. 7, no. 6, pp. 603–608, 2014.

56. H. Gupta, "Role of nanocomposites in agriculture," *Nano Hybrids Compos.*, vol. 20, pp. 81–89, 2018.

57. X. Zhong, R. Yuan, and Y.-Q. Chai, "Synthesis of chitosan-Prussian blue-graphene composite nanosheets for electrochemical detection of glucose based on pseudobienzyme channeling," *Sens. Actuators B Chem.*, vol. 162, no. 1, pp. 334–340, 2012.

58. K. M. Mitchell, "Acetylcholine and choline amperometric enzyme sensors characterized in vitro and in vivo," *Anal. Chem.*, vol. 76, no. 4, pp. 1098–1106, 2004.

59. G. E. Uwaya and O. E. Fayemi, "Enhanced electrocatalytic detection of choline based on CNTs and metal oxide nanomaterials," *Molecules*, vol. 26, no. 21, p. 6512, 2021.

60. F. Yang et al., "Heterostructure of $MnCo_2O_4$ intercalated graphene oxide coated with Ni-V-Se nanoparticles for supercapacitors with high rate capability," *J. Alloys Compd.*, vol. 926, p. 166762, 2022.

61. Y. Ni, C. Li, J. Gao, Y. Guo, and T. Li, "Preparation and electrochemical properties of Cu_3P/rGO nanocomposite protection strategy for lithium-ion batteries," *J. Solid State Electrochem.*, vol. 26, no. 12, pp. 2873–2881, 2022.

62. S. O'Donnell, K. Sprong, and B. Haltli, "Potential impact of carbon nanotube reinforced polymer composite on commercial aircraft performance and economics," in *AIAA 4th Aviation Technology, Integration and Operations (ATIO) Forum*, American Institute of Aeronautics and Astronautics, Chicago, Illinois.

63. A. Volodin, D. Buntinx, M. Ahlskog, A. Fonseca, J. B. Nagy, and C. Van Haesendonck, "Coiled Carbon Nanotubes as Self-Sensing Mechanical Resonators," *Nano Lett.*, vol. 4, no. 9, pp. 1775–1779, 2004.

64. S. J. Eichhorn et al., "Review: current international research into cellulose nanofibres and nanocomposites," *J. Mater. Sci.*, vol. 45, no. 1, pp. 1–33, 2010.

65. G. Siqueira, S. Tapin-Lingua, J. Bras, D. da Silva Perez, and A. Dufresne, "Mechanical properties of natural rubber nanocomposites reinforced with cellulosic nanoparticles obtained from combined mechanical shearing, and enzymatic and acid hydrolysis of sisal fibers," *Cellulose*, vol. 18, no. 1, pp. 57–65, 2011.

66. R. James, M. Deng, C. T. Laurencin, and S. G. Kumbar, "Nanocomposites and bone regeneration," *Front. Mater. Sci.*, vol. 5, no. 4, pp. 342–357, 2011.

67. C.-G. Sanporean, Z. Vuluga, and J. D. C. Christiansen, "Bioactive nanocomposites with applications in biomedicine," in *Innovative Strategies in Tissue Engineering*, M. Prasad and P. Di Nardo, Eds. River Publishers, 2015, pp. 1–24. Accessed: Sep. 26, 2022. [Online]. Available: http://www.riverpublishers.com/book_details.php?book_id=250.

68. J. Wróblewska-Krepsztul, T. Rydzkowski, G. Borowski, M. Szczypiński, T. Klepka, and V. K. Thakur, "Recent progress in biodegradable polymers and nanocomposite-based packaging materials for sustainable environment," *Int. J. Polym. Anal. Charact.*, vol. 23, no. 4, pp. 383–395, 2018.

69. A. R. Mclauchlin and N. L. Thomas, "Biodegradable polymer nanocomposites," in *Advances in Polymer Nanocomposites*, F. Gao, Ed. Woodhead Publishing, Cambridge, 2012, pp. 398–430.

70. S. I. Yun, G. E. Gadd, B. A. Latella, V. Lo, R. A. Russell, and P. J. Holden, "Mechanical properties of biodegradable polyhydroxyalkanoates/single wall carbon nanotube nanocomposite films," *Polym. Bull.*, vol. 61, no. 2, pp. 267–275, 2008.

71. 14:00–17:00, "ISO/TS 27687:2008," *ISO.* https://www.iso.org/cms/render/live/en/sites/isoorg/contents/data/standard/04/42/44278.html (accessed Jul. 03, 2022).

72. E. P. Giannelis, "Polymer layered silicate nanocomposites," *Adv. Mater.*, vol. 8, no. 1, pp. 29–35, 1996.

73. J. W. Gilman et al., "Flammability properties of polymer–layered-silicate nanocomposites. Polypropylene and polystyrene nanocomposites," *Chem. Mater.*, vol. 12, no. 7, pp. 1866–1873, 2000.

74. J. W. Gilman, T. Kashiwagi, and J. D. Lichtenhan, "Nanocomposites: A revolutionary new flame retardant approach," *NIST*, vol. 33, no. 4, pp. 40–46, 1997.

Index

For Product Safety Concerns and Information please contact our EU
representative GPSR@taylorandfrancis.com
Taylor & Francis Verlag GmbH, Kaufingerstraße 24, 80331 München, Germany

9 781032 381978